THE PROOF STAGE

The Proof Stage

HOW THEATER REVEALS THE HUMAN TRUTH OF MATHEMATICS

STEPHEN ABBOTT

PRINCETON UNIVERSITY PRESS

PRINCETON & OXFORD

Published by Princeton University Press
41 William Street, Princeton, New Jersey 08540
99 Banbury Road, Oxford OX2 6JX

press.princeton.edu

Library of Congress Cataloging-in-Publication Data

Names: Abbott, Stephen, 1964– author.
Title: The proof stage: how theater reveals the human truth of mathematics / Stephen Abbott.
Description: Princeton : Princeton University Press, 2023. | Includes bibliographical references and index.
Identifiers: LCCN 2022040785 (print) | LCCN 2022040786 (ebook) | ISBN 9780691206080 (hardback) | ISBN 9780691243368 (ebook)
Subjects: LCSH: Mathematics—History. | Theater. | Mathematics and literature. | BISAC: PERFORMING ARTS / Theater / General | MATHEMATICS / History & Philosophy
Classification: LCC QA21 .A24 2023 (print) | LCC QA21 (ebook) | DDC 808.8/036—dc23 / eng20230110
LC record available at https://lccn.loc.gov/2022040785
LC ebook record available at https://lccn.loc.gov/2022040786

British Library Cataloging-in-Publication Data is available

Editorial: Diana Gillooly, Kiran Pandey
Jacket Design: Chris Ferrante
Production: Danielle Amatucci
Publicity: Matthew Taylor, Kate Farquhar-Thomson
Copyeditor: Kelly Walters

This book has been composed in Arno Pro

Printed on acid-free paper. ∞

Printed in the United States of America

10 9 8 7 6 5 4 3 2 1

In memory of my father, Anthony S. Abbott (1935–2020);
the man who loved not wisely but too well.

CONTENTS

THE PROOF STAGE

Prologue

ANINDA: (*In an Indian accent.*) You're probably wondering at this point if this is the entire show. I'm Aninda, this is Al and this is Ruth. (*Pause. His accent changes.*) Actually that's a lie. I'm an actor playing Aninda, he's an actor playing Al and she's an actress playing Ruth. But the mathematics is real. It's terrifying, but it's real.

I REMEMBER the moment when the actor playing Aninda broke character and delivered the above confession in the opening scene of *A Disappearing Number*. I was at the University of Michigan in the 1,300-seat Power Center theater, and the house was full. At the top of the show, a mathematician entered what looked like a university classroom and, after nervously thanking us for coming, she started in on a lecture about infinite series. A bit flummoxed to find themselves back in math class, the audience responded with nervous laughter that grew in proportion to the complexity of the notation being scribbled on the chalkboard. When at last Aninda arrived to confirm for the anxious playgoers that the mathematician was fake—and that the mathematics was terrifying indeed—the response was a full-bodied cathartic guffaw.

Having spent my adult life studying and teaching mathematics, my personal reaction to Aninda's joke was limited to a wry smile. I had long ago made my peace with math's maligned public reputation and took the gibe as a backhanded compliment, which is how it was meant. Rather than being put at ease like my fellow audience members, what I most recall from this early moment in the show was the strange, unsettling feeling of coming face-to-face with the deeply contradictory nature of theater and mathematics. As Aninda makes plain, theater is chimeric by design. At its core, theater is an ephemeral event

1

consisting of live actors creating a momentary illusion for a live audience. Mathematics, by contrast, has an air of authentic, if sometimes forbidding, certainty. Along the wide spectrum of intellectual endeavors humans have engaged in since getting up on our hind legs, mathematics is arguably the most austere, enduring, and inevitable.

A Disappearing Number was first performed in 2007, and although the play vigorously acknowledges the stark incongruity of mathematics and theater—largely at theater's expense—it undercuts these arguments as quickly as it makes them. By every measure, this play full of mathematicians and their mathematics is a successful and compelling piece of art. It opened at the Theatre Royal in Plymouth and played to enthusiastic houses at the Barbican in London as well as touring worldwide. The play was performed with the original cast in Paris, New York, Barcelona, Milan, Sydney, and Mumbai, and was broadcast on National Theatre Live. It nearly ran the table of England's major theater awards.

Rather than emerging fully formed from the imagination of a solitary playwright, *A Disappearing Number* was the end result of a long improvisational process by the London-based theater company Complicité, working under the direction of its creative director and cofounder Simon McBurney. McBurney is not a mathematician, and he is not reticent about pointing that out. "Time for school," he writes about his childhood, "where I would understand nothing about maths except that I got the wrong answer."[1] Trained in physical and improvisational theater methods at the Jacques Lecoq School in Paris, McBurney had been making innovative theater with Complicité for two decades when a friend gave him a copy of *A Mathematician's Apology*, by Cambridge professor G. H. Hardy. Written in 1940, toward the end of his life, Hardy's essay begins with the following lament:

> It is a melancholy experience for a professional mathematician to find himself writing about mathematics. The function of a mathematician is to do something, to prove new theorems, to add to mathematics, not to talk about what he or other mathematicians have done.[2]

Hardy's essay is his attempt to justify a life dedicated to mathematics, and it is full of the wistful echoes of an artist past his creative prime. McBurney did not know much mathematics, but he knew a great deal about the imagination of an artist, and he immediately recognized a kindred spirit in the aging Cambridge don. The heart of Hardy's argument is that, despite its pragmatic nature, mathematics is an aesthetic discipline. "Beauty is the first test," he

Arcadia; Jessie Cave (Thomasina), Dan Stevens (Septimus), Samantha Bond (Hannah), Ed Stoppard (Valentine); Duke of York's Theatre, 2009.
© Donald Cooper/photostage.co.uk.

writes, "there is no permanent place in the world for ugly mathematics."[3] This viewpoint—of mathematics as a creative art form—cracked open a doorway for McBurney to forge a meaningful connection between his chosen craft and Hardy's, but the math-phobic McBurney must have felt a stronger force compelling him to push the door all the way open. One reason to suspect this is the sheer magnitude of the commitment that came with deciding to create a play about mathematics. McBurney's method of devised theater would require the total immersion of his company in the ideas of the play. The other reason I suspect that McBurney fell prey to a potent force linking theater to mathematics is that I experienced it myself, pulling me in from the opposite direction.

Twenty years ago, a friend of mine was directing a new play by Tom Stoppard called *Arcadia* that she reported was full of mathematics. She asked if I would come talk to her cast about the play and generously loaned me a copy of the script. I never returned it. It sits, even now, in a familiar place on my office shelf, worn, full of margin notes and missing its cover. The excitement I felt in my first read through the script was something I had only experienced a handful of times in my academic life. Although I was reading a play, it reminded me of the mix of agitation and awe I felt years before when, as an undergraduate, I first learned of the theorems of non-Euclidean geometry. My graduate school encounter with the work of Kurt Gödel was another such occasion. These moments are accompanied by a sense that tectonic changes are afoot. I like to imagine this is how Galileo felt looking through his telescope at the moons of Jupiter. At first, it seems like a portal to a fantastic new world, but the revelations I was experiencing in each case were really about a world I thought I already knew. The familiar was suddenly full of mystery.

I came to my colleague's rehearsal and told her cast that math was doing something in their new play that I had never seen it do—and that, possibly, theater was too. In the ensuing years I returned again and again to the intersection of mathematics and theater, and eventually I set up camp there, reading, attending, and eventually directing plays in an attempt to ferret out why these distinct entities were infecting me with the same visceral excitement. That journey led me through the many collaborations of math and theater that make up the contents of this book. To find a thesis for the book, however, I would have to get to the heart of the reason why a self-respecting mathematician like myself was spending so much time at the playhouses.

Mathematics as Art

Collaborations between mathematics and art more generally have taken myriad forms. The rich connection between mathematics and music is evident in the recursive structure of Bach's canons and the deftly patterned rhythms of classical Indian tabla music. One can sense mathematical principles at work in Palladian architecture, in Islamic tiling patterns, and even in the strict rules governing the Renaissance sonnet writer. But in each of these examples there is a natural affinity between the mathematics and the medium where the collaboration is taking place. Music theory, like architecture, requires a high level of mathematical fluency, while the tiling and wallpaper patterns created by artisans from different cultures over the centuries are a major inspiration for the mathematical study of tessellations and symmetry.

But how do we explain the curious conjunction of mathematics and theater? Other than falling under the same broad heading of "artistic endeavors," it is not at all clear how this particular collaboration might be carried out. Of the many forms of artistic expression, theater is distinctively human, performed by human actors telling human stories. Putting aside the applied mathematics of bankers and bridge-builders, as Hardy would insist we do, pure mathematics is a search for abstract truths. These truths are of a specific nature, and usually take the form of relationships that exist between the numerical and geometric objects that populate so-called mathematical reality. Even so, Hardy finds useful parallels between his craft and the more traditional arts. "A mathematician, like a painter or a poet, is a maker of patterns," is how he describes it.[4] Having aligned himself with visual and literary artists, Hardy goes on to explicitly include playwrights in his analogy, albeit in an equivocal way that echoes the sentiment in Aninda's direct address from *A Disappearing Number*. Hardy writes:

> Archimedes will be remembered when Aeschylus is forgotten because languages die and mathematical ideas do not.[5]

Hardy's intent here is to make the case for the superior nature of mathematical truth over poetic truth based on the former's potential for immortality, but this audacious sentence comparing mathematics specifically to theater implicitly draws attention to a compelling counterpoint that suggests there may be less of a difference than Hardy anticipates. Specifically, why did Hardy choose Archimedes and Aeschylus to make his point? No doubt the alliteration appealed to his poetic ear, but at the heart of this comparison is Hardy's desire to take us back to the roots of mathematics and the roots of theater. In my search to understand the kinship between these very distant cousins, ancient Greece seemed like a reasonable place to start.

Proofs and Plays

In what surely cannot be simple coincidence, the early Greek culture that developed and refined the core principles that characterize pure mathematics as it is currently practiced did much the same thing for theater, and at essentially the same time (Figure P.1). The hidden variable at the heart of this correlation is the Greek's unwavering allegiance to reasoned enlightenment. Whereas other ancient civilizations wanted to control the world, the Greeks wanted to understand it. Achieving this required getting below the surface to find the root causes for why things worked as they did. More fundamentally, it required a belief that there were root causes to find. To the Greeks, the universe was not random or capricious—there were laws at work. This was not a premise that applied only to the mathematical universe—it extended to the moral universe as well. The forces responsible for the choreography of the heavenly sky as well as the human condition were not arbitrary or subject to

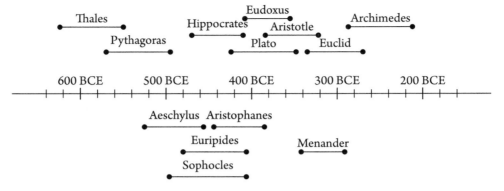

FIGURE P.1. Timeline of Greek mathematics and theater.

whim. There was a rational scaffolding underneath events, and the Greek faith in reason meant that truths about this scaffolding could be discerned by the well-trained mind.[6]

The manifestation of this conviction is at its most transparent in Euclid's *Elements* where the discerned truths—mathematical propositions like the Pythagorean Theorem—are inextricable from their proofs. Proofs in the *Elements* are referred to as demonstrations, and a proposition divorced from its demonstration is an inert platitude. Following the steps of a proof requires the active mental participation of the reader who, using the roadmap Euclid has laid out, reconstructs the argument to his or her own satisfaction. Demonstrations are not simply read, they are experienced, and the effect is that the validity of the proposition in question is essentially rediscovered by every cogent mind that engages the *Elements* in this active way.

The early Greek playwrights had something similar in mind in terms of how they intended their audience to interact with their art. The moral universe, like the mathematical one, was governed by laws. It was the job of the playwright to create a compelling demonstration of these laws. In this sense Greek tragedies are like Greek proofs—presentations crafted to reveal the existence of fundamental truths governing human nature while justifying their legitimacy at the same time. Now, to be sure, theater did not garner the same unquestioned respect that mathematics enjoyed as a vehicle for enlightenment. Plato, in fact, banishes theater from his model Republic on the grounds that its appeal to emotion is at odds with a properly rational pursuit of truth. But Aristotle comes to the poet's defense. In his *Poetics*, Aristotle makes the case that poetry is a perfectly valid means for accessing and communicating truth provided that it adheres to a set of formal guidelines that sound eerily similar to descriptions of mathematical proofs. In Aristotle's theory of Greek theater, plot is the "heart and soul, as it were, of tragedy,"[7] and it is of paramount importance that the plot, or action, of the play

> ... be both unified and complete, and the component events ... so firmly compacted that if any one of them is shifted to another place, or removed, the whole is loosened up and dislocated; for an element whose addition or subtraction makes no perceptible extra difference is not really a part of the whole.[8]

Just as in a mathematical proof, each event in a properly constructed plot leads inextricably to the next so that the sequence of scenes is not episodic but logically dictated.[9] "There is a great difference," Aristotle writes, "in whether the events happen because of those or merely after them."[10] Aristotle's clear favorite among the early masterpieces is Sophocles's *Oedipus Tyrannus* which, quoting *Poetics* again, obtains its artistic appeal by having its tragic revelations "come about contrary to one's expectations yet logically, one following from the other." Two millennia later, G. H. Hardy would unwittingly echo this sentiment in his *Apology* by describing a beautiful mathematical proof as possessing a "high degree of unexpectedness, combined with inevitability."[11]

A distinctive by-product of their austere architecture is the potential for universality possessed by Greek proofs and Greek plays. When Euclid includes a geometrical figure for his reader, the lines and circles he draws are stand-ins for idealized geometric objects that exist in a perfected mathematical reality all their own. Like the lines and circles sketched on the page, the masked actors in Greek tragedies are conduits to vivid mental constructions that are specific in name but general in their potential to shed light on human questions. Borrowing Aristotle's insights one last time, "The historian speaks of what has happened, the poet of the kind of thing that *can* happen ... for poetry speaks more of universals, history of particulars."[12] The characters and settings of Greek tragedy may seem to be specific—e.g., Oedipus sits on the throne in Thebes—but setting them in the mythical past frees the playwright from the constraints of contemporary social discourse.[13] Shedding these constraints is crucial for getting beyond how things appear to how they really are. Greek drama—like Greek mathematics—takes its cues from the empirical world but is carried out, unencumbered, in an abstract one. The irony is that by exiting physical reality, the discoveries that are recorded become arguably more real, especially in the way they transcend the circumstances of their origin.

On this last point, the results speak for themselves. Over two millennia later, the *Elements* is still in use as a geometry textbook, and Greek tragedies resonate with an eerily modern voice.[14]

Stages of Uncertainty

Assessing Hardy's claim for the immortality of mathematics, the best we can say at this point is that, two thousand years later, neither Aeschylus nor Archimedes has lost any luster. Meanwhile, more modern developments in the interactions between mathematics and theater, which were in their infancy during Hardy's lifetime, have since muddied the water around the question. Strange as it sounds, wading around in this muddy water is where I eventually found a sense of clarity about why theater had taken such a firm grasp on my imagination.

Although the common origins of mathematics and theater can be linked to the Greek quest for certainty, modern collaborations between these two art forms are characterized by negotiations around a rising tide of uncertainty. At the turn of the twentieth century, mathematics made a disguised appearance on the experimental theater stage in the form of a foil for a more avant-garde approach to art. Theater may have been incubated in a culture devoted to reasoned enlightenment, but by this point in its evolution, rational thinking was the enemy of dramatists intent on pushing against the realism prevalent in theater at the time. Because Euclid authored the rulebook that codified geometric space as the perfection of physical reality, or perhaps simply because it represented deductive thinking at its most austere, Euclid's authority had to be subverted along with the rest of establishment orthodoxy. This frontal assault on the *Elements* by artists advocating for an unleashed imagination is, on its own, not too surprising. Where the story takes its unexpected turn is in the work of a distinguished handful of playwrights who ventured deeply enough into the mathematical woods to discover that mathematicians had already dismantled Euclid, along with the rest of the foundations of mathematics.

Unlike, say, Einstein's theory of relativity, revolutions in mathematics have taken place largely out of the public eye, but they have come with a similar potential to reconfigure the philosophical landscape. The discovery of non-Euclidean geometries in the mid 1800s called into stark question whether the propositions in the *Elements* applied to the physical world, and this skepticism within geometry would spread to the laws of arithmetic in the decades ahead. Soon enough, the very process of deductive reasoning, so beloved by the Greeks, would become the subject of its own intense scrutiny. These were the paradigm shifts in the history of mathematics that had captivated me as a student, and they instilled in me a more ambiguous conception of mathematical certainty than the one Euclid had established.

If mathematical truth is not immutable or perfect, or even perfectible, then nothing is. This profound and simple fact becomes more humbling the more mathematics one knows, and it points to the core of what this book is really about. Initially at least, my fascination with the intersection of theater and mathematics was focused on the apparent tension between the two art forms. Was it really possible to embody the abstract nature of mathematics in the concrete, tangible world of live theater? Eventually, however, I found my way to the more fundamental question: To what end? Given theater's intrinsic compulsion to tell human stories, what meaningful relevance could pure mathematics have to the flesh-and-blood business of a life on Earth?

A great deal, as it turns out, when the emphasis shifts from transcendent certainty to the more equivocal relationship to truth that characterizes the mathematics of the previous century. In the journey since my initial encounter with *Arcadia*, the portrait of mathematics that has emerged out of its collaborations with theater has routinely caught me up short, most notably in the way that theater reveals the soulful insights embedded between the lines of math's beloved proofs. "Extraordinary how mathematics helps you to know yourself," says one of Samuel Beckett's characters, as though this statement might be perfectly obvious. It's not, of course, but filtered through the lens of twentieth-century playwrights this turns out to be very much the case.

Evolution of the Math Play

Bringing mathematics to bear on human questions requires focusing on its philosophical nature more than the content of its theorems. When Aninda says, "But the mathematics is real," he gets a hearty laugh from his math-anxious audience, but he is playing a semantic game because it is not clear what "real" means in this case. What are the theorems of mathematics really about? Is geometry about physical reality, or about some abstract mathematical world, or is it just an elaborate formal dance of symbols about nothing

$$2^3 \times 3^2 \times 5 \times 7 \times 11 \times 13 \times 97$$

A Disappearing Number; David Annen (G. H. Hardy), Saskia Reeves (Ruth), Paul Battacharjee (Aninda), Firdous Bamji (Al); Barbican Theatre, 2007. Stephanie Berger © 2022.

at all? This quest for meaning in mathematics is a regular source of inspiration for playwrights grappling with analogous questions of how their chosen medium acquires meaning and how its representations—i.e., the worlds of the plays themselves—do or don't correspond to reality. The function of language is a vibrant flash point for collaboration between math and theater, but the common DNA of these seemingly disparate disciplines is most vividly on display in the dexterity with which each is capable of investigating its *own* limitations.

In pure mathematics, proofs are paramount. This is why Bertrand Russell, David Hilbert, and others committed themselves to bringing the full power of mathematical reasoning to the study of axiomatic systems. Essentially they were asking whether there was a way to prove statements about the efficacy of mathematical proofs—and there was, although the conclusions turned out to be different from what anyone predicted. Just as there are proofs about the capabilities of proofs, there are plays about the capabilities of plays, and it is no coincidence that playwrights attuned to the dramatic potential of metatheatrical constructions are often ones who are otherwise engaged with mathematics. Tom Stoppard is a prime example here, and Samuel Beckett another, but Aninda's fourth wall–shattering monologue ("You're probably wondering at this point if this is the entire show") hints at how ubiquitous this phenomenon really is.

By the time *A Disappearing Number* premiered in 2007, the kinship between mathematics and theater was widely acknowledged among working playwrights. Simon McBurney's play was a high-profile example of exceptional merit; but it was, in fact, part of a trend of plays exploring different aspects of mathematics. David Auburn's *Proof* (2000) is probably the most well known of these, in part because it was made into a feature film in 2005. G. H. Hardy, whose essay was the seed for *A Disappearing Number*, had already appeared as a character in Ira Haupton's *Partitions* (2003), and Hardy's essay was an obvious source of inspiration for *The Five Hysterical Girls Theorem* (2001), a boisterous comedy by Rinne Groff featuring a dozen highly caffeinated mathematicians at a fictional math conference at the English seaside. There are a handful of plays about mathematician Alan Turing (e.g., *Breaking the Code*, 1986; *Lovesong of the Electric Bear*, 2003), and an armful about Isaac Newton (e.g., *Calculus*, 2003; *Newton's Hooke*, 2003; *Leap*, 2004; *Let Newton Be*, 2011; *Isaac's Eye*, 2012). There are plays about Cantor's Theorem (*Infinities*, 2002), plays about the Traveling Salesman Problem (*Completeness*, 2011), and there is even a catchy musical about Fermat's Last Theorem (*Fermat's Last Tango*, 2001), full of bawdy allusions to elliptic curves and modular forms.

This robust list of more recent plays engaging mathematics stands in contrast to the sparse landscape of such collaborations from the previous century. Mathematics was not always so welcome among artists—at least those who wanted to reach a broad audience—but it did prove to be potent inspiration for a few brave playwrights willing to push past math's sterilized classical facade. The last few chapters of this book take up a subset of plays from recent decades, but the majority of the attention is focused on the dramatic pioneers who employed mathematics in various degrees to impact the arc of modern theater: Alfred Jarry, Stanislaw Witkiewicz, Samuel Beckett, Bertolt Brecht, Friedrich Dürrenmatt, Tom Stoppard, Michael Frayn, and Simon McBurney. Implicit in the literary eminence of this list is an argument for the distinctive and unacknowledged contribution that mathematics has made to the evolution of theater. But this is by no means a one-way street, and it is the traffic going the other way—theater's humanizing contributions to a richer understanding of mathematics—that provides the narrative hook in the pages ahead.

The Human Truths of Mathematics

To tell this story, chapter 1 starts where I did—with Tom Stoppard's *Arcadia*. Stoppard has done more than any other playwright to bring mathematics from the experimental stage to the mainstage, although it took the Czech-born playwright three decades of experimenting with mathematical ideas to finally accomplish the feat. After chronicling Stoppard's journey from struggling beat reporter to *Arcadia*, chapter 2 goes back to the turn of the previous century and initiates a historical journey with the explosive debut of Alfred Jarry's *Ubu Roi*. Jarry's wrecking ball approach to Euclid in the form of his antihero Père Ubu is a storied part of early twentieth-century avant-garde theater lore, but it is the lesser-known Stanislaw Witkiewicz, writing in Poland in the 1920s and 1930s, who recognized that the non-Euclidean revolution in mathematics could be co-opted for a radical new agenda for theater. Chapter 3 charts the influence of mathematics on Samuel Beckett's singular career, from his richly literary novel *Murphy* to the wordlessly austere *Quad*, created when Beckett was in his 70s. Running in parallel through this chapter is an account of the attempts of Gottlob Frege, Bertrand Russell, and David Hilbert to establish a perfected formalized language for mathematics. This mathematical story acquires rich new overtones when juxtaposed with Beckett's aspirations for a formalized approach to language that he believed would get him closer to the deepest aspects of human identity.

Continuing the historical thread, chapter 4 explores three math-related plays by three important playwrights—Friedrich Dürrenmatt, Michael Frayn, and Simon McBurney. The influential Bertolt Brecht is given a cameo appearance for putting a mathematician center stage in *Life of Galileo*. Brecht's name is not included in the chapter title, however, because it was only Dürrenmatt, Frayn, and McBurney who fused the content of mathematics to the architecture of the plays they were creating. Tom Stoppard returns in chapter 5, but this time the focus is on how Stoppard's propensity for exploiting metatheatrical devices aligns him with the metamathematical discoveries of Kurt Gödel. At least it is initially. When Stoppard includes an explicit nod to Gödel's Theorem in a play about the mystery of the human ego, the chapter is transformed from one about the parallels between self-conscious theater and self-conscious mathematics to one about consciousness itself. The sparks arising at the end of this chapter from the interactions of Gödel's mathematics with contemporary theories of mind are brought to a full flame in chapter 6, which delves into the life, mathematics, and theater of Alan Turing.

Chapter 7 is a short epilogue that uses David Auburn's popular play full of mathematicians—but no mathematics—to ask one last time how mathematics and theater complement each other in their respective searches for truth.

One consequence of focusing on plays about mathematics—and giving the majority of the attention to the previous century—is the dearth of women playwrights discussed.[15] Curiously, widening the lens to science more generally reveals a generous number of notable plays by women—*An Experiment with an Airpump* (1998) by Shelagh Stevenson, *After Darwin* (1998) by Timberlake Wertenbaker, and *A Number* (2002) by Caryl Churchill being three important examples. More recently, Lauren Gunderson has penned a host of scientifically themed scripts (*Émilie du Châtelet*, 2009; *Silent Sky*, 2013; *Ada and the Engine*, 2018) that are regularly produced to broad acclaim. Fashioning an explanation for why women playwrights have not engaged more robustly with pure mathematics is complicated, but it has not helped that women continue to be underrepresented in the field itself. Gioia De Cari's one-woman play, *Truth Values: One Girl's Romp Through the MIT Male Math Maze* (2011) offers some vivid first-person insights into why this problem persists.

Real but Not Terrifying

No mathematical expertise is required to read this book, nor is a familiarity with any of the plays or playwrights discussed. Keeping the prerequisites to a minimum was part of an early decision to tell a self-contained story, accessible to someone who has never looked inside the *Elements* or attended a Beckett play. This doesn't mean there isn't some effort required in a few places. The mathematics is real, but it is by no means terrifying.

When the story slows down to take a deeper look into the mathematical weeds, an event that occurs about once a chapter, it is because there is some revelation lurking in the details that connects the mathematics to the art in a meaningful way. A look inside the *Elements*, long enough to appreciate the significance of Euclid's fifth postulate, provides the best vantage point from which to make sense of the self-described non-Euclidean theater of Stanislaw Witkiewicz. A firsthand encounter with formal languages like the one in Russell and Whitehead's *Principia Mathematica* provides a novel way to assess Beckett's career-long quest to exact more control over the intended function of language. It also provides the proper context for understanding Gödel's Incompleteness Theorem, a result that Beckett scholars have

Breaking the Code; Mark H. Dold (Turing); Barrington Stage Company, 2014 (photo: Kevin Sprague).

consistently associated with the Irish playwright. The fact that Gödel's Theorem also makes a cameo in a Stoppard play about self-identity is further
evidence that Gödel's mathematics has implications far beyond the arc of
twentieth-century logic. The same can be said about the intellectual achievements of Alan Turing. Even as Turing's name has become synonymous with
computing machinery, playwrights have used the dexterity of their craft to
reveal the deep, and sometimes painful, humanity lying beneath the surface
of Turing's mathematical ideas.

One can certainly appreciate all the plays discussed in these pages without any knowledge of mathematics—indeed, most audience members do just
that—but there are substantial rewards for those willing to bridge the disci-
plinary divide. This applies to travelers crossing in either direction, as I can
personally testify. In the two decades since my fateful encounter with Stoppard, what I have come to recognize is that playwrights, by virtue of their
chosen medium, possess a distinctive means for bringing mathematics to bear
on understanding who we are. What started as an attempt to align Gödel with
Godot in a way that explained why the music they make is in the same key gradually produced an entirely new song about the humanity that mathematics
acquires when actors engage mathematical ideas for an audience. Mathematics has the reputation of being the purest and most abstract form of knowledge,
but on stage all knowledge becomes self-knowledge. The great upheavals
in the history of mathematics that appear as limitations or imperfections—
the discovery of irrational numbers, alternate geometries, paradoxes of the
infinite, incompleteness, and chaotic dynamics—are transformed by gifted
playwrights into clarifying insights about the human journey.

Extraordinary how theater and mathematics help you to know yourself.

1

Stoppard: The Incline from Thinking to Feeling

I keep trying to find a play about mathematics. There is one somewhere but I can't find it.

—TOM STOPPARD, FROM A 1985 LETTER

THE CURTAIN rises in the opening scene of *Arcadia* on the drawing room of the stately Coverly manner. The year is 1809. Thirteen-year-old Thomasina Coverly scribbles in her lesson book while her tutor, recent Cambridge graduate Septimus Hodge, sits at a distance ignoring his pupil. Thomasina is supposed to be tending to her daily algebra assignment, but stronger forces are at play and she finally breaks the silence.

THOMASINA: Septimus, what is carnal embrace?
SEPTIMUS: Carnal embrace is the practice of throwing one's arms around a side of beef. (2)[1]

Thomasina reports that she has overheard the house staff gossiping that "Mrs Chater was discovered in carnal embrace in the gazebo," and she has come to her trusted tutor for enlightenment. Septimus, meanwhile, is eager to have the morning to himself. To that end, he has given Thomasina the task of finding a proof for Fermat's Last Theorem, "a problem that has kept people busy for 150 years," but he quickly finds out that mathematics is poor competition for carnality in the battle for the attention of a thirteen-year-old. As clever as she is curious, Thomasina is not completely satisfied with Septimus's various explanations and, detecting his evasions, she eventually backs him into

16

"Septimus, what is carnal embrace?" *Arcadia*; Emma Fielding (Thomasina), Rufus Sewell (Septimus); Royal National Theatre, 1993. © Fritz Curzon / ArenaPAL.

a rhetorical corner. "I don't think you have been candid with me Septimus," Thomasina insists. "A gazebo is not, after all, a meat larder."

SEPTIMUS: I never said my definition was complete.

THOMASINA: Is carnal embrace kissing?

SEPTIMUS: Yes.

THOMASINA: And throwing one's arms around Mrs Chater?

SEPTIMUS: Yes, now, Fermat's last theorem—

THOMASINA: I thought as much. I hope you are ashamed.

SEPTIMUS: I, my lady?

THOMASINA: If *you* do not teach me the true meaning of things, who will?

SEPTIMUS: Ah. Yes, I am ashamed. Carnal embrace is sexual congress, which is the insertion of the male genital organ into the female genital organ for purposes of procreation and pleasure. Fermat's last theorem, by contrast, asserts that when x, y, and z are whole numbers each raised to power of n, the sum of the first two can never equal the third when n is greater than 2.

(*Pause.*)

THOMASINA: Eurghhh!
SEPTIMUS: Nevertheless, that is the theorem.
THOMASINA: It is disgusting and incomprehensible. (3)

A Truly Marvelous Proof

Arcadia premiered at the National Theatre in London on April 13, 1993. Two months later, at the Isaac Newton Institute in nearby Cambridge, Andrew Wiles went public with an actual proof of Fermat's Last Theorem. This was almost certainly the first time a mathematics conference had any direct bearing on events at the National, but Wiles's surprise announcement was international news, and it sent playwright Tom Stoppard scrambling to update the program for his new play.

In 1993, Fermat's Last Theorem was the most famous unsolved problem in mathematics. Its fame was due in part to the ease with which it could be posed—Septimus's one sentence summary is perfectly accurate—and in part to its storied history. Around 1630, the great French mathematician Pierre de Fermat was studying a personal copy of an ancient text called *Arithmetica*, written in about 250 CE by Diophantus of Alexandria. The text is essentially a collection of exercises and examples illustrating properties of integers, and one in particular caught Fermat's attention. Problem 8 in Book II asks the reader "to divide a given square into two squares." Fermat, like most of us, was familiar with the Pythagorean Theorem, which states that if x and y are the lengths of the legs of a right triangle and z is the length of the hypotenuse then

$$x^2 + y^2 = z^2.$$

Keeping in mind that numbers in *Arithmetica* were implicitly understood to be whole numbers; i.e., 0, 1, 2, 3, . . . , Diophantus's challenge was equivalent to asking whether there existed a right triangle with integer length sides. Such examples were well known; e.g., $3^2 + 4^2 = 5^2$ or $5^2 + 12^2 = 13^2$. The real interest originated from Fermat's generalization of Problem 8. Appended to this problem in his copy of *Arithmetica* is Fermat's now infamous marginalia:

On the other hand it is impossible to separate a cube into two cubes, or a biquadrate into two biquadrates, or generally any power except a square into two powers with the same exponent. I have discovered a truly marvelous proof of this, which however the margin is not large enough to contain.[2]

"Biquadrate" refers to a fourth power. Thus, what Fermat claims is that the equation

$$x^n + y^n = z^n$$

has no integer solutions when n is any value bigger than two.

Needless to say, Fermat never supplied this "truly marvelous proof" in some other more spacious forum, and so resolving Fermat's claim, with either a proof or a counterexample, became an open question for the mathematical community that increased in stature every year it went unanswered. By 1809, when the fictional Thomasina was assigned the problem, the great Swiss mathematician Leonard Euler had provided a mostly satisfying proof for the case when $n = 3$ and Sophie Germain, a rare example of a nineteenth-century female mathematician, was solidifying her reputation by making significant progress on a large number of other cases. It is widely assumed that Fermat did not have the general proof he boasted about in his margin note, but he is credited with providing the argument for the case when $n = 4$ before he died. For three and a half centuries, mathematicians from around the world chipped away at Fermat's riddle, which is why Princeton mathematician Andrew Wiles decided it was best to work on the problem in secret to avoid the inevitable admonition he would have received from his colleagues that he was wasting his time. Even the title of his Cambridge lecture did not give away the surprise. When Wiles finally made the announcement that he had found a proof for Fermat's Last Theorem, the shock waves that rippled out from the Newton Institute were powerful enough to prompt the *New York Times* to run the front-page headline, "At Last, Shout of 'Eureka' in Age-Old Math Mystery."

You Cannot Stir Things Apart

Stoppard does not center the plot of *Arcadia* around the premise that Thomasina miraculously discovers a proof for Fermat's Last Theorem, but he does make her into a prodigious mathematician in a very different way. Later in the opening scene we get some additional evidence that *Arcadia* is more than just a comedy about the carnal escapades of the British aristocracy:

THOMASINA: When you stir your rice pudding, Septimus, the spoonful of jam spreads itself around making red trails like the picture of a meteor in my astronomical atlas. But if you stir backward, the jam will not come together again. Indeed, the pudding does not notice and continues to turn pink just as before. Do you think this is odd?

SEPTIMUS: No.

THOMASINA: Well, I do. You cannot stir things apart. (5)

This image of increasing entropy—of the inevitable and irreversible rise of disorder in any closed system—serves as a compelling illustration of the Second Law of Thermodynamics, a principle of physics that did not get fully articulated until the middle of the century. From this exchange we get the sense that Thomasina possesses not only wisdom beyond her years but insight beyond her era. The classically trained Septimus makes a sincere attempt to be a proper tutor to Thomasina, but she quickly grows restless with his Greek geometry and Fermatian number theory and sets off to invent a new kind of mathematics that can more accurately capture the jagged and unpredictable contours of nature. "Mountains are not pyramids and trees are not cones," Thomasina complains when faced with the prospect of learning more of Euclid's *Elements*. "If there is an equation for a curve like a bell," she says on a different occasion, "there must be an equation for one like a bluebell, and if a bluebell, why not a rose? Do we believe nature is written in numbers?" (37)

What Stoppard has in mind for his young prodigy is the twentieth-century branch of mathematical science that falls under the general heading of chaos theory. As her particular project, Thomasina sets herself a task in the subfield of fractal geometry. Taking a leaf from an onstage apple, Thomasina declares to Septimus that she "will plot this leaf and deduce its equation. You will be famous for being my tutor when Lord Byron is dead and forgotten." (37) A century and a half before mathematician Benoit Mandelbrot coins the term "fractal," Thomasina fills her lesson books with strange looking equations that will ultimately require a computer to be properly realized, along with an endearing note to future scholars:

I, Thomasina Coverly, have found a truly wonderful method whereby all the forms of nature must give up their numerical secrets and draw themselves through number alone. This margin being too mean for my purpose, the reader must look elsewhere for the New Geometry of Irregular Forms discovered by Thomasina Coverly. (43)

There Is One Somewhere

Euclidean geometry, fractal geometry, chaos theory, the Second Law of Thermodynamics, Fermat's Last Theorem. Before 1993 there were no examples of broadly successful plays that explicitly engaged this much mathematical content, but *Arcadia* was immediately recognized as something extraordinary. Having authored over forty scripts during the previous three decades, Stoppard was viewed as one of the leading playwrights of his generation, and many people who knew his work well were claiming that *Arcadia* was his best.[3] "A perfect marriage of ideas and high comedy," wrote the *Times*. The *Daily Telegraph* called it "a masterpiece." The success of Stoppard's play put mathematics onto the pages of the arts section, and two months later Andrew Wiles moved it to the front page, above the fold. Pinpointing cause and effect is difficult, but this one-two punch of public acclaim coincided with a shift in the relationship between mathematics and popular culture. In the years after *Arcadia*, writers were increasingly amenable to incorporating mathematics in plays, novels, film, and television.

Responding to the rush of positive reviews in 1993, Stoppard acknowledged that something special had occurred with his latest play. "I feel for once that I stumbled onto a really good narrative idea. *Arcadia* has got a classical kind of story and, whether we are writing about science or French maids, this whole thing is about storytelling first and foremost."[4] What Stoppard was hinting at is that *Arcadia* was groundbreaking, not so much because it contained a great deal of mathematics, but because it made the mathematics integral to the play's emotional arc. What is missing from Stoppard's comments is any indication about how this merging of mathematics and storytelling was carried out. Why had no one done this before?

The attempt to weave mathematics into the plot of *Arcadia* was not a one-off idea for Stoppard. Early in his career, he acquired a reputation as a playwright of ideas. These ideas were in no way confined to science and mathematics. History, art, and politics are ubiquitous themes in Stoppard's writing, but Stoppard was perfectly willing to include mathematics in his palette alongside these other more traditionally acceptable motifs. This was true from the outset. The plot of the early radio play *Albert's Bridge* turns decisively on the details of a simple optimization problem. Probability makes an appearance in Stoppard's breakthrough play *Rosencrantz and Guildenstern Are Dead*. In *Jumpers*, the highly anticipated follow-up to *R&G*, Stoppard again mined mathematics to give the play some intellectual breadth.

As Stoppard's career progresses, the nature of the mathematics grows more sophisticated, as does the way he leverages it within the framework of the script. While early on, mathematics appears whimsically, it eventually becomes a point of focus for the playwright. "I keep trying to find a play about mathematics," he wrote to a friend in 1985. "There is one somewhere but I can't find it."[5] While on the surface it may seem as though the central challenge is transporting technical ideas onto the stage without the burden of too much didactic luggage, the revelation of Stoppard's journey is that success depended on something else entirely. The intellectual heft he might give his plays by accessing the largely untapped terrain of mathematics would all be for show unless he could find a beating heart somewhere in the austere headiness of the theorems. He had to make the mathematics accessible, he had to make it authentic, and then he had to make it matter just as authentically on a personal level. This last trick was the crucial one and the most elusive. For many, including Stoppard, the allure of mathematics is its propensity for certainty, which is why a search for the humanity of mathematics sounds so incongruous. It's also why wedging a description of Fermat's Last Theorem up against the definition of carnal embrace generates such a hearty laugh. Sex is the opposite of math, or so their respective reputations would suggest. To find his math play, Stoppard would have to flip this public perception on its head, a challenge that required venturing beyond the predictable certitude of traditional algebra and geometry to find a more romantic incarnation of mathematics in the service of paradox and unpredictability.

So what kind of education might lead to an artistic sensibility like Stoppard's—one that saw as much dramatic potential in fractals and Fermat's Last Theorem as it did in poetry and politics? Given his rogue multidisciplinary disposition, it's a bit stunning and also, somehow, perfectly obvious that Stoppard never attended university. Dropping out of school at age seventeen, this highly decorated English playwright did not start out as an intellectual by any traditional definition. In fact, he did not start out English either.

A Bounced Czech

Tom Stoppard was born Tomáš Sträussler in July 1937. His parents, Marta and Eugen Sträussler, were settled in the town of Zlín in what was then Czechoslovakia. The year Stoppard was born, the Nazi threat was looming, and the Munich agreement of 1938 to give Hitler the Sudetenland fully unleashed the forces of anti-Semitism. Stoppard's father was an unobservant Jew and his mother's family had long ago converted to Catholicism. These details were irrelevant in the current political climate, and Stoppard's parents began to search for a way out. Somewhat miraculously, the Sträussler family managed to escape Czechoslovakia for Singapore in April 1939 under the pretense of a job transfer. All four grandparents stayed behind and would die in the Holocaust.

Singapore provided only a temporary respite. Over the next two years, Stoppard and his older brother Petr (who would later become Peter) learned some rudimentary English while living among the British in what was a tropical trading crossroads of Southeast Asia. By late 1941, however, Japanese bombs began to fall on the city, and in January of the following year, Stoppard boarded a ship in the dark of night with his mother and brother bound for Australia. Eugen Straüssler was a doctor who nobly stayed behind in Singapore to volunteer his services as part of the local defense effort. When he finally managed to secure passage on a departing ship some two weeks later, his vessel was attacked and destroyed by Japanese aircraft. The three surviving members of the Straüssler family, meanwhile, transferred midjourney to a ship bound for India and arrived in Bombay in February 1942.

From the time Stoppard was five until he was eight, his family lived in the city of Darjeeling, safely tucked up against the Himalayan Mountains. The boys attended an English-speaking school while their mother took a job as a clerk in a local shoemaking factory. In 1945, Marta Straüssler married a British army officer—Major Kenneth Stoppard—and soon the family was once again on the move, this time to England. Tomáš Straüssler legally changed his name to Tom Stoppard, and the young boy who had already lived in three countries finally found one he was able to call home.

Partly at his traditional stepfather's insistence, and partly of his own volition, Stoppard tried valiantly to shed his outsider status. In many ways he succeeded. He immediately took to the pastoral British landscape, and he has always maintained that English is his first language.[6] His mother took pains to sequester the family's refugee history from her boys, and both received fairly typical English prep school educations at Pocklington Grammar School in Yorkshire. Stoppard was a good student, although not distinguished in any particular way. He focused mostly on classics and history, taking some standard courses in mathematics and science but nothing beyond what was required. He did not participate in theater other than trudging through the requisite Shakespearean texts in his English classes, an experience he claims was uninteresting to him at the time.[7] His two passions at Pocklington were the debate society and the cricket team, both of which would become regular motifs in his playwriting career.

When it came time to decide between the university-bound track or wrapping up his education and heading into the professional world, Stoppard eagerly chose the latter. Still three years short of his twentieth birthday, he moved to Bristol, the town where his mother and stepfather were currently living, and took a job as a journalist writing for various local newspapers about everything from car accidents to art openings. Stoppard enjoyed the writing but would later admit to some significant journalistic shortcomings. "I wasn't much use as a reporter," he said in a 1967 interview, "I felt I didn't have the right to ask people questions." This hesitancy led to some early experience writing fiction. "It was OK when [the articles] didn't use a photograph," he confessed. "I just sat in the canteen and made up quotes from people who always lived in one of Bristol's longest streets."[8] Among his many journalistic duties in the early years was serving as the second string theater reviewer, a task he enjoyed. Theater was undergoing something of a heyday in England at the time. Harold Pinter, John Osborne, and Samuel Beckett were contributing to a vibrant atmosphere that captured more and more of Stoppard's imagination. Around 1960, the twenty-three-year-old journalist made the decision that a career critiquing plays wasn't enough—he wanted to write them.

Albert's Bridge

Hear that, George? The City Engineer's figures are a model of correctitude.

—CHAIRMAN, FROM *ALBERT'S BRIDGE*

The early 1960s for Stoppard were a balancing act between freelance journalism to avoid poverty and hocking his scripts for stage, radio, and television. Stoppard's first modest success was late in 1961 when his stage play, *A Walk on the Water*, garnered enough interest that a production agency paid him 100 pounds for an option on the work. Although the agency did not ultimately produce it, a performance was eventually filmed for British Independent Television in 1963. In the meantime, Stoppard published a few short stories and authored several short radio plays that managed to find their way into the BBC's programming rotation. In 1964, Stoppard received a grant to attend a five-month playwriting workshop in Berlin during which he gave most of his attention to a work in progress with the title *Rosencrantz and Guildenstern Meet King Lear*. A public reading of this twenty-five-minute, one-act pastiche did not generate much enthusiasm among those who saw it, including its author. It was also around this time that Stoppard signed a contract to produce a novel. The common denominator in all of these projects was a sense of general ambition to secure a career as a writer in some form and a legitimate need for financial security. Stoppard was routinely in debt to anyone who would lend him money.

At this early stage in his career, the majority of Stoppard's work that made it to production was written for radio. With the limitations of the fifteen- and thirty-minute time slots of the BBC sound stage serving as catalyst for his creative imagination, Stoppard honed his dramatic voice with pieces like *The Dissolution of Dominick Boot* (1964), *M Is for Moon among Other Things* (1964), and *If You're Glad I'll Be Frank* (1965). In 1965, the BBC commissioned a longer radio drama from Stoppard that resulted in *Albert's Bridge* the following year. With a full hour to fill, Stoppard revealed his early willingness to entrust his play's narrative to mathematics, even if all the high school classics major knew at the time was a little basic algebra.

I. E. B + E. G. Q

The eponymous central character of *Albert's Bridge* is a newly minted university graduate with a chronic disinterest in the real world. Having majored in philosophy, the best Albert can imagine for himself is an entry-level position at a retail philosophy boutique. "Of course, a philosopher's clerk wouldn't get the really interesting work straight off," he muses. "It'll be a matter of filing the generalizations, tidying up paradoxes, laying out the premises before the boss gets in—that kind of thing." (61)[9]

In the meantime, Albert takes a temporary job painting the Clufton Bay Bridge alongside three other blue-collar types. The fact that there are four total painters working on the bridge is significant. The silver paint the city uses lasts precisely two years before it requires repainting, which is exactly the amount of time required for four men to complete the job. Thus, when the four painters finish up the last few steel girders at the end of the span, they return the next morning to the other side of the bridge and begin all over again. Oppressive to the other three, the Sisyphean task is pure solace to Albert, who sings while he wiggles his brush into corners he knows no one else will ever see. There is no doubt, then, that it is going to be Albert who volunteers for the lonely duty required in a money-saving plan being hatched in the city below at a special meeting of the Clufton Bay Bridge Sub-Committee.

Mr. Fitch is the "clipped, confident, rimlessly-eyeglassed" town accountant, obsessed with efficiency and possessed of just enough mathematical intuition to do a great deal of damage:

FITCH: The cycle is not a fortuitous one. It is contrived by relating the area of the surfaces to be painted—call it A—to the rate of the painting—

B—and the durability of the paint—C. The resultant equation determines the variable factor X—i.e. the number of painters required to paint surface A at speed B within time C. For example—
CHAIRMAN: E.g.
FITCH: Quite. Er, e.g. with X plus one painters the work would proceed at a higher rate—i.e. B, plus, e.g. Q. (59)

In 1967, the only established way to talk about mathematics within a play was through farce, which this certainly is, but there is some logic driving Fitch's presentation. Fitch is proposing that the city switch from using the current brand of paint, which lasts two years, to a new brand of paint which lasts eight years. What is causing the confusion is that this new brand of paint is more expensive; in fact, it costs exactly four times as much as the current brand and thus on the surface of things, so to speak, it seems like a wash. Fitch's pseudo-math jargon manages to baffle everyone at the meeting, including Fitch, until the Chairman finally calls the question:

CHAIRMAN: Pull yourself together, Fitch—I don't know what you're drivillin' about.
GEORGE: In a nutshell, Fitch—the new paint costs four times as much and lasts four times as long. Where's the money saved?
FITCH: We sack three painters.
(Pause.)
CHAIRMAN: Ah . . . . (60)

If four men can paint the bridge in two years, then one can paint it in eight. The rigidity of the algebra of this eighth-grade word problem is accented by the steel beams of the bridge itself. The bridge, like the algebra, is functional, definitive, and unambiguous in its purpose. Albert, by contrast, is an amorphous soul. He unwittingly stumbles into a tryst with his mother's house cleaner and dutifully marries her when she gets pregnant. But a wife, a baby girl, and a job prospect at his father's firm are no match in Albert's mind for the allure of Fitch's unique brand of optimizing mathematics:

FITCH: I'm the same. It's poetry to me—a perfect equation of space, time and energy—
ALBERT: Yes—
FITCH: It's not just slapping paint on a girder—
ALBERT: No—

FITCH: It's continuity—control—mathematics.
ALBERT: Poetry.
FITCH: Yes, I should have known it was a job for a university man. . . .
    You'll stick to it for eight years will you?
ALBERT: Oh, I'll paint it more than once. (65)

The Bridge Man

Some of the most entertaining and thought-provoking moments in the play come when Albert is joined on the bridge by Fraser. Convivially neurotic, Fraser is convinced that society has exceeded its capacity to stay ordered and senses that the apocalypse is at hand. He makes a habit of climbing up the bridge with the intent of throwing himself off, but each time he does so the perspective of the city as a gently humming arrangement of dots and squares makes his anxieties go away. "Yes, from a vantage point like this, the idea of society is just about tenable." (78) Albert, meanwhile, is morphing into "the bridge man" who lords over the toy town below. Having his solipsistic daydreams routinely interrupted, Albert is annoyed rather than moved by Fraser's suicidal agenda. "Aren't you going to try to talk me out of it?" Fraser pleads to Albert. "You know your mind. And you're holding me up," is Albert's reply. "I've got to paint where you're standing." (78)

The laugh lines and barroom philosophy bantered about between Albert and Fraser are good fun, but the real hook for *Albert's Bridge* turns out to be the mathematics. What Stoppard instinctively knew, even at this early stage, was that the sure-footedness of mathematics was most potent when it was deployed to *undermine* common sense. At its artistic best, mathematics could be a tool for creating uncertainty. Although Fitch's algebra seems airtight, it becomes clear as the play progresses that there is a fatal oversight in his calculations. (As a challenge, take a moment to try to spot it before reading on.) Although the paint Albert is applying to the bridge lasts eight years, the paint he is covering up is only meant to last for two. Working on his own, Albert cannot keep up with the rate of decay. After two years, Albert is only a quarter of the way across the span, meaning that three-fourths of the bridge are now in various stages of unsightly disrepair. The Chairman of the Clufton Bay Bridge Sub-Committee is undone. Broken and in a panic, Fitch crafts a solution that holds up on paper but is anathema to Albert:

FITCH: I have made arrangements.
ALBERT: What arrangements?

FITCH: Eighteen-hundred painters will report for work at seven o'clock tomorrow morning. By nightfall the job will be done. I have personally worked it out, and my department has taken care of the logistics.

ALBERT: Eighteen-hundred?

FITCH: Seventeen-hundred and ninety-nine. I kept a place for you. I thought you'd like that. (82)

For what it's worth, Fitch's calculations are fairly reasonable. Working 300 days out of the year, it is going to take Albert roughly

$$(6 \text{ years}) \cdot (300 \text{ days/year}) = 1{,}800 \text{ days}$$

to completely refinish the old paint on his own. By compressing the totality of this labor into a single workday, Fitch sets the stage for the play's denouement, which includes one final nod to mathematics.

As the army of painters marches inexorably toward the bridge, Albert and Fraser each see an incarnation of their own worst nightmare. For Fraser, it is the chaos of society pushing outward and upward into his last place of refuge. For Albert, the painters are an angry mob coming to take the imperious bridge man away. Both end up being partially correct. Forgetting to break stride as they march onto the surface of the bridge, the resonating frequencies of the 1,800 collective footsteps bring the bridge, Fraser, Albert, and the play to a crashing end.

Albert's Bridge was completed in 1966. Greeted with modest approval after it ran several times on BBC radio, the play went on to enjoy an interesting afterlife. In 1968 it won an international prize awarded to plays for radio and was performed as a stage play a number of times. One memorable series of three performances was held in the girders of the Royal Exchange Theatre in Manchester. In 1969, Stoppard expanded *Albert's Bridge* into a full-length screenplay.[10] Although the film was never produced, Stoppard's collaborator on the project—a friend and writer named Anthony Smith—eventually adapted the original version of *Albert's Bridge* into an operetta which was performed in 1999.

Rosencrantz and Guildenstern Are Dead

Well, it was an even chance, if my calculations are correct.

—GUILDENSTERN, FROM
ROSENCRANTZ AND GUILDENSTERN ARE DEAD

At the same time that he was working on *Albert's Bridge* as well as his contracted novel, Stoppard returned to the *Hamlet* pastiche he had started in Berlin the year before. It would take several years and many drafts, but the end result would transform Stoppard's career. First performed at the Edinburgh Fringe Festival in August 1966, *Rosencrantz and Guildenstern Are Dead* vaulted Stoppard from struggling writer and journalist to the upper echelons of active playwrights. A half century and some fifty plays and screenplays later, *Rosencrantz and Guildenstern Are Dead*—or *R&G* for short—is still Stoppard's most iconic and identifiable piece of writing.

In Shakespeare's original play, Rosencrantz and Guildenstern are school friends of Hamlet who appear briefly in a number of scenes, always in tandem and with their respective identities somewhat conflated. They first arrive when Claudius summons them to court to tease out what is afflicting Hamlet, but their old friend has no trouble recognizing Rosencrantz and Guildenstern for the spies that they are. Later, Claudius gives Rosencrantz and Guildenstern the task of taking Hamlet to England where Hamlet is to be executed upon his arrival. Shakespeare arranges for Hamlet to escape at sea, but not before Hamlet has altered the contents of the sealed order so that it now requests that Rosencrantz and Guildenstern be the ones executed by the English king.

The title of Stoppard's 1964 Berlin sketch—*Rosencrantz and Guildenstern Meet King Lear*—reflects his original idea to explore what would happen if Lear were the king that Hamlet's two friends encountered when they arrived, without Hamlet, on English soil. Although the various plot possibilities ultimately proved unsatisfying, the characters of Rosencrantz and Guildenstern held Stoppard's interest. Rather than invent elaborate biographies for them, the interesting challenge was to explore who they would be if their lives consisted of no more than the scant bits of action that Shakespeare's script requires of them. Following this logic, Stoppard removed Lear from the story and shifted his attention back in time to when Rosencrantz and Guildenstern are at Elsinore. In fact, he followed them back in time to a moment just before Elsinore, before anything has happened to them—before any choices have been made or any memories have accrued. "Two Elizabethans passing the time in a place without any visible character," is how the stage directions of this new account of Rosencrantz and Guildenstern begin. What might such a place be like? How could Stoppard make it nondescript yet different from any place we had ever been? What could the two be doing to pass the time? What could there be to talk about if nothing had happened yet?

Stoppard solved all these problems with a single mathematical device.

A Multipurpose Coin

The curtain rises in *R&G* to find the two misplaced Elizabethans betting on the flip of a coin; heads and the coin goes to Rosencrantz, tails and it belongs to Guildenstern. It is immediately clear that something is amiss. Each flip we witness turns up heads, and Rosencrantz's heavy bag of coins indicates that this has been happening for quite some time. Rosencrantz is slightly embarrassed to be taking so much money from his friend but seems uninterested in considering the matter much further. Guildenstern couldn't care less about the money but is clearly disturbed by the implications.

> GUIL: It must be indicative of something, besides the redistribution of wealth. (*He muses.*) List of possible explanations. One: I am willing it. Inside where nothing shows, I am the essence of a man spinning double-headed coins, and betting against himself in private atonement for an unremembered past. (*He spins a coin at Ros.*)
>
> ROS: Heads.
>
> GUIL: Two: Time has stopped dead, and the single experience of one coin being spun has been repeated ninety times. (*He flips a coin and tosses it to Ros.*) On the whole doubtful. Three: divine intervention . . . Four: a spectacular vindication of the principle that each individual coin spun individually (*he spins one*) is as likely to come down heads as tails and therefore should cause no surprise each individual time it does. (*It does. He tosses it to Ros.*) (16)[11]

In between flips, Rosencrantz and Guildenstern scour their essentially nonexistent memories for scraps of information that might help them determine what they are about. Rosencrantz is the instinctive, emotional member of the pair—the tail, so to speak. Guildenstern is the more cerebral one. Conveniently, mathematics is universal, present in a world devoid of empirical experiences, and Guildenstern is enough of a mathematician to sense that a run of ninety heads is highly suspect in any conception of nature. He is also enough of a mathematician to instinctively know some Aristotelian logic, and

"It must be indicative of something, besides the redistribution of wealth."
Rosencrantz and Guildenstern Are Dead; Simon Russell Beale (Guildenstern),
Adrian Scarborough (Rosencrantz); National Theatre, 1995. © Donald
Cooper/photostage.co.uk.

he begins organizing his arguments into the form of the logical syllogisms that
Aristotle championed.

> GUIL: One, probability is a factor which operates within natural forces.
> Two, probability is not operating as a factor. Three, we are now within
> un–, sub– or supernatural forces. Discuss.

But the heady Guildenstern is not done yet. Moments later, he attempts to
turn his own logic back on itself.

GUIL: If we postulate, and we just have, that within un–, sub– or supernatural forces the probability is that the law of probability will not operate as a factor, then we must accept that the probability of the first part will not operate as a factor, in which case the law of probability will operate as a factor within un–, sub– or supernatural forces. And since it obviously hasn't been doing so, we can take it that we are not held within un–, sub– or supernatural forces after all; in all probability, that is. (17)

The coin has one other important function to fulfill. In addition to setting the existential tone in its refusal to obey the law of averages, it also points to the symbiotic relationship between Stoppard's script and the script of *Hamlet*. After muddling about on their own at the beginning of the play, Rosencrantz and Guildenstern encounter a ragged group of traveling actors on their way to Elsinore. These turn out to be the players that Hamlet recruits to perform for Claudius in order to "catch the conscience of the king." The Player is the shifty spokesperson for the troop, and he does his best to tempt Rosencrantz and Guildenstern with a private performance of their own choosing. "We do on stage the things that are supposed to happen off," he says suggestively, "which is a kind of integrity, if you look on every exit being an entrance somewhere else." (28) On cue, the coin does finally come up tails—at precisely the moment when Hamlet and Ophelia swoon on stage and the action is taken over by *Hamlet*, act II, scene 2.

Stoppard manages to sustain this complementary relationship between his play and Shakespeare's throughout the evening. It is as though there is a full production of *Hamlet* taking place just offstage. When they are required, Rosencrantz and Guildenstern are swept into the action. They earnestly recite their scant few lines and just as abruptly are left alone again to pass the time and ponder their predicament. The comic opportunities are substantial, but there is still the question of whether this arrangement can support the weight of a full-length play in three acts. Recounting the story of the original performance of *Rosencrantz and Guildenstern Are Dead*, Stoppard explained that, in point of fact, the whole production nearly collapsed before the curtain even went up.

Clean What Afflicts Him

The Edinburgh Fringe Festival is an open access showcase for the performing arts that has been running annually since 1947. It was spontaneously created as an offshoot of the established Edinburgh International Festival when a handful of uninvited theater companies crashed the more formal festival, taking advantage of the large crowds who had gathered by performing in smaller venues around the "fringe" of the city. By 1966, the Fringe was more coordinated and growing quickly, but it still featured unvetted and innovative work performed in nontraditional venues. Eager to get *R&G* on its feet, Stoppard gladly consented to letting a reputable student group mount a production as part of the Fringe at the Cranston Street Hall. When Stoppard arrived to sit in on the last stretch of rehearsals, he encountered a show in disarray. Here is how Stoppard described what he found, in a program note he wrote fifteen years later:

> I had arrived in Edinburgh a few days previously to be shown the fruits of rehearsal. The Oxford Theatre Group had been laboring under certain disadvantages. The director had abandoned ship before we had left port. The actors were using scripts typed by somebody who knew somebody who could type. And the first thing that struck me was that there were a few unfamiliar cadences and some curious repetitions in the text they were using. . . . It turned out that such was the Oxford Theatre Group's touching faith in my play that they were faithfully rehearsing the typographical errors. The authentic Shakespearean phrase 'Glean what afflicts him' was coming out as 'Clean what afflicts him'. So we stopped and tidied all that up.[12]

One reason that Stoppard gives for remaining so sanguine throughout the chaos was that his novel, *Lord Malquist and Mr. Moon*, was due out from the publisher in the same week that *R&G* was to open. As odd as it sounds to say it now, Stoppard was confidently banking on the book to establish his writing credentials with the larger public.

Reports of the size of the opening night audience of *Rosencrantz and Guildenstern Are Dead*, which occurred on Wednesday, August 24, 1966, range from a handful to a couple of dozen. Stoppard recalls their response was more affable than awed. The following Sunday, when he boarded his train back to London carrying a copy of the *Observer* under his arm, Stoppard still had no idea his life as a playwright was about to radically change. A number of the local reviewers who reported on the show were skeptical about whether *R&G* was much more than a clever sketch that had gone on for too long, but an influential critic named Robert Bryden who knew something of Stoppard's BBC work had been among the sparse early audiences and recognized the full potential of both the play and its author. "Behind the fantastic comedy," Bryden wrote in his review,

> you feel allegoric purposes move: is this our relation to our century, to the idea of death, to war? But while the tragedy unfurls in this comic looking-glass, you're too busy with its stream of ironic invention, metaphysical jokes and linguistic acrobatics to pursue them. Like *Love's Labour's Lost* this is erudite comedy, punning, far-fetched, leaping from depth to dizziness. It's the most brilliant debut by a young playwright since John Arden.[13]

The last line in the review was also the headline in the *Observer* that Stoppard saw next to his picture when he finally opened the paper on the train. When he arrived back in London he was greeted by a telegram from Kenneth Tynan at the National Theatre inquiring about the play.

A polished production of *Rosencrantz and Guildenstern Are Dead* played to sold-out houses at the National Theatre starting in April 1967. It opened on Broadway in October 1967, where it earned a sizable haul of commendations, and soon after, translated versions started appearing across the major cities of Europe.

Newton's Apple or Eve's?

While *R&G* effectively launched Stoppard's career, it is overstating matters to say it represents a popular breakthrough for mathematics on stage. The nods to mathematics in *R&G* are modest and safely ensconced in the bantering word play. Guildenstern has a mathematician's sensibilities, leaning on his analytical skills to make sense of the world he inhabits. Rosencrantz, meanwhile, is the empirical philosopher, content to draw his conclusions from his experiences, however scant those may be.

Much of the comedy in *R&G* stems from the classical arrangement of having the audience know more than the actors on stage. Our knowledge of the plot of *Hamlet* gives us a vantage point from which to enjoy Rosencrantz and Guildenstern's metaphysical struggles without feeling any threat to our own existential security. But for *R&G* to work as allegory—to borrow Bryden's word—that security needs to be eroded. *R&G*'s darker overtones become more audible as Stoppard aligns Rosencrantz and Guildenstern's situation with our own by granting them just enough freedom to wonder at their predicament but not enough to see it for what it is. "Intrigued without ever quite being enlightened," is how Guildenstern summarizes it. (41) The script of *Hamlet* that spawned the two Shakespearean extras looms omnipotently

"Intrigued without ever quite being enlightened." *Rosencrantz and Guildenstern Are Dead*; Simon Russell Beale (Guildenstern), Adrian Scarborough (Rosencrantz), Paul Rattigan (Hamlet), Claudie Blakley (Ophelia); National Theatre, 1995. © Donald Cooper/photostage.co.uk.

over their hollow lives, but in their copious offstage time Stoppard manages to turn Rosencrantz and Guildenstern into flesh and blood seekers of truth whose fears and uncertainties sound more and more familiar as the evening wears on.

The major contribution of mathematics to establishing this uncertainty is in the mischievous coins from the opening scene, but when he was offered the chance to adapt *R&G* into a screenplay, Stoppard saw that there were more untapped dramatic possibilities lying in wait at the scientific end of the intellectual spectrum. In the film, which Stoppard wrote and directed, Rosencrantz's empirical disposition brings him face to face with a number of the laws that govern the physical world. En route to his predestined demise, the cinematic version of Rosencrantz almost manages to discover Galileo's principle of falling objects, Newton's principle of the conservation of momentum, and Archimedes's law of floating bodies. With each of these encounters, the film draws attention to the natural forces at play in the universe, implicitly asking whether the future is controlled by providence or probability. With each fumble by Rosencrantz, the answer gets more obscure. When an apple falls on Rosencrantz's head, he has a fleeting vision of the law of universal gravitation, an idea that will one day usher Western civilization into the Enlightenment. He stares at the apple:

> ROS: I say . . .
> GUIL: What?
> ROS: Well . . .[14]

But no. Rosencrantz abandons the thought and instead takes a bite out of his apple. In the blink of an eye, we have gone from Newton under the tree to Adam in the garden, whimsically illustrating the tug-of-war between science and religion as competing systems for making sense of the human condition.

A Play-within-Itself

The extensive critical attention directed at Stoppard's work has produced an array of fascinating interpretations of R&G as an existential, allegorical, absurd, and even postmodern play. Endorsing this diversity of opinion, Stoppard scholar and theater historian John Fleming compares R&G to Shakespeare for the way that "critics often find what they bring to it; their own values are reflected back." Eschewing labels, Fleming celebrates R&G as "contradictory and expansive, [a play] that raises as many questions as it offers tentative answers."[15] An innovative way that Stoppard's play achieves this universal sense of relevance is by harnessing the full potential of the gift Shakespeare bequeathed with the traveling players' production of the *Murder of Gonzago*. Although not so obvious, there is a mathematical aspect to this particular component of R&G which is central to the play's "contradictory and expansive" aura.

The way it is arranged in Shakespeare's play, Hamlet requests some edits to *Gonzago* that essentially transform it into a portrayal of Claudius's murder of Hamlet's father. Not content with the single scene Shakespeare provided, Stoppard has Rosencrantz and Guildenstern stumble onto a dress rehearsal of the troupe's *Gonzago* which Stoppard extends into a full rendition of the story of *Hamlet*. Among other things, this leads to the provocative moment of Rosencrantz and Guildenstern unknowingly witnessing a portrayal of their own executions at the hands of the English king.

In the history of mathematics, the reflexive arrangement of having some object (e.g., a set, a function, a logical sentence) contain or refer to itself has led to powerful new constructions and as many controversies. R&G represents an intriguing artistic translation of this phenomenon. While the play-within-a-play device is a familiar theatrical trope, the more distinctive play-within-*itself* structure that *R&G* inherits from *Hamlet* is what enables Stoppard's play to toggle back and forth between a music hall slapstick and an exegesis on the human condition. For his part, Stoppard is adamant that he was intent on writing a comedy.[16] Taking the playwright at his word, it is still evident that the logically attuned Stoppard was enticed by the introspective overtones that emerged as he moved the focus of his story from England back to Elsinore where the Player and his tragedians resided. By the time he makes *R&G* into a movie, Stoppard is acutely aware of the paradoxical echoes that result from embedding a copy of a play inside the original. To highlight this in the film version, Stoppard adds in a puppet show of *Gonzago* as part of the troupe's rendition of *Hamlet* happening within Stoppard's *R&G* which is conjoined with Shakespeare's *Hamlet*. All of this is, of course, taking place in front of an audience, who, judging from the fact that Rosencrantz and Guildenstern are caught unaware in this strange loop, should not assume that the recursive levels end with the theater in which they sit.

Jumpers

If rationality were the criterion for things being allowed to exist, the world would be one gigantic field of soya beans!

—GEORGE, FROM *JUMPERS*

The mathematics in *R&G* is significant but subtle, largely appearing in disguise in the form of statistical anomalies and Guildenstern's Lewis Carroll-like musings. On this same point, most audience members would not associate the self-referencing structure of *R&G* with being explicitly mathematical. Stoppard probably didn't either. The mathematics in *R&G* is not so much a conscious artistic decision as it is a by-product of Stoppard's logical instincts. In terms of mathematics, logical instincts and some algebra were all the high school–educated playwright had to go on at this early point in his career, but this was about to change. In 1968, Stoppard wrote to a friend about his ongoing self-education:

I'm on a ridiculous philosophy/logic/maths kick. I don't know how I got into it, but you should see me trying to work out integral calculus with one hand, while following Wittgenstein through 'Tractatus Logico-philosophicus' with the other. I shall end up writing an unsatisfactory play by preparing just enough ground to reveal the virgin and impenetrable tract(atus).[17]

As Stoppard progressed further into the intellectual weeds, an opportunity appeared for bringing mathematical insights to bear on human ones, albeit in a backhanded way. Ludwig Wittgenstein is a prime example of a philosopher who attempted to export the tools of formal logic from the mathematical realm, where they had proved inordinately effective, to other fields of inquiry. Stoppard, like Wittgenstein, became fixated on the tension between the clarity that a strictly logical approach to knowledge offers in the abstract and the largely unsatisfying results it yields when applied to issues like language and morality. This is the fulcrum on which Stoppard perched *Jumpers*. His next full-length play after *R&G*, *Jumpers* opened in February 1972 at the National Theatre in London to great fanfare. It is a philosophy play, not a math play; there is more Wittgenstein in the script than there is integral calculus. Logical reasoning, however, is a significant motif, and the roots of calculus in the form of Zeno's paradoxes play a pivotal role.

The title of *Jumpers* refers to its acrobats. As Stoppard was piecing together the intellectual arguments of the play, he had a vision of a large human pyramid. In his imagination, one of the acrobats on the bottom row gets shot, bringing the whole structure tumbling down. "I had this piece of paper with this dead acrobat on the floor and I didn't know who he was, who shot him or why."[18] In search of a metaphor for his collapsing pyramid, Stoppard combined his skepticism of higher education with his newfound infatuation with philosophy. The result, many drafts later, was an elaborate staging of a philosophical disputation in which mathematics is invoked in support of a position one might not anticipate: the existence of God.

Mental Gymnastics

Jumpers is centered, literally and figuratively, around an academic debate. In one corner is George Moore, professor of moral philosophy. Throughout the play, George has a number of extended monologues during which he dictates a speech he is scheduled to deliver at an upcoming symposium titled "Man: Good, Bad, or Indifferent." The central topic of the debate is the question of the existence of moral absolutes, with George taking the affirmative. This antiquated position puts George squarely at odds with the prevailing empiricist movement, which requires that statements be experimentally verifiable to be meaningful. George's adversary, and overall nemesis, is the university vice-chancellor Sir Archibald ("Archie") Jumper.[19] For Archie and his followers, the moral code that undergirds a functioning society does not originate from a benevolent creator, nor does it enjoy any privileged universal status. Instead, notions of ethical behavior are the result of practical norms and therefore subject to change. Assessing this point of view with equal parts contempt and fascination, George summarizes his adversaries's position by saying that in their minds "good and bad aren't actually *good* and *bad* in any absolute or metaphysical sense, . . . [they are] categories of our own making, social and psychological conventions which we have evolved in order to make living in groups a practical possibility, in much the same way as we have evolved the rules of tennis without which Wimbledon Fortnight would be a complete shambles." George concedes that empiricists like Archie are not advocating that murder and lying become commonplace, as anarchy would likely ensue, but it does allow them "to conclude that telling lies is not *sinful* but simply anti-social." (48)[20]

To turn this philosophical debate into a play, Stoppard surrounds George with an array of ethically ambiguous scenarios. The curtain rises on a raucous party taking place in George and Dotty's apartment. The celebration is for the recent victory of the Radical Liberal party, a political incarnation of the morally relativistic tenants of Archie's logical positivist worldview. Archie is the party's MC, Dotty its songstress, but the main attraction of the evening is a performance by the "INCREDIBLE—RADICAL!—LIBERAL!!—JUMPERS!!" Consisting generally of the members of the philosophy department (without George, of course), the Jumpers' show ends tragically when an unseen gunshot takes out Duncan McFee, professor of logic and the colleague George is scheduled to debate at the upcoming symposium. By morning, Dotty has concealed the body by hanging it on the back of her bedroom door. She may or may not have fired the shot. She also may or may not be having an affair with Archie who, in addition to being vice-chancellor and gym coach, is also a coroner, lawyer, and Dotty's psychiatrist.

> GEORGE: (*reckless, committed*) I can put two and two together, you know. Putting two and two together is my *subject*. I do not leap to hasty conclusions. I do not deal in suspicion and wild surmise . . . Now let us see. What can we make of it all? Wife in bed, daily visits by gentleman caller. Does anything suggest itself?
> DOTTY: (*calmly*) Sounds to me he's the doctor. (32)

"Somewhere there is a domino which was *nudged*." *Jumpers*;
Michael Hordern (George); Old Vic Theatre, 1972. © Donald
Cooper/photostage.co.uk.

This exchange illustrates the central asset that theater brings to a debate about whether knowledge should be restricted to what can be empirically confirmed. Over and over again in *Jumpers*, Stoppard allows his audience to experience firsthand how the same collection of observations can be explained by a variety of different realities. A compelling anecdote recounted in the play makes clear how fundamental this phenomenon is in any clear-minded pursuit of truth:

GEORGE: (*facing away, out front, emotionless*) Meeting a friend in a corridor, Wittgenstein said: 'Tell me, why do people always say it was *natural* for men to assume that the sun went round the earth rather than the earth was rotating?' His friend said, 'Well, obviously, because it just *looks* as if the sun is going around the earth.' To which the philosopher replied,

'Well, what would it have looked like if it had looked as if the earth was rotating?' (75)[21]

Saint Sebastian Died of Fright

Stoppard's budding interest in higher mathematics enters the play through George's long monologues in which he dictates his symposium lecture to his secretary. George has an ambitious agenda, and Stoppard does too. George's intention is to establish a universal moral compass by arguing for the existence of a benevolent God; Stoppard's job is to find the right balance between philosophy and farce so that the academic lecture does not actually feel like one but the play doesn't collapse into frivolity.

Postponing his argument for benevolence, George's initial goal is to make the case for the existence of some kind of cosmic deity in general. "Does, for the sake of argument, God, so to speak, exist? . . ."

> GEORGE: We see that a supernatural or divine origin is the logical consequence of the assumption that one thing leads to another, and that this series must have had a first term; that, if you like, though chickens and eggs may alternate back through the millennia, ultimately, we arrive at something which, while perhaps no longer resembling either a chicken or an egg, is nevertheless the first term of that series and can itself only be attributed to a First Cause—or to give it its theological soubriquet, God. (27)

This line of reasoning puts George on sound theological ground alongside St. Thomas Aquinas, who in George's words argued for the existence of God based on "the simple idea that if an apparently endless line of dominoes is knocking itself over one by one then somewhere there is a domino which was *nudged*." (29) In anticipation of counter arguments, George brings mathematics more explicitly into the story by invoking its extensive experience negotiating with the infinite.

> GEORGE: Mathematicians are quick to point out that they are familiar with many series which have no first term—such as the series of proper fractions between naught and one. What, they ask is the first, that is the smallest, of these fractions? A billionth? A trillionth? Obviously not: Cantor's proof that there is no greatest number ensures that there is no smallest fraction. There is no beginning. (27)

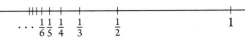

FIGURE 1.1. A sequence with no beginning.

The name George drops here is that of Georg Cantor, a late nineteenth-century mathematician whose contributions to the understanding of the infinite were nothing short of revolutionary. Cantor's ideas will garner their share of attention in discussions to come, but with all due respect to George, Cantor is not really required here. Long before Cantor, it was a commonly understood notion that the set of natural numbers—1, 2, 3, 4, 5, 6, . . .—has no largest element and increases without bound. This is all one needs to prove that the sequence of proper fractions

$$1, \frac{1}{2}, \frac{1}{3}, \frac{1}{4}, \frac{1}{5}, \frac{1}{6}, \dots$$

decreases indefinitely with no candidate for a smallest one. Looking at it from left to right along the number line (Figure 1.1), the collection of proper fractions has no beginning; and so, the logic goes, maybe the universe of falling dominos does not either.

Does this example deal a fatal blow to George's argument for God as First Cause? Not necessarily! George is prepared to defend himself—with props, no less. Continuing his dictation, George provocatively produces a quiver of arrows and a bow, and deftly notches an arrow in the string:

GEORGE: But it was precisely this notion of infinite series which in the sixth century BC led the Greek philosopher Zeno to conclude that since an arrow shot towards a target first had to cover half the distance, and then half the remainder, and then half the remainder after that, and so on *ad infinitum*, the result was, as I will now demonstrate, that though an arrow is always approaching its target, it never quite gets there, and Saint Sebastian died of fright. (28)

With the bow taut, George is suddenly startled by his wife's cry for help and proceeds to fire the arrow over the wardrobe in his study.

Zeno of Elea

Historical knowledge of Zeno is scant. He appears in one of Plato's dialogues as a student and defender of Parmenides, which is essentially how Zeno has come to be viewed. Parmenides was a philosopher who argued

"And Saint Sebastian died of fright." *Jumpers*; Michael Hordern (George);
National Theatre, London, 1976. © Donald Cooper/photostage.co.uk.

for the unorthodox idea that all of reality consisted of a single changeless
unity. This position is significantly at odds with the plurality of distinct things
that common observation would naturally suggest. In defense of Parmenides,
Zeno reportedly wrote a book of paradoxes designed to discredit the senses by
demonstrating logical contradictions that arise from sensory observation such
as physical motion. The book did not survive, but references to a handful of
these paradoxes appear in later texts, most notably in Aristotle's *Physics*.

Keeping in mind that everything we know about Zeno is secondhand,
Stoppard, through the character of George, is participating in what is a long

tradition of adapting Zeno to his particular purpose. That said, some moderate tidying up of the details is probably in order. One of Zeno's paradoxes is indeed about an arrow, but it says something a bit different. Zeno argues that at a fixed instant in time, a flying arrow occupies an amount of space exactly equal to its shape and size. In other words, the arrow is not moving. Stepping back and noting that any interval of time is composed of a collection of such instances, it follows that at no time is the arrow in motion and thus it cannot in actuality progress from the bow to its target.

Another of Zeno's paradoxes argues for the impossibility of motion along the lines that George describes.[22] George's conflation of these two paradoxes causes no real harm, but it is worth noting that Zeno is being more thorough than George gives him credit for by considering the problem of infinite divisibility of both space and time. Is a line segment a collection of points? Is a time interval a collection of instances? In either case, Zeno points to a logical contradiction that arises from uncritically accepting the evidence of our eyes.

In citing Zeno, George's agenda is different from the one Zeno originally intended. Zeno attempted a radical attack on empiricism, an agenda to which George is generally sympathetic. But at the moment, George is laden with proving the existence of God, and standing in the way of his First Cause argument is his unease with Zeno's infinities. George's other problem is the chaos taking place in his apartment that routinely scrambles his already addled train of thought. "Everything has got to begin somewhere," he dictates to his secretary over the hum of the political celebration going on outside,

> and there is no answer to *that*. Except, of course, why does it? Why, since we accept the notion of infinity without end, should we not accept the logically identical notion of infinity without beginning? My old— Consider the series of proper fractions. Etcetera. (*To Secretary.*) Then Cantor, then no beginning, etcetera, then Zeno. Insert: But the fact is, the first term of the series is not an infinite fraction but *zero*. It exists. God, so to speak, is nought. Interesting. Continue. (29)

Aha! God is zero, if only George can get there along his sequence of decreasing fractions. Thus, all George has to do is show that Zeno erred and that it *is* possible to complete an infinite number of tasks in a finite amount of time. Putting his own particular dramatic spin on another of Zeno's paradoxes, George pulls out two small animal cages, explaining that, just as Zeno had argued that the arrow could never reach its target, he similarly asserted that a "tortoise given a head start in a race with, say, a hare, could never be overtaken." (29)

The tortoise and the hare are borrowed from the race in Aesop's famous fable. In Zeno's formulation it is Achilles, the fleetest of all mortals, who gives a head start to the tortoise. The reason Achilles cannot overtake the tortoise, Zeno says, is because each time Achilles reaches the place where the tortoise was, the tortoise has meanwhile moved ahead ever so slightly, and this process repeats itself ad infinitum. Substituting a rabbit for Achilles makes the point just as well—better in fact, because once again George is going to reveal the fallacy of Zeno's argument with a live demonstration. But just as his archery demo went awry, this one does too. Sadly for philosophy, Thumper the hare has escaped from his cage and is nowhere to be found.

More than Meets the Microscope

Throughout the play, the jokes and the philosophy swirl around the mystery of who killed Professor Duncan McFee. Dottie is a suspect, as is Archie, especially when it is revealed that McFee planned to resign his position as professor of logic and join a monastery. Even George's secretary, who never speaks, is dragged into the zone of suspicion as McFee's disgruntled mistress. Stoppard does not provide a definitive answer, which is wholly appropriate. Rather than a whodunit, *Jumpers* is an exploration of the limits of rational certainty in human affairs, and mathematics is invoked as a point of contrast to George's persistent befuddlement, often with language itself.

GEORGE: Do I say 'My friend the late Bertrand Russell' or 'My late friend Bertrand Russell'? They both sound funny.
DOTTY: Probably because he wasn't your friend.

While Russell does not appear, Stoppard contrives events so that this iconic mathematician's spirit is hovering just offstage.[23] The three large volumes of *Principia Mathematica* that Russell cowrote in an attempt to perfect the language of mathematical certainty cast a daunting shadow on George's struggle with his own language's imprecision with words like "good" and "bad." After two acts of trying to logically infer a benevolent God from self-evident principles while murders, infidelity, and self-interest encroach from all sides, George the moral philosopher abandons reasoned debate. "All I know is that I think that I know that I know that nothing can be created out of nothing," George confesses in desperation, "that my moral conscience is different from the rules of my tribe, and that there is more in me than meets the microscope." (68)

George's instinctive faith in an altruistic universe comes off as the more sympathetic alternative to Archie's overzealous rationalism, but even if it is closer to what Stoppard believes, that doesn't mean George emerges victorious in any way.[24] Quite the opposite, in fact. The portrait Stoppard paints of the future is bleak in its lack of inspiration: "More eat than starve, more are healthy than sick, and one of the thieves was saved" is Archie's pragmatic assessment. (87)[25] Meanwhile, George ends the play racked with grief, although not for his murdered colleague. Spotting blood on the back of his secretary's coat as she exits in the final scene, George peers over the wardrobe in his study to find his beloved Thumper fatally impaled by Zeno's errant arrow from the first act. Undone, he staggers backward and crushes his tortoise under the weight of his boot.

Hapgood

Who needed God when everything worked like billiard balls?

—KERNER, FROM *HAPGOOD*

The success of *Jumpers* enhanced Stoppard's reputation as an exceedingly clever and acrobatic writer. Over the next decade, that reputation continued to grow. His next major stage play after *Jumpers* was *Travesties*, another comic high-wire act inspired by the historical fact that Vladimir Lenin, James Joyce, and Tristan Tzara were all simultaneously in Zürich at the conclusion of World War I. Mathematics is one of the few subjects that does not make an appearance in the pastiche of *Travesties*, but it does resurface several times—in a metaphorical way—as part of Stoppard's evolving interest in writing about the political realities of the '70s and '80s. The 1977 play *Every Good Boy Deserves Favour* began as a riff on mathematics and music before being pulled in a political direction. Invited by renowned composer André Previn to cocreate a play that encompassed a full orchestra, the eager playwright was initially flummoxed by his complete lack of musical experience. Stoppard's solution was to create a protagonist who is a lunatic triangle player with delusions of being part of an orchestra. "Music and triangles led me into a punning diversion based on Euclid's axioms," Stoppard noted, "but it didn't belong anywhere."[26] Instead, current events compelled him to situate his delusional percussionist in a Russian mental institution with a cellmate who is a political prisoner. The original Euclidean inspiration survives, however, in a few scenes involving Sacha, the political prisoner's son. In one of them, Sacha is asked to read from his geometry book while he gets a lesson in Soviet mental health policy:

SACHA: Will I be sent to the lunatics prison?

TEACHER: Certainly not. Read aloud.

SACHA: 'A point has a position but no dimension.'

TEACHER: The asylum is for malcontents who don't know what they're doing.

SACHA: 'A line has length but no breadth.'

TEACHER: They know what they're doing but they don't know it's anti-social.

SACHA: 'A straight line is the shortest distance between two points.'

TEACHER: They know it's anti-social but they are fanatics. . . .

SACHA: 'A polygon is a plane area bounded by straight lines.'

TEACHER: And it's not a prison, it's a hospital.[27]

Like the tonal answer in a two-voice fugue, geometry serves as a subversive counterpoint to Communist orthodoxy in this dialogue, which eventually concludes with Sacha's retort that "a plane area bordered by high walls is a prison, not a hospital."[28]

Stoppard invokes geometry for a related purpose in *Squaring the Circle*, a 1984 docudrama for television about Lech Walesa and the Solidarity movement in Poland. In the opening scene the narrator explains:

NARRATOR: Between August 1980 and December 1981, an attempt was made in Poland to put together two ideas that wouldn't fit, the idea of freedom as it is understood in the West, and the idea of socialism as it is understood in the Soviet empire. The attempt failed because it was impossible, in the same sense as it is impossible in geometry to turn a circle into a square with the same area—not because no one has found out how to do it but because there is no way in which it can be done.[29]

A catalyst for Stoppard's attention to political oppression, especially in Eastern Europe, was the plight of fellow playwright Václav Havel. An internationally acclaimed writer, Havel was at the forefront of human rights protests in Czechoslovakia, the country of Stoppard's birth. Havel was arrested and jailed multiple times in the '70s and '80s, and when Communism finally collapsed in 1989 he was elected president of Czechoslovakia. Stoppard was not only a strong public advocate for Havel during his various imprisonments, but the two artists eventually formed a meaningful friendship. Stoppard's 1977 television play *Professional Foul* is dedicated to Havel.

The modest nods to geometry in these political dramas notwithstanding, Stoppard's ambition to find a play about mathematics remained largely on hold for the decade and a half after *Jumpers*, during which time he continued to demonstrate his creative dexterity in other ways. Stoppard's bawdy political farce *Dirty Linen* set attendance records on the West End in the mid-70s, and his liberal adaptation of Johann Nestroy's *Einen Jux will er sich Machen*, translated as *On the Razzle*, was also enormously popular. In the early '80s, Stoppard accrued some notable screenplay credits, including a shared byline on the highly acclaimed dystopian comedy *Brazil* and another on Steven Spielberg's *Empire of the Sun*. When Stoppard was criticized it was usually along the lines of being too popular to be regarded as an important artist or too cerebral to create truly compassionate characters. Whether or not it was a conscious effort, Stoppard always seemed to have an answer for his critics. When it was pointed out that his female characters were lacking in depth, he created journalist Ruth Carson in *Night and Day*, a play that dives into the politically thorny issues surrounding freedom of the press. Accused of being unemotional, Stoppard answered with a modern love story called *The Real Thing* that was enormously popular with audiences and critics.

The Real Thing premiered in London in 1982. It would be six years before Stoppard came out with his next major stage play, and it was with this new script that Stoppard finally discovered a mathematical hook he felt was sturdy enough to support the weight of a full play. When Stoppard said that "Euclid's axioms . . . didn't belong anywhere" with respect to *Every Good Boy Deserves Favour*, this was a tacit acknowledgment that the austerity of classical geometry had proved too rigid to be shaped into something poetically useful, but when his mathematical education reached the twentieth century, Stoppard finally found what he was looking for:

> I was aware that for centuries mathematics was considered the queen of the sciences because it claimed certainty. It was grounded on some fundamental certainties—axioms—which led to others. But then, in a sense, it all started going wrong . . . The mathematics of physics turned out to be grounded on *un*certainties, on probability and chance. And if you're me you think—there's a play in that.[30]

After a twenty-year flirtation, Stoppard was ready to cast mathematics—or mathematical science at least—in a leading role. The question was whether the audience was ready.

Homework Lying around the Stage

Hapgood opened at the Aldwych Theatre in London on March 8, 1988. On March 9, the headline in the *Times* review read, "Going Nowhere Fast." Making an analogy with the launch of a new British sports car that is all bells and whistles but short on performance, reviewer Irving Wardle summarized *Hapgood* as "a spy thriller of a complexity that reduces Len Deighton and John le Carré to the narrative simplicity of Little Red Riding Hood." Other overnight reviewers used phrases like "convoluted," "user-hostile," and "thoroughly incomprehensible," asserting that "it would need a seeing eye dog with A-level physics to guide most of us through what was going on." "Stoppard leaves his homework lying around the stage," said another. One particularly disgruntled critic went so far as to proclaim that "with *Hapgood*, Mr. Stoppard is signaling his intention to give up any pretence of being a serious playwright."[31]

Apparently, Stoppard had overstepped his bounds. A coin violating the law of averages is provocative; trained pets reenacting Zeno's paradoxes has a certain charm. But in *Hapgood*, a line had been crossed. Long monologues in *Jumpers* contrasting idealism with logical positivism were embraced as erudite comedy. Political speeches about art lifted verbatim from Lenin's collective writings were acknowledged as bringing gravitas to *Travesties*. This time, however, Stoppard asked his reviewers to wade in more technical waters, and they were having none of it. In his *Times* review Wardle offered a clue for why this unwillingness bordered on hostility. "Whereas in his earlier plays he scored by flattering the audience into feeling intelligent," Wardle wrote, "this time he makes them feel stupid."

On the surface, *Hapgood* looks like familiar entertainment. The narrative form fits into the established spy-thriller genre, pitting British intelligence against Russian KGB in a game of Cold War espionage. Central to the story is a Russian scientist named Kerner. Originally sent to England as a spy by the KGB, Kerner has since become a double agent working for British intelligence, only now there is suspicion that he may be secretly leaking his defense-related research back to Moscow. When confronted for the truth, Kerner responds with a lesson in quantum mechanics that is meant to be the central metaphor of the play. "A double agent is more like a trick of the light," Kerner says to British agent Paul Blair, and launches into a description of a famous experiment first carried out in 1801 by Thomas Young. (9)[32]

"The experimenter makes the choice. You get what you interrogate for."
Hapgood; Roger Rees (Kerner), Nigel Hawthorne (Blair); Aldwych Theatre, 1988. © Donald Cooper/photostage.co.uk.

The Double-Slit Experiment

In published editions of *Hapgood*—and there are several different versions—Stoppard includes the following quotation by Nobel laureate Richard Feynman as a preamble to the text:

> We choose to examine a phenomenon which is impossible, *absolutely* impossible, to explain in any classical way, and which has in it the heart of quantum mechanics. In reality it contains the *only* mystery . . . Any other situation in quantum mechanics, it turns out, can always be explained by saying, 'You remember the case of the experiment with the two holes? It's the same thing.'[33]

Feynman's acknowledgment of the difficulty in making sense of the quantum world is reason to be sympathetic to reviewers who found Stoppard's play baffling. The behavior of nature at the atomic level is not at all like the classical picture Newton painted where apples fall deterministically from trees and planets fall deterministically around the sun. The world of the atom, for those who have not experienced it, is a place where the conclusions of hard science outpace the imagination.

At stake in the "experiment with the two holes" is the question of how light propagates through space; specifically, whether it does so (i) as a particle, like a stream of tiny bullets, or (ii) as a wave, like a ripple across the water or sound through the air. A phenomenon that unambiguously distinguishes waves from particles is *interference*. A consideration of waves on the surface of water is sufficient to make the point. Waves have a peak (a high point) and a trough (a low point). When two traveling waves of equal amplitude cross paths and overlap, the net effect one sees at each moment is the addition of the respective amplitudes of the two overlapping waves. If the peaks of the two waves line up, they make a peak that is twice as high. In the same way, if a peak lines up with a trough, the net effect is that the two cancel out and the surface of the water is perfectly flat. The term for this place where a peak and a trough cancel is called a *node*. A node for water waves is a flat spot; a node for sound waves is a quiet spot; a node for light waves is a dark spot.

Interference nodes are at the core of what Young's double-slit experiment revealed. By shining a monochromatic light through two narrowly spaced holes, Young created two perfectly synchronized light waves emanating out in semicircular arcs on the other side (Figure 1.2). The resulting image on a screen opposite the two holes is an unmistakable pattern of alternating nodes and bright spots. The bright spots appear where the waves arrive in phase, with peaks, lining up with peaks, while the dark nodes appear at places on the screen where the waves arrive out of phase, with peaks lining up with troughs.[34] Details aside, the crucial point is that this interference pattern of alternating light and dark patches can only be explained by thinking of light as a wave. If one of the holes is covered up then the alternating pattern disappears. There is no interference taking place with the light coming through only one hole, and the pattern on the screen becomes an uninterrupted swath of light, brighter in the middle and gradually fading out toward the edges[35] (Figure 1.3). The unambiguous conclusion of this experiment is that light must propagate as a wave. With two holes open, the pattern we see can only be explained by interference, and interference is a phenomenon exclusively associated with waves.

Where then is the mystery that Feynman refers to?

The mystery is that one can show just as conclusively that light propagates as a discrete stream of particles called *photons*. In *Hapgood*, Kerner uses the more audience-friendly term "little bullets." In fact, it is not hard to arrange Young's experiment to see these individual photons. Harking back to our discussions of Zeno's paradoxes, it turns out that light does not propagate as a

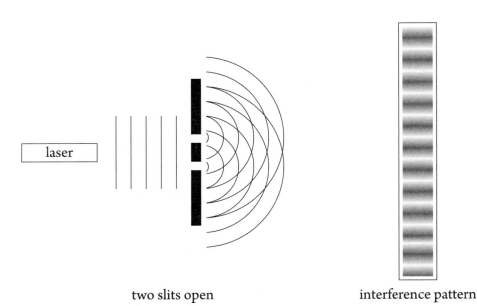

laser

two slits open   interference pattern

FIGURE 1.2. When monochromatic light passes through the two slits, it
emanates out the other side in two synchronized semicircular waves that
generate a striped interference pattern on the screen.

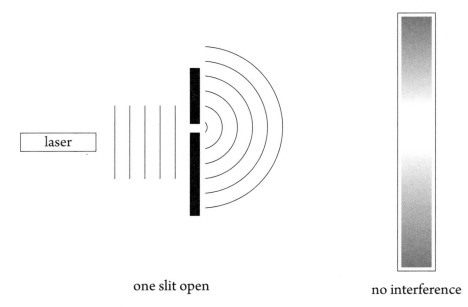

laser

one slit open   no interference

FIGURE 1.3. When the light passes through only one slit, the interference pattern
on the screen is replaced by a single swath of light that is brightest in the middle
and fades toward the edges.

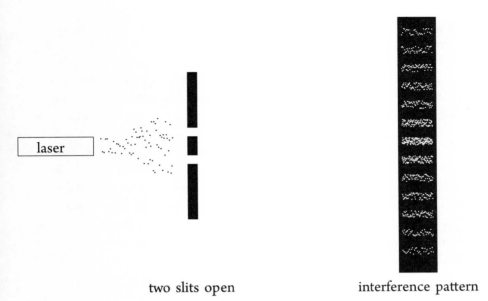

two slits open   interference pattern

FIGURE 1.4. Turning down the laser intensity so that the light arrives
on the screen one photon at a time still generates an interference pattern
when both slits are open.

continuous, infinitely divisible flow of energy but is actually more like sand
in the way that it becomes grainy at a small enough scale. Turning down
the intensity of the laser may at first appear to dim the light, but eventually,
when the intensity is low enough, it becomes apparent that the light is arriving in discrete, indivisible packets. One at a time, these photons arrive at the
screen, each possessing the exact same energy as the next. There is no such
thing as half a photon; lowering the intensity of the laser just spaces out the
time gaps between each photon's arrival. A useful image is to take the screen
where the light is absorbed to be a piece of film with each photon that hits
creating a small dot on the film. When the laser is at its normal intensity,
we do not notice the individual dots because there are a hundred trillion
photons arriving every tenth of a second. With the laser intensity extremely
low, the dots appear one at a time, and the film keeps a running record of
where the photons hit. Over a suitably long length of time, this generates
the same final image we see instantaneously when the laser intensity is high
(Figure 1.4).

But now Feynman's mystery is at hand. The image we see in the two-slit
experiment is the wave interference pattern—alternating spots of light and
dark—*regardless of whether the laser is at regular intensity or at its low, one*

photon at a time intensity. The pattern makes sense when we conceive of light as a wave, but to borrow Feynman's words, it is absolutely impossible to explain conceiving of light as a particle. Particles don't exhibit interference; i.e., they don't "cancel out." Recall that if only one hole is open there is no interference pattern. In this case, the dark nodes disappear and the light shows up everywhere across the screen. Another way to say this is that each individual photon is free to land anywhere on the screen, and indeed this is what happens. When *two* holes are opened, however, the nodes reappear. This means that there is no light at these particular points on the screen. In this case, an individual photon *never* winds up hitting the screen at one of these dark spots. But why not? Presumably, the photon must travel through one hole or the other, and as this is happening it seems nonsensical to assume the photon is somehow aware of whether the other hole is open or closed. But it matters. If the other hole is open, our photon is forbidden from hitting the screen at a node. If the other hole is closed, that restriction is rescinded. How does a photon passing through the slide "know" whether one hole or both holes are open and thus what trajectory it is allowed to take on the other side?

The problem here is that we are bringing a classical Newtonian mindset into the quantum world where the same rules do not apply. A Newtonian view is that light has to be either wave or particle, but a pillar of quantum mechanics is that light—and atomic particles more generally—manifest properties of both. Fully embracing this viewpoint is hard because it requires us to overhaul our conception of how reality works. Wave interference is the only way to explain the nodal pattern we see when two holes are open, but when we send a single photon through on its own, what does the photon interfere with? The poetic answer is actually much like the quantum mechanical answer—the photon interferes with itself; it goes through both holes.

One way to call quantum mechanics' bluff for proposing such a fantastical explanation is to watch the two holes and observe what happens as the photons pass through. Is it really the case that a single photon emanating from the laser appears in both slits as it passes? No. One might even say, of course not. There is only one photon, and it gets detected in only one of the two slits. But now quantum mechanics gets the last laugh. When the experiment is arranged this way, so that each photon is observed passing through one of the two holes, the interference pattern vanishes. The dark nodes disappear, and the photons are free again to land anywhere on the screen, just as though only one hole were open. Unobserved, the photons live a double life, passing through both slits and proceeding along a path that is a superposition of these two possibilities and exhibiting wave interference. Observed, however, the photon commits to precisely one of these two options and the wave interference pattern is no longer present.

This is the story Kerner tells Paul Blair in response to Blair's question about whether Kerner is working for Britain or the KGB, and Blair, like the audience, is understandably puzzled.

> BLAIR: Joseph. I want to know if you're ours or theirs, that's all.
> KERNER: I'm telling you but you're not listening. Now we come to the exciting part. We will watch the bullets to see how they make waves. This is not difficult, the apparatus is simple. So we look carefully and we see the bullets, one at a time. Some go through one gap and some go through the other gap. No problem. Now we come to my favourite bit. The wave pattern has disappeared. It has become particle pattern again.
> BLAIR: (*Obligingly*) All right—why?
> KERNER: Because we looked. Every time we don't look, we get wave pattern. Every time we look to see how we get wave pattern we get particle pattern. The act of observing determines what's what.
> BLAIR: How?
> KERNER: Nobody knows. Somehow light is continuous and also discontinuous. The experimenter makes the choice. You get what you interrogate for. (10)

This education in nuclear physics takes place early in *Hapgood*, and although it is employed here as a means of explaining the ambiguity of whether Kerner is a Russian "sleeper" or a British "joe," Stoppard's real agenda is to bring the physics to bear on the more intimate parts of human personality.

Elizabeth Hapgood is a tough and classy agent who plays chess without a board and runs an otherwise all-male British intelligence office. The Russian physicist Joseph Kerner was "turned" by Hapgood years earlier, and we learn that not only did she fall in love with her scientist but that he is also the father of her twelve-year-old son, Joe. To engage in Stoppard's language games, Joseph Kerner is Hapgood's British joe and the young Joe's missing parent. Hapgood's current feelings for Kerner include a great deal of trust but, from a romantic viewpoint, are not so clear. Whatever fondness she may still have for Kerner is obscured because she is currently in a somewhat strained relationship with Blair, who is her superior. Amid these personal dynamics is the question of who in Hapgood's professional circle is the mole smuggling classified information back to the KGB. Staying true to the spy genre—for the moment at least—Stoppard offers up one more potential suspect, an abrasive and somewhat crude fellow agent named Ridley. Midway through the first act, the suspicion around Ridley gains momentum when it is discovered that there are two of them.

The Bridges of Königsberg

Contributing to *Hapgood*'s negative reception might be that Stoppard inserts an additional mathematics lesson on top of the physics. Having already treated his audience to an introduction to quantum mechanics, Stoppard pushes his luck by revealing the existence of Ridley's twin via a famous problem in graph theory. Stoppard decides to have Kerner hail from Kaliningrad. Before being annexed by the Soviet Union, Kaliningrad was the Prussian city of Königsberg, and in Königsberg there were seven bridges spanning a confluence of two rivers (Figure 1.5).[36] "An ancient amusement of the people of Königsberg was to try to cross all the seven bridges without crossing any of them twice," Kerner tells Hapgood. "It looked possible but nobody had solved it." (38) As Kerner explains this bit of fabled mathematical history to Hapgood, he is in possession of a diagram of an intricately choreographed drop-off that took place in the men's locker room of a public gym facility earlier that week. The audience has seen this drop-off, in fact. It is the opening scene of the play. In that scene, Ridley circumnavigates in and out of view, revealing the various passageways leading to the showers, lockers, and pool. With the diagram in hand, Kerner deduces what the audience surely did not:

> KERNER: Leonard Euler took up the problem of the seven bridges and he presented his solution in the form of a general principle based on vertices . . . When I looked at Wates's diagram I saw that Euler had already done the proof. It was the bridges of Königsberg, only simpler.
> HAPGOOD: What did Euler prove?
> KERNER: It can't be done, you need two walkers. (39)

Euler's "general principle based on vertices" is clever in its simplicity. By replacing the map of the bridges with a graph like the one in figure 1.6, the problem becomes whether it is possible to trace every edge of the graph exactly once without lifting the pen. Euler's principle states that such a walk is possible exactly when the number of edges emanating out from each vertex is even.[37] Considering the graph in figure 1.6, all four vertices have an odd number of

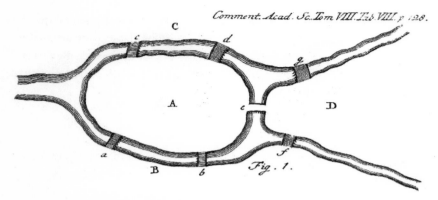

FIGURE 1.5. Euler's original drawing of the seven bridges of Königsberg.

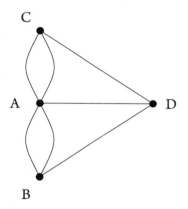

FIGURE 1.6. This graph is derived from figure 1.5 by substituting a vertex for each lettered land mass and an edge for each bridge. Note, in particular, that each vertex is met by an odd number of edges. This observation is the cornerstone for Euler's proof that the Königsberg walk is impossible.

edges. This is how Euler concluded the Königsberg walk could not be done. Observing that Ridley performed this impossible feat in the opening scene is how Kerner concludes Ridley must have an identical twin.

Stoppard does not include many details about Euler's argument, and that is because the mathematics is not as important as the physics. The graph theory provides a pivotal clue to the spy plot, but the physics goes to the center of the play. Like a photon whose location is not uniquely specified by the laws of quantum mechanics, Ridley has for years been existing simultaneously in two places at once, essentially serving as his own alibi. Ridley's twin, however, is not as interesting as Hapgood's twin, and it is via this latter conceit that the play achieves a moment of distinction. After two decades of experimenting, Stoppard creates a scene where the full capabilities of theater are utilized to reveal a human epiphany embedded in a mathematical idea, albeit one employed in service of science.

We Meet Our Sleeper

Complaining that it is rude for authors of spy novels to withhold crucial information until the end, Kerner makes a case to Hapgood for the superior literary manners of scientists. "This is why a science paper is a beautiful thing," he argues. "First, here is what we will find; now here is how we find it." (40) Applying this lesson to Stoppard's play, act I of *Hapgood* presents the hypothesis of Ridley as traitor; act II is the experiment designed to test the hypothesis.

The test involves an elaborate trap designed to trick Ridley into employing his secret twin in another sting. The ruse is rich with the details necessary to make their story credible, but the broad strokes are that the Russians have pegged Kerner as the traitor. In the process, they have discovered Kerner fathered a child with Hapgood and have kidnapped the boy as ransom to be exchanged for research. To spring the trap, however, is going to require one more ingredient—Celia, Hapgood's foulmouthed, chain-smoking twin sister.

To be clear, there is no such person; Hapgood must invent her and play both parts. Kerner's long soliloquies about the double life of photons come to theatrical life as Hapgood jumps between her established personality and Celia's, depending on whether she is being observed by Ridley. The harder Stoppard pushes on his physics metaphor, the more the play's human story overtakes the espionage plot. The play's emotional crescendo is not when the guns are fired at the end of the play but comes just before. Ridley and Hapgood (as Celia) are waiting in a hotel room to swap a computer disk of science research for Hapgood's son, and as they wait, the irony of their situation grows steadily thicker. The "bad guy" Ridley is there because he instinctively knows that the international spy game is a racket and that the value of Kerner's physics pales in comparison to the life of a twelve-year-old kid. He is there because some latent fondness for Hapgood makes him willing to take the risk while Blair, Hapgood's supervisor and lover, shows no scruples about using young

Joe as bait. Stoppard inverts the spy-thriller template by revealing Ridley as the traitor early in the play, but then he inverts his inversion by revealing that Blair is the emotional traitor while Ridley is something of a hero.

Although impersonating the invented Celia was originally part of the ruse to catch Ridley, Hapgood begins to see the world very differently through Celia's eyes. Why has she been so willing to deny her son a father just to preserve Kerner's usefulness to the agency? Has Blair been running her? Short and powerful, the hotel room scene between Hapgood and Ridley is full of stunning emotional shifts.[38] Like light that is unambiguously a wave and also unambiguously a particle, Hapgood embodies her own dual nature. Onstage, the performance of Celia is indistinguishable from viewing Celia as an autonomous, cogent entity. To borrow Feynman's phrasing, it is absolutely impossible to explain in any classical way. Hapgood is in the act of betraying Ridley and falling for him at the same time, and the paradox is too much for Ridley to grasp:

RIDLEY: Shut up.

HAPGOOD: (*Cranking up*) You'd better be sure, she plays without a board. You haven't got a prayer.

RIDLEY: *Shut up!*

HAPGOOD: If you think she's lying, walk away. If you think bringing back her son will make you her *type*, walk away. You won't get in the money, women like her don't pay out—take my advice and open the box.

RIDLEY: (*Grabbing her*) *Who the hell are you?*

HAPGOOD: I'm your dream girl, Ernie—Hapgood without the brains or the taste.

(*She is without resistance, and he takes, without the niceties; his kiss looks as if it might draw blood.*)[39] (71)

When the time comes to spring the trap, Hapgood—the original one, let's say—is forced to shoot Ridley. After the stretcher is taken away there is a brief moment when she—the Hapgood from the hotel, perhaps—asserts, "it was the shoulder," even though the shot was fatal.

There are other tender surprises as the dust settles out, and they point conclusively to what this play full of spies and physics is really about—that we all live the superimposed double life of a photon. Hapgood as secret agent and mother, Kerner as friend and lover. It is not schizophrenia or having multiple personalities; it is that this kind of duality is an inherent part of what human

"At night—perhaps in the moment before unconsciousness—we meet our sleeper." *Hapgood*; Roger Rees (Kerner), Felicity Kendal (Hapgood); Aldwych Theatre, 1988. © Donald Cooper/photostage.co.uk.

personality entails.[40] As Kerner says early in the play, "The one who puts on the clothes in the morning is the working majority, but at night—perhaps in the moment before unconsciousness—we meet our sleeper."

Evidently Satisfied Customers

With regards to how well *Hapgood* ultimately holds up as theater, the London theater critics should not be given the last word. In fact, they probably should not have been given the first word except that their hostile tone says something important about where mathematics and science stood as subjects for popular entertainment in 1988. Reviews that came later took a more measured tone, describing the play with mixed phrases like "enervating and entertaining, taxing and stimulating."[41] This is a fairer assessment. The profundity of *Hapgood*'s blending of science and psychology grows richer with reflection, but in real time the play asks a great deal of its audience. Initially, Stoppard was of the opinion that his new play was misunderstood. "We have a problem with the way the play is being perceived," he wrote in a letter to producer Michael Codron two months after *Hapgood* opened, what has surfaced from the reviews in general is the idea that it is "difficult," and what has submerged is the idea that it is entertaining. Rightly or wrongly, and I think rightly, this is the gist of our backstage ruminations. We are all a little bewildered and certainly frustrated by the combination of disappointing audiences and evidently satisfied customers. For my part I receive more and nicer mail for Hapgood than for anything I've written including The Real Thing.[42]

In later interviews, Stoppard did come to acknowledge the play had flaws, although in his mind it was more to do with the intricacies of the spy plot than with the physics. Major revisions accompanied a production of *Hapgood* in Los Angeles in 1989, and still more were necessary when the play was brought to New York five years later for a limited but relatively successful run at Lincoln Center. By this point the physics had been pruned somewhat, and significant edits were made to clarify the narrative mechanics. But something else had transpired in the meantime—*Arcadia* had been packing out the National Theatre for months and a proof of Fermat's Last Theorem had been announced. For mathematics and theater, it was a whole new world.

Arcadia

The decline from thinking to feeling, you see.

—HANNAH, FROM ARCADIA

Any cautionary lessons *Hapgood* might have suggested about the compatibility of mathematics and theater went unheeded by Stoppard, who increased the quantitative demands on his audience in his follow-up play a few years later. Whereas *Hapgood* is an acquired taste, the dominant response to *Arcadia* from audiences and critics was euphoric from the outset. One explanation is that *Hapgood* was simply ahead of its time. Post-*Arcadia* revivals of Stoppard's spy thriller fared reasonably well, as have swaths of more recent plays engaging mathematical ideas. But this is really evidence that Stoppard achieved something distinctive, and transformative, with his 1993 play full of fractals and chaos.

The first difference between Stoppard's two scientifically themed plays is narrative structure. There is a more organic flow to the sequence of events in *Arcadia* that extends to the science as well. The quantum mechanics in *Hapgood* functions effectively as metaphor, but its content is imported into the story by wedging in a physicist among the spies. The mathematics in *Arcadia* has a metaphorical role to play as well, but it serves a more primal function by being central to the passions and motivations of several protagonists. Running underneath the action of *Arcadia* is an unabashed embrace of the pursuit of knowledge. Rather than feeling stupid, the audience feels grateful for what they understand, and they are rewarded by having their new mathematical knowledge synthesized into the romantic tragedy that unfolds. *Hapgood* without its physics would still at least be melodrama; *Arcadia* without its mathematics would be nonsense.

In terms of *Arcadia*'s intellectual heft, the math is only part of the story. Filling out the academic raw material are substantial references to Lord Byron, the history of English landscape gardening, and some art history in the form of Salvator Rosa and Nicolas Poussin. Having gathered together a critical mass of painting and poetry in one corner and a similar arsenal of math and physics in the other, it may seem like *Arcadia* is set up to be another installment in the ongoing debate between art and science. Although there are some notable flare-ups along this familiar fault line, overlaying this paradigm onto Stoppard's play misses the mark. A better framework from which to view the overarching tension in *Arcadia* is classical versus romantic, or perhaps rational versus passionate.[43] The opening scene of *Arcadia* is set in the year 1809, a time when the Western world is on the cusp of the Romantic movement. The historical Byron is a major figure in the reaction within literary circles to the rigors of Enlightenment thinking. Byron does not explicitly appear in Stoppard's play, but he is a guest at Sidley Park, the Coverly family's pastoral estate, and his romantic tendencies, in art and otherwise, figure heavily in the events of the play.

The most interesting domain in which Stoppard explores the classical/ romantic duality is within the characters themselves. When the play opens, the thirteen-year-old Thomasina is exhibiting the telltale signs of her own passionate awakening. When Stoppard arranges for Septimus to wedge a definition of carnal embrace up against a description of Fermat's Last Theorem, the playwright not only gets his first big laugh of the night, but he also puts up a signpost for what is to come. Math and sex incongruously sharing the same short speech represents the first of many encounters between the head and the heart.[44]

Romantic Mathematics

With the battle lines established, it is important that mathematics not exist solely on one side of the fight. Euclid, Fermat, and Newton are the subjects that Septimus teaches Thomasina, but Thomasina is no ordinary student.

"Each week I plot your equations dot for dot, xs against ys in all manner of algebraic relation," she complains, "and every week they draw themselves as commonplace geometry. . . . Armed thus, God could only make a cabinet." (37) Thomasina's "New Geometry of Irregular Forms" advertised in the margin of her lesson book is a fictional precursor to the twentieth-century field of fractal geometry. Fractal geometry offers a visually stark contrast to the familiar arcs and angles of Euclid, which makes it an ideal candidate to be the romantic alternative to classical Greek mathematics that Stoppard needs. In fact, it was one of the initial seeds from which his play was created.[45]

A central influence cited by Stoppard for *Arcadia* is James Gleick's 1987 bestseller *Chaos: Making a New Science*. One chapter of Gleick's book is dedicated to Thomasina's fractals; another, late in the book, discusses the relevance of chaos theory to the Second Law of Thermodynamics. The opening chapter describes Edward Lorenz's fabled encounter with what he eventually coined as the "butterfly effect." At MIT in the early sixties, Lorenz was using an unwieldy computer called a Royal McBee to build a model for forecasting the weather. The computer took initial inputs for parameters like temperature, pressure, and wind speed, and then updated them several thousand times via a series of programmed equations. For this early generation of machines, a simulation like this took hours, with the progressively updated results printed in hard copy on long rolls of paper. As the story goes, one of the simulations was halted prematurely by a power outage. In possession of the hard copy record of the output, Lorenz restarted the run using the same initial conditions as he had before, only to witness an entirely different output from the original run. The mystery was solved when Lorenz realized that the hard copy he used for the restart only listed three decimal places while the computer stored six.

Having found the source of the discrepancy, Lorenz was still unsettled by what he saw. A tiny difference in the initial inputs—one part in a thousand—was being magnified within the computations of his model so that after relatively few iterations, the model was predicting an unrecognizably new weather pattern. This "sensitive dependence on initial conditions," to use its formal name, became a core principle of chaotic dynamics and a daunting hurdle to the predictive aspirations of deterministic models. To Stoppard, who had demonstrated a keen interest in challenges to deterministic thinking in *Hapgood*, it was fuel for the romantic fire.

The Coverly Set

To portray the butterfly effect on stage, Stoppard sets *Arcadia* across two time periods. The first scene of the play is the early nineteenth-century setting previously discussed. This features Thomasina Coverly and her tutor Septimus Hodge who, when he is not teaching his talented student the classics, is engaging in carnal embrace in the gazebo with the wife of Mr. Chater the poet. The second scene of the play takes place in the same room, but in the present, where three academics are trying to re-create what has taken place in the house two centuries earlier. The most caricatured of the three is Bernard, a blustery and impulsive Byron scholar. Bernard traces Byron to the Coverly house at Sidley Park, but his tiny errors in initial conditions lead him to erroneously conclude that Lord Byron fought a duel and killed the poet Chater before fleeing England.[46]

Hannah is the second academic. She is also a historian but is interested in the evolution of the Sidley Park gardens. Hannah is more serious and disciplined than Bernard, a characteristic that applies to her overall personality as well as her research methods. The actor playing Hannah has a fine line to walk. Strong, smart, and dedicated to her field, Hannah is driven by the same kind of gut instincts as Bernard but is inherently suspicious of unreasoned passion and stiffens at the first sign of intimacy. This inner tension extends to the content of her current project. As the audience gets to see, the Sidley Park gardens during Thomasina's time were being transformed from pastoral meadows full of carefully arranged sheep to the "picturesque style" characterized by gloomy forests, craggy ruins, and a hermitage in place of the infamous gazebo. Of particular interest to Hannah is the unidentified occupant of the hermitage. Reported in a local magazine to be a "a sage of lunacy," the Sidley Park hermit lived out his days in the hermitage, filling it with thousands of pages of incoherent mathematics in an attempt "to save the world through good English algebra." For Hannah, the hermit is "a perfect symbol [of] the whole Romantic sham . . ."

> HANNAH: It's what happened to the Enlightenment, isn't it? A century of intellectual rigour turned in on itself. A mind in chaos suspected of genius. In a setting of cheap thrills and false emotion. The history of the gardens says it all, beautifully. . . . The decline from thinking to feeling, you see. (27)

The third academic is Valentine, his name making it clear that he resides on the romantic side of the arc from thinking to feeling. Valentine is a mathematician and a member of the Coverly family who is using the data in the game books to model the grouse populations as a chaotic system. In addition to explaining the rudiments of chaos theory to Hannah and the audience, his main service is to translate Thomasina's lesson books. When Thomasina sets herself the task of finding an equation for the leaf from her apple, she is a bit like a deaf composer. Without a computer there is no practical way to graph the results. "You couldn't see to look before," is how Valentine explains it to Hannah. "The electronic calculator was what the telescope was for Galileo." (51) Valentine eventually takes it on himself to run Thomasina's equations through his laptop, and the results catch Hannah up short:

HANNAH: Oh!, but . . . how beautiful!
VALENTINE: The Coverly set.
HANNAH: The Coverly set! My goodness, Valentine!
VALENTINE: Lend me a finger.
 (*He takes her finger and presses one of the computer keys several times.*) See? In an ocean of ashes, islands of order. Patterns making themselves out of nothing. I can't show you how deep it goes. Each picture is a detail of the previous one, blown up. And so on. Forever. Pretty nice, eh? (76)

The audience cannot see the image on Valentine's screen but the one in figure 1.7 is a reasonable candidate. There are two observations to make about this graph. The first is the detailed complexity of the image compared to the parabolas and ellipses characteristic of "good English algebra." Figure 1.7 is idealized, but it just looks like a better representation of the jagged mixture of order and disorder possessed by natural objects. As Thomasina says, "Mountains are not pyramids and trees are not cones," a quotation lifted nearly verbatim from Benoit Mandelbrot's 1982 book *The Fractal Geometry of Nature*.[47] The second observation is that the equations responsible for figure 1.7 are *not* complicated, nor is the graphing method. There are four equations, and each one is quite simple. There are no squares, no roots, no sines and cosines, and yet still we get this finely detailed image. Instead of graphing in the traditional way—"*x*s against *y*s in all manner of algebraic relation"—Valentine's computer does what Lorenz's computer did. It takes a given point as input and iterates it, alternating randomly through the four equations. Each

FIGURE 1.7. A rendering of a fractal leaf.

output becomes the new input for the next iteration, and the computer keeps
a record of where the dot lands each time.

> VALENTINE: You'd never know where to expect the next dot. But grad-
> ually you'd start to see this shape, because every dot will be inside the
> shape of this leaf. It wouldn't *be* a leaf, it would be a mathematical object.
> But yes. The unpredictable and the predetermined unfold together to
> make everything the way it is. It's how nature creates itself, on every
> scale, the snowflake and the snowstorm. (47)

One of the biases we develop from an overexposure to classical Newtonian
thinking is the assumption that simple systems lead to simple behavior and
complicated systems lead to complicated behavior.[48] The Coverly set does
not fit this expectation. It is an example of complex structure arising out of

"The unpredictable and the predetermined unfold together to make everything the way it is." *Arcadia*; Ed Stoppard (Valentine), Samantha Bond (Hannah); Duke of York's Theatre, 2009. © Donald Cooper/photostage.co.uk.

the repeated implementation of very simple rules, and what Valentine is suggesting is that nature is engaged in a similar trick. But nature, it turns out, is fighting a losing battle, and it is around this point that the mathematics begins to mesh with the storytelling of *Arcadia* in its most compelling way.

The Attraction That Newton Left Out

Thomasina is as passionate as she is prodigious, at one point crumbling at the thought of the losses incurred when the ancient library of Alexandria was burned. Septimus consoles her with the argument that "we shed as we pick up, like travelers who must carry everything in their arms, and what we let fall will be picked up by those behind." Septimus believes in the perpetual mechanical wheels of the Newtonian universe, and what is clear is that Stoppard is fascinated by the challenges Newton's deterministic system has sustained over the centuries. In *Arcadia*, determinism is discussed repeatedly, once even by Valentine's flirtatious sister Chloë:

> CHLOË: The future is all programmed like a computer—that's a proper theory isn't it?

VALENTINE: Yes, there was someone, forget his name, 1820s, who pointed out that from Newton's laws you could predict everything to come—I mean, you'd need a computer as big as the universe but the formula would exist.

CHLOË: But it doesn't work, does it?

VALENTINE: No, it turns out the maths is different.

CHLOË: No, it's all because of sex. (73)

"Ah," Valentine concedes. "The attraction that Newton left out." Mirroring this exchange is a similar one between Thomasina and Septimus.

THOMASINA: Well! Just as I said! Newton's machine which would knock our atoms from cradle to grave by the laws of motion is incomplete. Determinism leaves the road at every corner, as I knew all along, and the cause is very likely . . . the action of bodies in heat. (84)

The pun conflating heat and sex is certainly intended, but Thomasina is actually referring to the Second Law of Thermodynamics. It was three years earlier that an adolescent Thomasina was enthralled by the red jam turning her rice pudding an even shade of pink. Now, nearly seventeen, she has a sharper vision of what is at stake. In every physical process there is an inherent inefficiency—energy lost in the form of heat—and the disorder in the system increases. One implication of this is that time is a one-way street. Heat goes to cold, and never the other way. Simple and profound, this law of the universe takes on added significance in a play where so many of its characters are attempting to look back in time to reconstruct history.

One way *Arcadia* succeeds with its science, most notably the thermodynamics, is through its staging. "In performance the play becomes a working demonstration of the ideas at its core," writes theater scholar Kirsten Shepherd-Barr, articulating a beautiful insight about why *Arcadia* "must be theater and no other genre."[49] Through most of the play, the two separate casts alternate scenes in a drawing room whose sparse furnishings do not anachronistically offend either time period. The large central table, however, acts as something of a portal where an apple brought on in the present might get picked up by someone in the past. Various parallels and repeated lines create echoes across time, and when the modern characters begin dressing up for a nineteenth-century theme party it feels as though the periods are about to collide. Then, at the point where the Second Law of Thermodynamics enters the story to tell us that this is impossible, Stoppard puts both casts on stage together, theatrically doing what nature cannot.[50] The effect is literally breathtaking.

Without the mathematical vocabulary to express her insight that the equations of heat flow are not reversible, Thomasina draws a diagram illustrating the principle of increasing entropy which, through Stoppard's staging, is studied by both Septimus and Valentine doubled in time.

> SEPTIMUS: So the Improved Newtonian Universe must cease and grow cold. Dear me.
> VALENTINE: The heat goes into the mix.
> THOMASINA: Yes, we must hurry if we are going to dance.
> VALENTINE: And everything is mixing the same way, all the time, irreversibly . . .
> SEPTIMUS: Oh, we have time, I think.
> VALENTINE: . . . till there's no time left. That's what time means.
> SEPTIMUS: When we have found all the mysteries and lost all the meaning, we will be alone, on an empty shore.
> THOMASINA: Then we will dance. Is this a waltz? (94)

Amid the mathematical and scientific fireworks, *Arcadia* works as art because the fireworks are pivotal to the play's heartbreaking denouement. The more one understands the mathematics, the more heartbreaking it becomes. Thomasina and Septimus do finally dance, gracefully and to their mutual delight, passionately. But we have learned midway through the play that, like the library in Alexandria, Thomasina dies in a fire that night, on the eve of her seventeenth birthday. When Septimus lights a candle for Thomasina, we recognize it as the source of the entropy that engulfs her. Watching the lovers dance as the curtain falls, it is finally clear that it is the devastated Septimus who is Hannah's hermit, living out his years in the hermitage stuffing it full with pages of mathematical scribbling. Perhaps he is trying to finish what Thomasina started with her new geometry, or maybe he is trying to rescue his perfectly tuned Newtonian universe from the inexorable heat death that her insights foresaw. In actuality, these two quests are two sides of a single coin. The question of how complex structures arise in the face of the downhill current caused by the Second Law of Thermodynamics has long been an issue at the forefront of science, and the mathematics of chaos theory has offered some compelling avenues forward. This is at the heart of what Valentine is referring to when he describes Thomasina's apple leaf as an "island of order" in an "ocean of ashes . . . patterns making themselves out of nothing."

"So the Improved Newtonian Universe must cease and grow cold.
Dear me." *Arcadia*; Dan Stevens (Septimus), Samantha Bond (Hannah),
Ed Stoppard (Valentine); Duke of York's Theatre, 2009. © Nigel
Norrington / ArenaPAL.

"Do you mean the world is saved after all?" Hannah asks him later.

"No, it's still doomed," he says. "But if this is how it started, perhaps it's how the next one will come." (78)

Uncertainty Principles

"I keep trying to find a play about mathematics," Stoppard said in the mid-1980s. "There is one somewhere but I can't find it."

Assessing Stoppard's long journey to *Arcadia*, scholar William Demastes observes that the British playwright was unwittingly invoking the principles of chaos theory long before he learned the mathematical details. "All along Stoppard has been onto the trail of orderly disorder," Demastes argues, making the case that this phenomenon starts with the early radio plays. "[Stoppard] was groping for a system of thought that would confirm that things do *cause* other things to happen but that there's always the chance that unanticipated reactions undermine life's predictabilities."[51] Endorsing Demastes's thesis, the revelation that emerges from Stoppard's search for his math play is that what ultimately sparked his creative imagination was not the austere authority of mathematics but the challenges to its claims on absolute certainty.

In *Albert's Bridge*, Albert's fate turns on the subtle error in Fitch's calculations. In *Jumpers*, Bertrand Russell's historic attempt at a perfected form of mathematical truth via *Principia Mathematica* is relegated to the wings while Zeno's paradoxes about the infinite are performed center stage. The central metaphor of *Hapgood* is the collapse of objective reality at the heart of quantum mechanics. When Stoppard finally finds his math play in *Arcadia*, his agenda is once again an assault on certainty—this time in the form of Newtonian determinism's demise at the hands of chaos theory and thermodynamics.

Chloë's conversation with Valentine about sex being "the attraction that Newton left out" is an echo of an earlier one between Septimus and Thomasina who, in typical fashion, pushes the limits of her tutor's lessons and patience:

> THOMASINA: If you could stop every atom in its position and direction, and if your mind could comprehend all the actions thus suspended, then if you were really, *really* good at algebra you could write the formula for all the future; and although nobody can be so clever as to do it, the formula must exist just as if one could.
>
> SEPTIMUS: (*Pause*) Yes. (*Pause*) Yes, as far as I know, you are the first person to have thought of this. (5)

Contrary to what Septimus says, these ideas were in fact a staple of Enlightenment thinking. When Valentine says to Chloë, "There was someone, forgot his name, 1820s," he is referring to Pierre Simon Laplace. A major figure in the development of calculus, Laplace wrote an essay in 1814 proposing the existence of "an intelligence" that is eerily similar to Thomasina's mythic algebra wizard. "[This intelligence] would embrace in the same formula the movements of the greatest bodies of the universe and those of the lightest atom; for it, nothing would be uncertain and the future, as the past, would be present to its eyes."[52]

What Laplace and the young version of Thomasina have in common is a zealous faith in the authority of Newton's laws of motion, but as Valentine indicates to Chloë, this theory has taken on water over time because "it turns out the maths is different." The butterfly effect—the phenomenon where tiny variations in initial conditions are magnified in chaotic systems—combined with the quantum mechanical assertion that perfect precision in initial conditions is a theoretical impossibility creates a significant roadblock to any vision of a deterministic universe. Is it a death knell for this theory? That argument continues, and by the closing moments of *Arcadia*, Thomasina is firmly on the other side of the debate.

In the journey to find his math play, Stoppard cast mathematics in progressively more prominent roles. Mathematics provides a rich accent to the philosophical banter in his early scripts, but the clever scene shifts and prevalence of twins in *Hapgood* allow the audience to experience quantum doubling in a tangibly palpable way. In *Arcadia*, the theoretically impossible collapsing time periods literally bring the mathematical ideas to life. Having fully assimilated the mathematics into the storytelling, Stoppard executes a final artistic feat in *Arcadia* by bringing the mathematical discoveries to bear on soulful ones—ones where mathematics offers genuinely new insights. With an assist from Valentine, Thomasina's romantic vision of fractal geometry and the mathematics of heat flow offer a portrait of the human journey that sneaks up on an audience occupied by an unfolding love story that ends in tragedy. Thomasina identifies the chink in the armor of Newton's laws that dispels the myth of determinism, liberating the future to become a continuum of undetermined possibilities. She also uncovers the geometry that holds the clue to how nature exploits this newfound freedom "to create itself, on every scale." The price of liberation is the recognition that the future is finite and unpredictable, and the universe is in decline. Septimus is undone by this revelation. Whereas Newton famously described himself as a boy playing by the ocean "finding a smoother pebble or a prettier shell than ordinary," the despondent Septimus recasts this hopeful image of discovery into a desolate vision where "we will be alone on an empty shore."[53]

"Then we will dance," Thomasina replies, invoking the power of the free will she has restored and bequeathed to the rest of us.

———

In one sense, Stoppard is a pioneer. He is the first playwright to put mathematics squarely on the popular mainstage, and in the decades following *Arcadia* there is a paradigm shift in the depth and frequency with which playwrights engage mathematics and science.[54] In another sense, however, Stoppard's accomplishments are the culmination of a century of innovative writers using mathematics to help them challenge the status quo in their respective eras.

This sounds odd because, to the uninitiated, the immutability of mathematics might seem to establish it as part of the status quo of any era. Indeed, this was Stoppard's early mindset. "I was aware that for centuries mathematics was considered the queen of the sciences because it claimed certainty," he commented when explaining the origins of *Hapgood*. "It was grounded on some fundamental certainties—axioms—which led to others. But then, in a sense, it all started going wrong."

The twentieth-century discoveries of quantum mechanics and chaos theory are the weapons Stoppard chose to undercut the authority of Euclid's *Elements*, but the preeminence of Greek geometry was first challenged in the nineteenth century by the emergence of alternative geometries that threatened to downgrade the propositions in the *Elements* from absolute truths to useful conventions. This upheaval in the foundation of mathematics was a source of inspiration for a few early-century playwrights whose agenda was more ambitious than Stoppard's. Whereas Stoppard employed modern mathematics to create theater that engaged new kinds of ideas, decades earlier, the insurgency to overthrow Euclid inspired entirely new kinds of theater.

2

Jarry and Witkiewicz: Geometry of a New Theater

In our opinion, the breaking of certain inessential bad habits touching the real-life aspect of works of art opens new horizons for formal possibilities.

—STANISLAW WITKIEWICZ, FROM
HIS *THEORETICAL PREFACE*[1]

AT THE beginning of the twentieth century, theater and mathematics were each engaged in their own version of a philosophical debate centered on a common theme. At stake, in both cases, was how to understand the relationship between the real world and the artistic endeavor in question.

At the heart of the matter on the mathematical side was the following innocuous-sounding statement that appears among Euclid's postulates in the opening pages of the *Elements*:

> Let it be postulated . . . that, if a straight line falling on two straight lines make the interior angles on the same side less than two right angles, the two straight lines, if produced indefinitely, meet on that side on which are the angles less than the two right angles.

This statement is the last on a list of five postulates, or axioms, that Euclid selected to be the foundation for his system of geometry. These five postulates were *not* propositions requiring a demonstration. They were intended to be self-evident truths from which the propositions could be derived. They represented the building blocks of the *Elements*, so fundamental in nature that their validity was to be accepted, without proof, as an intrinsic quality of the mathematical world.

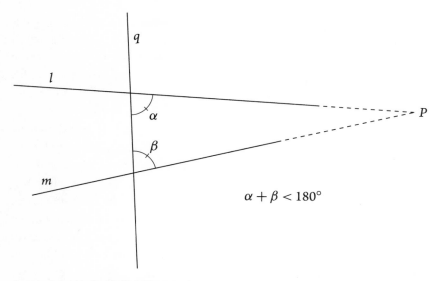

FIGURE 2.1. Euclid's fifth postulate says that if $\alpha + \beta < 180°$, then the lines l and m intersect at some point P.

The topic of Book I of the *Elements* is geometry in the plane, and the meaning of this particular postulate becomes evident with the aid of a figure. Given two straight lines labeled l and m, draw a third straight line q that crosses the first two. Interior angles are the ones that fall between the lines l and m, and the postulate asks us to consider two of these angles that fall on the same side of q. Specifically, it says that if these two angles sum to less than two right angles (i.e., less than 180°), then at some point on that same side of q, the straight lines l and m will eventually cross at a point P (Figure 2.1).

After giving it the proper reflection, this postulate certainly begins to sound like a true statement, although whether it passes for self-evident is up for debate. If the interior angles on one side sum to less than 180°, the straight lines appear to be on a trajectory that brings them closer together. Our intuition tells us that the lines, set on a collision course, will eventually come to a point where they intersect if we extend them far enough.

The problem is that intuition is a fickle way to get at absolute truth—it can be hard to distinguish intuition from deep-seated bias. On what basis, then, can we be absolutely confident that Euclid's fifth postulate is really valid? If we attempt a short calculation, using some basic trigonometry perhaps, to determine exactly where the lines must intersect, that calculation will inevitably rely on the geometric machinery developed in the *Elements*. Because this machinery depends on the accepted truth of the fifth postulate, the argument is circular and therefore invalid. A proposal for moving forward is to consider the question from an empirical point of view. Can we simply declare Euclid's fifth postulate true because it describes a fundamental property about the nature of the space in which we live?

This was essentially the approach Euclid adopted, and it became the default view of most pre-twentieth-century mathematicians. This is not to say that there weren't significant philosophical hurdles to overcome. The dominant position, which falls under the general heading of Platonic realism, was that the geometric world of points and lines laid out in the *Elements* existed separately from the natural world in some unchanging and eternal domain beyond the reach of our five senses. The physical world of balls and boxes and diagrams constructed with a straightedge and compass, meanwhile, was an evolving approximation of this ideal realm. "[Geometers] are not reasoning, for instance, about this particular square and diagonal which they have drawn," Plato writes in the *Republic*,

> but about the Square and the Diagonal; and so in all cases. The diagrams they draw and the models they make are actual things, which may have their shadows or images in water; but now they serve in their turn as images, while the student is seeking to behold those realities which only thought can comprehend.[2]

Plato emphasizes his view that mathematical knowledge is acquired by rational thought, not observation, but rational thought alone kept coming up short as a means for establishing the validity of the fifth postulate. The natural alternative is to take advantage of Plato's "images in water" analogy and argue that abstract geometry should intrinsically correspond to physical space. Although only a shadowy reflection, physical reality can at least serve as an imperfect roadmap of mathematical reality. Because the thought experiment of extending the lines *l* and *m* in figure 2.1 produces a point of intersection, the reasoning goes, this same truth should hold in the idealized Platonic world as well.

If this sounds like an unsatisfying way out of the philosophical quandary, that's because it is. If it is not part of physical reality, then where does this mathematical reality reside? And why, and by what means, do its properties correlate with the observable ones of the physical world? Practically speaking, theorems of mathematics are indispensable to the science we use to describe nature, so it seems odd to quarantine mathematical objects to a strictly metaphysical existence. Aristotle's solution is not to completely abandon Plato's perfected mathematical reality, but to conflate it with the observable world. "[P]hysical bodies contain surfaces, volumes, lines, and points, and these are the subject matter of mathematics," Aristotle writes. He goes on to explain how these objects can be separated from their tangible origins via a process of mental abstraction so that, for instance, "geometry investigates physical lengths, but not as physical."[3] This solves the problem of the extraordinary usefulness of mathematics to natural science, but it generates its own nest of new issues. If we propose that the theorems of geometry are about physical space, then we have to be prepared to say precisely what a "line" is in the real world. The path that light travels? The shortest distance between two points? What, then, does "point" refer to in a purely physical model of geometry? Without unambiguous definitions for the primitive notions of point and line, it is not obvious whether the term "true" is even meaningful in relation to a statement like Euclid's fifth postulate, much less whether it applies.

As the philosophical wrangling continued over the centuries, what never wavered was the conviction that Euclid's geometry described space, and that intuition about space was hardwired into our cognitive capabilities. "The concept of [Euclidean] space is by no means of empirical origin," the influential Immanuel Kant wrote in 1781 in *Critique of Pure Reason*, "but it is an inevitable necessity of thought."[4]

Taking Place in Real Life

The fascinating story of the fifth postulate and its impact on mathematical realism was precursor, catalyst, and, in one remarkable case, direct inspiration for the early twentieth-century confrontations around the role of realism in the theater. The artistic sensibilities of the nineteenth century, at the outset at least, were characterized by the Romantic imagination. In theater, this manifested itself in the form of melodramas and operettas that featured intensely emotional acting and exaggerated scenic effects. A majority of mid-nineteenth-century theater performances were mounted without a central creative authority—a person we would refer to now as the director—in part because there was not a pressing need for such a person. Characters were types (e.g., heroes, villains) and acting these types involved mastering a standard array of voices and gestures.[5]

As the century wore on, this sense of spectacle was gradually replaced by more naturalistic theater. A major figure in this movement was Norwegian playwright Henrik Ibsen. *A Doll's House*, from 1879, stands out as a definitive example of the shifting emphasis toward realism in the way that it renders emotional conflicts in a familiar and unadorned setting. Ibsen's fellow Scandinavian August Strindberg was another important voice in this movement. Ibsen's female protagonist stunned audiences at the end of *A Doll's House* by walking out on her husband and children with a reverberating door slam. In *Miss Julie*, Strindberg tells the even more shocking story of an aristocratic daughter who initiates an affair with her father's valet that ultimately destroys her life. Written in 1888, *Miss Julie* was first performed in Berlin and Paris in the early 1890s before finally making it past the Scandinavian censors in 1905.

By taking up such deeply personal psychological subject matter, Ibsen, Strindberg, and others were asserting that theater had the potential to engage serious contemporary topics. But to do so effectively, the art form had to trade its flamboyant and overtly theatrical tendencies for a more colloquial and nuanced style. Writing to one of his directors in 1880, Ibsen insisted that the staging should give "the illusion that everything is real and that one is sitting and watching something that is actually taking place in real life."[6] In a preface to the 1888 published script of *Miss Julie*, Strindberg asserted his vision this way:

> I do not believe, therefore, in simple characters on the stage. And the summary judgements of the author upon men—this one stupid, that one brutal, this one jealous, and that one stingy—should be challenged by the naturalists who know the fertility of the soul-complex and who realize that vice has a reverse very much resembling virtue. . . . My souls (or characters) are conglomerates, made up of past and present stages of civilization, scraps of humanity, torn off pieces of Sunday clothing turned into rags—all patched together as is the human soul itself.[7]

Whether it was crafting characters on the page, or set design, or a new approach to acting pioneered by Konstantin Stanislavski, the goal was to probe the social and psychological realities of the day to render life as it really was, without cliché or euphemism.[8]

By every measure, the movement toward realism in the theater proved successful and lasting, so much so that it still characterizes mainstream theater more than a century later. Meanwhile, Euclid's geometry has dominated the intellectual landscape around mathematics and beyond. Authored in 300 BCE, the *Elements* was still in wide use as a textbook well into the 1900s, solidifying the consensus view that math, like art, is a reflection of nature and that we inhabit a Euclidean world. But the respective successes of this art-imitating-life approach in both theater and mathematics would not go unchallenged. In theater, this is not too surprising. The preeminence of realistic theater makes it an obvious target for pioneers interested in pushing back against the status quo. What is more surprising is that the realism attached to Euclid's geometry would be undermined in an irreparable way. In fact, by the time Ibsen and Strindberg were writing their groundbreaking plays, this insurrection in geometry had already happened, but the repercussions were only just starting to surface.

What is most surprising of all is where these two stories cross. When the non-Euclidean revolution in mathematics finally hit its stride, it helped inspire alternatives to theatrical realism by providing a template for how art could be liberated to pursue entirely new forms. What is more, these new forms in theater exploited the same biases that for centuries had flummoxed mathematicians who devoted their careers to resolving the puzzle of Euclid's fifth postulate.

If two straight lines are getting closer together, then they should eventually intersect; and if two old friends are talking under a streetlamp then we shouldn't expect one of them to level a shotgun and shoot the other—at least that's what intuition suggests if mathematics and theater are intent on imitating nature. The alternative is to ask what happens if we drop the requirement of naturalism and consider a "pure" form of mathematics and, likewise, a "pure" form of theater, free of the constraints imposed by any application to the observable world. The new universes that emerge from experiments like these, in both math and theater, have a fairy-tale quality in the way they violate familiar norms but yet don't collapse into nonsense. This preservation of sense-making is where the mathematics makes its richest contribution to early avant-garde theater. Endowed with their own inner logic, these fabricated worlds maintain a degree of integrity. They are strange and unfamiliar, but their internal consistency imbues them with the potential for being tangible places.

This leads to a provocative question: Having freed mathematics and theater to be defined on their own terms, what happens if we flip the art-imitates-nature relationship around and ponder the possibility that nature might imitate this new art? In the case of mathematics, this line of inquiry led to

FIGURE 2.2. The five Platonic solids.

revolutions in modern science that dismantled the Newtonian understanding of time and space. In the case of the playwright Stanislaw Ignacy Witkiewicz writing in Poland in the 1920s, it produced a harbinger of a descending darkness that no realist playwright could have imagined.

Euclid's *Elements*

> At age eleven, I began Euclid, with my brother as tutor. This was one of the great events of my life, as dazzling as first love.
>
> —BERTRAND RUSSELL, FROM HIS AUTOBIOGRAPHY

A full appreciation of the consternation surrounding Euclid's fifth postulate requires seeing it in its original context in Book I of the *Elements*. The *Elements* is divided into thirteen books, or chapters, that form a comprehensive survey of the fundamental results of mathematics that were known at the time. Roughly speaking, Books I–VI focus on geometry in the plane, Books VII–X deal with arithmetic, and Books XI–XIII take up the geometry of three-dimensional solids. Book XIII provides a grand finale for the *Elements* by undertaking a study of regular polyhedra—convex solids whose faces are identical copies of the same regular polygon joined together so that every corner looks the same as every other corner—and concluding that there are five and only five such beasts in the mathematical universe. The deep significance of the so-called Platonic solids in Greek philosophy and astronomy is likely what prompted Euclid to give them such an honored position in his compendium (Figure 2.2).

Far more than just a catalog of results, the *Elements* became the archetype of a deductive methodology for establishing domains of intellectual certainty beyond geometry and arithmetic. Isaac Newton wrote his *Principia* laying out the laws governing the universe in this style, and philosophers such as Thomas Hobbes and John Locke were heavily influenced by Euclid's axiomatic method. Baruch Spinoza subtitled his *Ethica* with "ordine geometrico demonstrata" (demonstrated in geometrical order) and organized it like a math text. When Russell and Alfred North Whitehead adopted the Euclidean template for *Principia Mathematica*, it was not a stylistic choice. This is just how mathematics was done by the end of the nineteenth century. The allure is in the compelling clarity of the method. Each proposition in the *Elements* is accompanied by a demonstration for why it is true, and each demonstration adheres to the strict policy that it only refers to propositions that appear earlier in the text. In this way, the *Elements* becomes a completely self-contained justification for the long list of theorems that it houses. The Pythagorean Theorem, for example, appears in Book I as Proposition 47. Euclid states it this way:

Proposition 47. *In right-angled triangles the square on the side opposite the right angle equals the sum of the squares on the sides containing the right angle.*

In the first paragraph of the demonstration for Proposition 47, Euclid sets up the notation shown in figure 2.3 and then instructs his reader to draw a straight line through the point *A* that is parallel to the vertical sides of the square with corners *BCED*. The figure makes it look obvious that such a line should exist—it's drawn in as the straight line from *A* to *L*—but figures do not constitute justifications.[9] So how does Euclid defend this step? By referencing an earlier result:

Proposition 31. *[It is possible] to draw a straight line through a given point parallel to a given straight line.*

This is precisely what we need in the first step of the proof of Proposition 47. Our given point is *A*, and our given straight line is the side of the square through *BD*. This hierarchical structure characterizes the entirety of the *Elements*. Euclid's proof of the Pythagorean Theorem in Proposition 47 contains references to Propositions 46, 41, 31, 14, and 4. The proofs of these propositions in turn rest on lower numbered propositions, but how does the process get off the ground? What does Euclid use in his demonstration of Proposition 1, for instance? This is where the postulates enter the picture.

Definitions and Common Notions

On the opening pages of Book I, before any propositions are stated, Euclid lays out the raw material for everything that is to follow. This raw material consists of twenty-three definitions, five postulates, and five common notions. The common notions are aptly named. They include assertions such as:

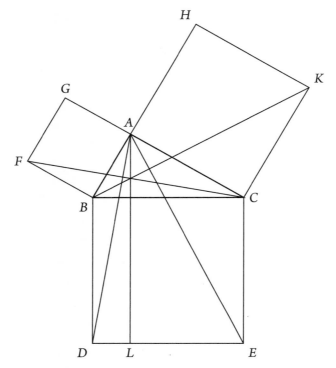

FIGURE 2.3. Euclid's diagram accompanying his proof of Proposition 47
(the Pythagorean Theorem).

- If equals are added to equals, then the wholes are equal.
- The whole is greater than the part.

For the most part, the common notions did not generate any significant debate. This was not so for the definitions and the postulates, however. The first four definitions—which are in fact the first four lines of the *Elements*—read as follows:

1. A *point* is that which has no part.
2. A *line* is breadthless length.
3. The extremities of a line are points.
4. A *straight line* is a line which lies evenly with the points on itself.

From this small sample, we can see Euclid wrestling with the ontological issues addressed earlier. In order to define the primitive notions "point" and "straight line," Euclid has to resort to using other poetic-sounding phrases such as "has no part," "breadthless length," and "lies evenly," which then beg for their own definitions. Trying to define these terms would just kick the can farther down the road of infinite regress. Euclid likely imagined he was describing preexisting entities that didn't require more clarification than he gave them, and, in any event, he never actually referred to these early definitions. The effect, then, is that terms such as "point" and "straight line" can be thought of as undefined, with their usage governed by the way in which they relate to each other in the statement of the postulates.

Although they are not useful for proving propositions, Euclid's first few definitions do clarify a few things about his terminology. Euclid employs the term "line" the way we would use the word "curve," and he employs the term "straight line" to refer to what we would now call a "line," or more precisely, a "line segment." Definition 3 implies that lines and straight lines can be, and perhaps by default are, finite in length. Note specifically that this is how the term is used in the proof of the Pythagorean Theorem and the applicability of Proposition 31.

Once Euclid muddles through the first few, the definitions on his list become progressively more meaningful and satisfying. One in particular is worth mentioning; it is the last definition on Euclid's list:

23 *Parallel* straight lines are straight lines which, being in the same plane and being produced indefinitely in both directions, do not meet one another in either direction.

Notice that this definition does not say that parallel lines are everywhere equidistant from each other. Whether this turns out to be a consequence of Euclid's definition remains to be explored, but we need to be absolutely clear that to say two lines are parallel means only that the two lines do not share a common point, no matter how far they are extended in either direction.

The Postulates

With his list of definitions complete, Euclid provides the following five bedrock assumptions on which to build his system of geometry.

Let the following be postulated:
1 To draw a straight line from any point to any point.
2 To produce a finite straight line continuously in a straight line.
3 To describe a circle with any center and radius.

4 That all right angles equal one another.

5 That, if a straight line falling on two straight lines make the interior angles on the same side less than two right angles, the two straight lines, if produced indefinitely, meet on that side on which are the angles less than the two right angles.

The first four postulates on Euclid's list certainly possess the air of self-evident truth. The first three describe the most basic functions of an unmarked straightedge and compass. These had become the natural tools of geometry, and Euclid was articulating in his postulates what he could do on a piece of papyrus or parchment. Exhibiting a self-evident truthfulness on par with the common notions, the fourth postulate emphasizes the universal quality of right angles as a standard for measuring the size of other angles.

This brings us to Euclid's fifth postulate, which we shall more formally designate as Postulate 5. Because parallel lines are defined as lines that do not intersect, another way to state the conclusion of Postulate 5 is that the lines l and m from figure 2.1 are not parallel. For this reason, the fifth postulate is also referred to as the parallel postulate. The original controversy surrounding Postulate 5 was not whether it was true—no one doubted its validity. The complaint was that it was not self-evident in the way that the other four postulates were, and that it was inappropriate to accept it without proof. The physical impossibility of extending the straight lines l and m "indefinitely" made this postulate tangibly distinct from the others. If nothing else, the sheer wordiness of it suggested something was amiss. Yes, Postulate 5 did describe something true and fundamental about the nature of space, but it belonged among the other propositions where demonstrations were required. This ugly duckling quality to the parallel postulate was evident early on. Proclus, the head of the Platonic Academy in Athens around 450 CE, wrote in his detailed commentary on the *Elements*,

[the fifth postulate] ought even to be struck out of the Postulates altogether; for it is a theorem involving many difficulties. . . . It may be that some would be deceived and would think it proper to place even the assumption in question among the postulates as affording, in the lessening of the two right angles, ground for an instantaneous belief that the straight lines converge and meet. [But] we have learned from the very pioneers of this science not to have any regard to mere plausible imaginings when it is a question of the reasonings to be included in our geometrical doctrine.[10]

Thus began the task of trying to find a proof that would enable mathematicians to move Postulate 5 out of the list of assumptions and into the list of demonstrated propositions. Allowed in such a proof would be any of the first four postulates and any of the propositions one could deduce from them. Incidentally, this includes the first twenty-eight propositions in Book I. Euclid apparently had his own reservations about his fifth postulate and delayed using it for as long as he could, finally succumbing with Proposition 29.

Attempts to deduce the parallel postulate from the other four continued through the centuries, coming to a dramatic conclusion in the mid-1800s. We will pick up the story later in the chapter, but we conclude this brief introduction to the *Elements* by pinpointing exactly how Postulate 5 finds its way into the proof of the familiar theorem that the angle sum of any triangle is 180°.

The *Elements* in Action

The first proposition Euclid proves is that it is possible to construct an equilateral triangle using a given finite line segment as one of the sides. Triangles in fact are the subject of many of the early propositions. Proposition 5 states that if a triangle has two equal sides then the angles opposite these two sides must also be equal. Proposition 6 is the converse statement. In Proposition 17, Euclid proves that for any given triangle, the sum of two of the angles is *less than* two right angles. Saying "two right angles" is Euclid's way of saying 180°, and the statement of Proposition 17 feels like a natural precursor to the fundamental fact that the sum of all three angles of a triangle is precisely 180°. But that is not the next proposition, nor is it among the next ten propositions. At this point in his organization of Book I, Euclid is still postponing the use of the controversial fifth postulate, and one gets the sense that he is doing his best to prove this seminal result about triangles without it. Finally, however, he relents.

Proposition 29. *A straight line falling on parallel straight lines makes the alternate angles equal to one another.*

Euclid's short demonstration is not hard to follow, and taking a quick look at the details provides a chance to see a master at work. As he does with most propositions, Euclid starts with a diagram similar to the one in figure 2.4. The lines *l* and *m* are assumed to be parallel, which means that they do not intersect, and the angles marked α and β are what Euclid is referring to when he says

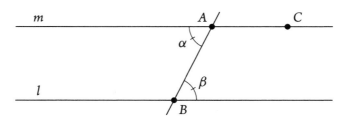

FIGURE 2.4. Diagram for Proposition 29.

"alternate angles," although the other two interior angles would serve just as well.

A favorite strategy of Euclid's is to argue by contradiction. In this proposition he is trying to show $\alpha = \beta$ and so he starts with the *negation* of that statement which is $\alpha \neq \beta$. The goal, as with all contradiction proofs, is to start from this premise and proceed along logical lines until we reach an unacceptable conclusion. At this point we will be forced to retrace our steps and reject the erroneous assumption that $\alpha \neq \beta$ in favor of $\alpha = \beta$.

From the hypothesis $\alpha \neq \beta$, Euclid points out that one of these angles is therefore larger than the other, and for the sake of the argument he takes $\alpha > \beta$. Now $\angle BAC$ when added to α makes 180°, or two right angles, because the combined angle lies along the line m. So, what happens when we take the sum $\alpha + \angle BAC$ and replace α with the smaller angle β? Well, we get that $\beta + \angle BAC$ must be *less* than two right angles. This is precisely the situation for which the parallel postulate was designed! Given that two interior angles on the same side sum to less than two right angles, Postulate 5 says that the straight lines l and m must intersect, which of course is a blatant violation of our hypothesis that they are parallel. This is the sought-after contradiction. Euclid then concludes that $\alpha = \beta$, as desired.

With this proposition established, it was finally possible to state and prove a fundamental truth about triangles.

Proposition 32. *In any triangle . . . the sum of the three interior angles of the triangle equals two right angles.*

The ease and elegance of the demonstration for this seminal proposition goes a long way toward illustrating the allure of the deductive structure of the *Elements,* and of pure mathematics in general. Starting from an arbitrary triangle *ABC* (Figure 2.5), the first step is to invoke Proposition 31 referenced

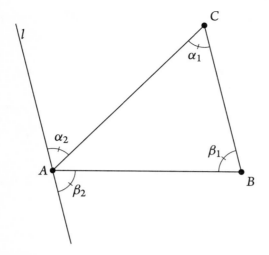

FIGURE 2.5. Diagram for Proposition 32.

earlier.[11] Taking A as our given point, Proposition 31 asserts that we can draw a straight line l through A that is parallel to the line CB. Then, because l and CB are parallel, Proposition 29 asserts that $\alpha_1 = \alpha_2$ and $\beta_1 = \beta_2$. Finally, we know

$$\alpha_2 + \angle CAB + \beta_2 = \text{two right angles}$$

because these three angles taken together lie along the line l. Substituting α_1 for α_2 and β_1 for β_2 yields

$$\alpha_1 + \angle CAB + \beta_1 = \text{two right angles,}$$

and the proposition is proved.

Ubu Roi and Ubu Cocu

But whichever sense is attributed to the piece, *Ubu Roi* is "hundred percent theatre," what we today would call "pure theater," synthetic and creating, on the margin of reality, a reality based on symbols.

—HENRI GHÉON, FROM *L'ART DU THÉÂTRE*[12]

The opening night of Alfred Jarry's first Ubu play has achieved legendary status in the annals of avant-garde theater. What is rarely discussed, however, is the relationship of this event to Euclid's *Elements*, or to mathematics more generally. This is understandable. The script Jarry ultimately chose to perform meant that there was no explicit mention of mathematics in the dialogue. But it was there—disguised, deformed, and under assault, like everything else on the stage that night.

On the evening of December 9, 1896, in Paris, the renowned comic actor Firmin Gémier, playing the lead role of Père Ubu, walked to the front of the stage of the Théâtre de l'Oeuvre to deliver the show's opening line. This was, in fact, the first public performance of any play written by the twenty-three-year-old Jarry. As a playwright, Jarry had built a controversial reputation for himself hosting private performances of his plays in his own home, most often using marionettes in place of live actors, and so the anticipation was thick as literary and art critics from across Paris gathered for the premiere. In his curtain speech, Jarry provocatively explained that although this was not a performance of puppets, the actors had nevertheless gone to great lengths to make themselves as inhuman as possible. "And the action, which is about to start, takes place in Poland," Jarry concluded, "that is to say, Nowhere."[13]

Indeed, the grotesque sight of Père Ubu ambling downstage with his huge, spiraled belly and carrying a toilet brush as a scepter conveyed an unmistakable sense of otherworldliness. And then the first line:

PÈRE UBU: Merdre!

This lyric twist on "merde"—the French word for "shit"—set the tone for a raucous evening whose factual details and significance are still being debated by historians and scholars of modern theater. Often described as a violent explosion of indignation from a shocked bourgeois audience, the riotous premiere is more appropriately characterized as a vociferous heckling match between assembled artistic factions who knew about Jarry's work and came prepared to play their part in a public fracas.[14] Reports on the intensity of the altercations vary with who is telling the story and on which night they attended, but Jarry unambiguously succeeded in generating the cultural spectacle he desired. When the performances concluded, the fight moved to the public papers where columnists debated whether Jarry's play was the precursor to an artistic revolution or the height of imbecilic folly. History now suggests that it was the former, although it would take several decades for the ripples from Jarry's work to be assimilated by the movements of surrealism and Dada and eventually become a touchstone of inspiration for the absurdist movement in twentieth-century theater.[15]

In terms of the implications for Alfred Jarry's life, the events he set in motion in December 1896 had the dual effect of making him a household name in the Paris art scene and simultaneously scaring off any future producers of his plays. The two scheduled performances of *Ubu Roi* would be the only public performances using live actors that Jarry would see of any of his plays during his truncated lifetime.

Schoolboy Pranks

Mathematics enters Jarry's artistic story via the fact that the root source for the infamous Père Ubu was his high school science instructor. When Jarry was fifteen years old he befriended Charles and Henri Morin at the lycée in Rennes, and the trio performed farcical skits primarily at the expense of the school's hapless physics teacher, Professor Hébert. Hébert was an easy target—excessively overweight, pretentious, ineffectual as a teacher—and Jarry and his coconspirators were relentless in their attacks. Jarry antagonized Hébert in class while the character of "Père Héb" became the subject of an ongoing torrent of theatrical mockery.[16] As Jarry's writing evolved, Père Héb morphed into Père Ubu, and Hébert's physics morphed into a new imaginary science that Jarry dubbed pataphysics. "Pataphysics . . . is the science of that which is superinduced upon metaphysics," Jarry wrote several years after Ubu's controversial debut, "whether within or beyond the latter's limitations, extending as far beyond metaphysics as the latter extends beyond physics."[17] The play performed at the Théâtre de l'Oeuvre in 1896 did not mention pataphysics. It was largely based on early material Jarry developed with the Morin brothers. Jarry's alternative scientific worldview was still nascent in Rennes, but it would become a central component of his artistic vision.

After Rennes, Jarry moved to Paris and began publishing poetry and short prose pieces, eventually finding his way into the society of the *Mercure de France*, the leading literary magazine of Paris. Stories of Jarry holding forth in the smoke-filled salons reveal his simultaneously erudite and irreverent nature. He routinely performed the character of Père Ubu in the *Mercure* salon to his audience's comic delight, but when *Ubu Roi* (*Ubu Rex*) was finally staged for the larger public, the reaction was divided and vitriolic. The profane opening line is just the beginning of a strange and obscene play. The first few lines (from here on translated) give a sense of the tone of what is to come:

"Watch out I don't bash yer nut in!" *Ubu Roi*; Andreas
Katsulas (Père Ubu); The Young Vic, 1978. © Donald
Cooper/photostage.co.uk.

PA UBU: Pschitt!

MA UBU: Ooh! What a nasty word. Pa Ubu, you're a dirty old man.

PA UBU: Watch out I don't bash yer nut in, Ma Ubu!

MA UBU: It's not me you should want to do in, Old Ubu. Oh no! There's
someone else for the high jump.

PA UBU: By my green candle, I am not with you. (21)[18]

The insults and violence are a staple throughout, as is the word "pschitt"—one of several imperfect attempts to translate "merdre" into English. The narrative arc of *Ubu Roi* is a simplified distortion of *Macbeth*. Ma Ubu needles Pa Ubu into assassinating the king of Poland, after which he begins wreaking havoc as the new king by executing judges and torturing his subjects. More significant than the farcical threads of plot is the utter depravity of Pa Ubu himself. Made to appear inhuman—Gémier wore an enormously stuffed costume and a mask in the original production—Pa Ubu is foul, dishonest, cowardly, devoid of morals, and quite stupid. He is also potentially funny, depending on whether one can get beyond his utterly offensive nature. Most people outside Jarry's immediate circle of supporters could not.

"It is a filthy hoax which deserves only contemptuous silence," wrote one critic who took heart in the heckling of the audience. "This is the beginning of the end," he went on. "For too long now these pranksters have laughed at us. Enough is enough." Another reviewer christened his comments with the quip, "In spite of the late hour, I have just taken a shower. An essential purgative measure after such a spectacle."[19] There was no real intellectual debate within Jarry's play to parse, but there was one around the play about the direction and purpose of theater more generally. The following firsthand report of the performance, recounted by author and Jarry supporter Catulle Mendès, conveys a palpable sense of what was at stake:

> Whistles? Yes! Screams of rage and ill-tempered laughter? Yes. Torn-up seats about to fly on stage? Yes. The boxes vociferous and shaking their fists? Yes, in a word, a crowd enraged at being tricked. . . . [A]llusions to the eternal imbecility of humanity, its eternal lust, its eternal greed, misunderstood? Yes, and the symbol of its base instincts which erect a tyranny, unnoticed? Yes . . . and more unfunny jokes, offensive absurdities, braying laughter that ends up resembling the macabre rictus of a corpse? Yes, and in fact, the whole piece tedious and lacking any explosion of joy, despite it being so keenly awaited? Yes, yes, yes, I tell you. . . . But nevertheless, make no mistake, there was nothing indifferent about these performances over the last two nights at the Oeuvre, nor were they devoid of significance. From out of the ruckus came a shout: "You would not have understood Shakespeare!" He was right. Let me make myself clear. I am not saying at all that Alfred Jarry is Shakespeare; in him everything Aristophanic has been turned into a repulsive puppet show or a squalid fairground attraction, but believe it or not, despite the nonsensical action and its mediocre structure,

a new type has emerged, created by an extravagant and brutal imagination that is more that of a child than a man.

Père Ubu *exists*.[20]

Ubu Cocu

Mendès admirably communicates what Jarry's inhuman incarnation of theater was dismantling, but he is more uncertain about what Jarry was offering in its place. In his own writing about drama, Jarry's hostility was not directed so much at theatrical realism—he clearly admired Ibsen for instance—as at the sentiment that naturalistic tendencies were superior or represented some kind of hierarchical destiny for art. What Jarry readily objected to was the "horrifying and incomprehensible" attempts by theater to imitate nature as the common person sees it because "it scandalizes those who see nature in an intelligent way."[21] Jarry put himself among a cohort who viewed the world differently—he would say intelligently, others might point out that he was a chronic alcoholic. This truth notwithstanding, Jarry had no patience for the view that he and his fellow artists were, in his own words, "madmen suffering from a surfeit of what [the public] regard as hallucinatory sensations produced in us by our exacerbated senses."[22] As incongruous as it sounds given the exotic landscape of his plays, Jarry was making the case for a personalized brand of realism. The observable world of nature could still be the guiding light of theater, he argued, as long as it was nature as he observed it.

To carry his project in subjective realism forward, Jarry recognized the need to address the objective character of mathematics. By the time *Ubu Roi* was performed, Alfred Jarry had penned a second play featuring the menacing Pa Ubu as protagonist which, unlike the collaboratively devised *Ubu Roi*, was very much his own creation. Jarry's original title for this new play was *Les Polyèdres* (*The Polyhedra*) which he later changed to *Ubu Cocu* (*Ubu Cuckolded*). It is here that he began to explore the role of mathematics and science in his reimagined universe. Instead of profanity, *Ubu Cocu* opens with the following curious speech by a professorial character named Achras.

> ACHRAS: Oh, but it's like this, look you, I have no grounds to be dissatisfied with my polyhedra; they breed every six weeks, they're worse than rabbits. And it's also quite true to say that the regular polyhedra are the most faithful and most devoted to their master, except that this morning the icosahedron was a little factitious, so that I was compelled,

look you, to give it a smack on each of its twenty faces. And that's the kind of language they understand. And my thesis, look you, on the habits of polyhedra—it's getting along nicely, thank you, only another twenty-five volumes!

Tying Euclid's celebrated Platonic solids to Achras in this campy opening speech has several effects: one is to identify Achras as a symbol of the intellectual status quo by associating him with the most well-established bastion of certainty in the Western canon. Another is to set him up to be mocked. In the very next line, Pa Ubu's arrival is announced, and in his usual belligerent fashion he smashes through the door and forces Achras out of his home so Ubu and his family can move in.[23] With Euclid banished, we are then presented with an alternative source of enlightenment.

ACHRAS: Oh but it's like this, excuse me. I was very far from expecting the visit of such a considerable personage . . . otherwise, you can be sure I would've had the door enlarged. But you must forgive the humble circumstances of an old collector, who is at the same time, I venture to say, a famous scientist.

PA UBU: Say that by all means if it gives you any pleasure, but remember that you are addressing a celebrated pataphysician.

ACHRAS: Excuse me, Sir, what did you say?

PA UBU: Pataphysician. Pataphysics is a branch of science which we have invented and for which a crying need is generally experienced. (78)

The crying need was, of course, Jarry's, whose evolving commitment to validating the reality he perceived in his brutal imagination meant that he had to do something about the objective rigidity of traditional science. The most lucid description of pataphysics that Jarry provides is in *Les Gestes et Opinions du Docteur Faustroll, Pataphysicien* (*Exploits and Opinions of Doctor Faustroll, Pataphysician*). In this self-described "neo-scientific novel," Jarry momentarily adopts a Euclidean affect, titling one of the chapters "The Elements of Pataphysics" and defining his central term this way:

DEFINITION. Pataphysics is the science of imaginary solutions, which symbolically attributes the properties of objects, described by their virtuality, to their lineaments.[24]

A bit more elucidating than this opaque statement is Jarry's declaration in the same chapter that pataphysics

... will examine the laws governing exceptions and will explain the universe supplementary to this one; or, less ambitiously, will describe a universe which can be—and perhaps should be—envisaged in the place of the traditional one, since the laws that are supposed to have been discovered in the traditional universe are also correlations of exceptions, albeit more frequent ones, but in any case accidental data which, reduced to the status of unexceptional exceptions, possess no longer even the virtue of originality.[25]

As tempting as it might be to critique Jarry's argument for the insignificance of Professor Hébert's version of science, what is more useful is to appreciate Jarry's motivation for inventing his own "science of imaginary solutions." These passages confirm that underneath Ubu's destructive rampages is Jarry's relatively sanguine conviction that the physical world of our senses—the "traditional one" mentioned above—ought not be given any special place in the hierarchy of possible worlds that we might wish to contemplate.

This optimistic interpretation of Jarry's pataphysics depends very much on its incarnation through Doctor Faustroll, a character who came later and is unconnected to Jarry's theater writing. Returning to the odd action of *Ubu Cocu*, Achras with his sixty years of studying Euclidean polyhedra must deal with Jarry's original pataphysician who does not have much patience for the classical approach to natural philosophy. When Pa Ubu learns that his wife has been cheating on him, he tests out his preferred method of execution by impaling Achras with a long spear that hoists him into the air where his deceased body is left for public display. Pa Ubu's Conscience—a character who is typically kept locked in a suitcase—rescues Achras and resuscitates him. Achras and Conscience then plot revenge which leads to the metaphorically rich moment of Pa Ubu throwing his Conscience into the barrel of sewage in the lavatory.

In the original *Ubu Roi*, the consensus view is that Jarry was taking aim at the whole of bourgeois society and "the eternal imbecility of its humanity," as Mendès intimates in his review. When Pa Ubu continues the destructive orgy in *Ubu Cocu*, Achras's professorial identity has the effect of focusing the target of the playwright's ire toward the established intelligentsia; Pa Ubu and his pataphysics on one side with Achras, Euclid, and Newton and his physics on the other. It's a back-and-forth duel. In act IV, Ma Ubu and Conscience are hiding in the sewage barrel when Pa Ubu comes in to use the

"Oh no! There's someone else for the high jump." *Ubu* adaptation; Anna Prucnal
(Mère Ubu), Georges Wilson (Père Ubu); Cloître des Célestines, Avignon,
1974 (photo: Suzanne Fournier/Gamma-Rapho via Getty Images).

lavatory and falls into the barrel himself. This leads to one of the moments
where the laws of old-fashioned physics have the upper hand. "He (Ubu)
emerges again," the stage directions read, "thanks to the Archimedean principle." (99)

Père Ubu Exists

A few years later, Jarry would complete the Ubu cycle with a third play called
Ubu Enchaîné (*Ubu Enchained*), but neither this nor *Ubu Cocu* were performed
until many years after Jarry died. One reason for this is that Jarry's death came
much sooner than it should have.

After the controversial production of *Ubu Roi*, Jarry penned a series of innovative and enigmatic novels, many of which were autobiographical in nature. A number of them were published but none achieved any commercial success, and Jarry was consistently in financial difficulty. He lived in a tiny Paris apartment, literally on the second-and-a-half floor, with a ceiling so low that few people besides himself had room to stand up straight. Later, Jarry moved out of the city into a small hut along the banks of the Seine where he happily subsisted, in large part, on fish that he caught himself. He also drank excessively—absinthe if it was available and eventually ether. Beyond his plays and novels, he pursued opportunities in journalism and marionette theater, but the steady decline in his physical health eventually made these endeavors untenable. Although the assumption was that Jarry's alcoholism was the primary cause of his demise, his fever-induced delirium and ultimate death in 1907 were the result of damage to his brain from undiagnosed tuberculosis. Jarry was just thirty-four years old when he died.

Although Jarry never saw another proper production of any his Ubu plays after the one at l'Oeuvre, one of the most curious facts surrounding his artistic journey is that the character of Ubu was performed with unwavering conviction every day for the last decade of his life—by Alfred Jarry himself.[26] This transformation into the persona of his artistic creation is perhaps the strongest piece of evidence for the argument that Jarry's assaults on theater and mathematics were part and parcel of a crusade for a new kind of realism of imagined possibilities. Jarry's style of daily dress, his manners, his home decor, all morphed into something consistent with Père Ubu, which not surprisingly became the name by which Jarry was universally known. This aspect of his biography makes trying to disentangle his literary life from his lived life exceptionally complicated and, to a large extent, not really a possibility.[27] Jarry did not become Ubu in some kind of delusional sense. He was not crazy or incapable of making rational judgments. He continued to have meaningful relationships with friends and fellow artists; he wrote some of his most traditional and accessible novels long after he had assumed the public identity of Père Ubu. With his own idiosyncratic flair, he managed to engage with the business of daily life, but he did so in a way that oscillated between optimistic eccentricity and self-destructive impulsivity. The choices he made did not always make sense to an outside observer, but in Jarry's mind they were carefully reasoned. Whether consciously or not, Jarry was using the example of his own life to make a compelling case for the legitimacy of alternate worlds created by the imagination.[28]

Mathematics and the Imagination

In 1899, Jarry wrote an essay entitled "Commentary and Instructions for the Practical Construction of the Time Machine." Adopting the tone and format of a scientific paper, Jarry set out an argument for the theoretical possibility of building the device at the center of H. G. Wells's book *The Time Machine*. Although intended as a work of experimental prose, it was coherent enough to catch the attention of several eminent scientists, including Samuel Langley, the future namesake of Langley Air Force Base, and William Crookes, a fellow and president of the Royal Society. Neither was convinced of the efficacy of Jarry's proposal, but both mistook Jarry's paper for a sincere scientific argument rather than the work of fiction that it was.[29]

This anecdote highlights the dilemma of how to approach Jarry's work. Up close and in isolation, much of his writing comes across as juvenile and nonsensical, but those judgments are often the result of bringing some preconceived notion of what writing was meant to represent. Juvenile or not, Jarry made a lasting impression on the artists that followed him. Pablo Picasso put himself among Jarry's followers, Antonin Artaud named his theater after Jarry, and the surrealists essentially made him into a patron saint. In 1948, the Collège de 'Pataphysique was founded in Paris which, among other things, became a center of support for the mathematically inspired literary movement of Oulipo.

Jarry's working knowledge of mathematics is difficult to quantify. He was certainly widely read. In *The Exploits and Opinions of Doctor Faustroll*, for instance, he finds reason to mention Kelvin, Navier, Poisson, Cauchy, and Cayley—but they are inevitably blended into something supernatural. In the very last section of *Faustroll*, entitled "Concerning the Surface of God," Jarry adopts the most explicitly mathematical syntax of anything he wrote. Taking inspiration from the theory of the Holy Trinity, he postulates that God can be represented as an equilateral triangle like the one in figure 2.6.[30] Jarry describes this figure very precisely, including the labeled variable names, and then sets off to calculate the triangle's area. Along the way, he properly invokes the Pythagorean Theorem as well as the formula for the area of a triangle, demonstrating that pataphysics, whatever else it entails, is not entirely at odds with Euclid's geometry. It's also not restrained by it. The pursuit of the divine usually requires some kind of negotiation with the infinite, and Jarry's calculation of the "surface of God" is no exception. By incorporating the infinity symbol ∞ into his algebra and taking advantage of its indeterminate nature,

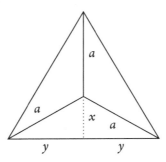

FIGURE 2.6. Alfred Jarry from *Faustroll*: "[L]et us suppose God to have the shape and symbolic appearance of three equal straight lines of length *a*, emanating from the same point and having between them angles of 120 degrees."

Jarry's calculation ends with the discovery that the variable *a* is equal to zero, and that the triangle must then collapse to a single point. "Therefore, *definitively*," Jarry poetically concludes with a nod to the three-in-one dogma of the Holy Trinity, "God is the tangential point between zero and infinity. Pataphysics is *the* science . . ."[31]

Jarry's uneasy relationship to mathematics is on full display in this pseudo-calculation, most notably in the way that mathematics represents both friend and foe. Sometimes he enlists math's sterling reputation to justify his alternative worldview; other times he attacks mathematics as part of the establishment thinking he is so eager to undermine. In this particular case, Jarry constructs a proof for a new ethereal universe using geometric tools that for two thousand years had been anchored to the structure of the old physical universe. Jarry resorts to some algebraic sleight of hand involving the mathematical infinite to make his argument, but at the time when Jarry was creating his new science of imaginary solutions there were more legitimate ideas afoot in mathematical circles very much in the spirit of Jarry's pataphysics. To borrow Doctor Faustroll's controversial phrasing, it is perfectly accurate to say that a handful of nonconforming mathematicians "describe[d] a universe which can be—and perhaps should be—envisaged in the place of the traditional one."

Jarry was in fact aware of the emergence of alternatives to Euclid's geometry, but he did not pursue them at any length.[32] In his defense, it is not obvious how the technical ideas of non-Euclidean geometry could be employed by a creative artist, particularly one focused on writing for the stage. But this is precisely what transpired a decade after Jarry died. To find the impact of the non-Euclidean revolution on the development of theater, we have to travel many miles from the vibrant theaters of Paris to a small city in what feels like the middle of nowhere—that is to say, Zakopane, Poland.

Tumor Brainiowicz

Papa! I don't want to hear any of that. What have all of you done to infinity?

—IRENE BRAINIOWICZ, FROM *TUMOR BRAINIOWICZ*

When Alfred Jarry died in 1907, Stanislaw Ignacy Witkiewicz was twenty-two years old. The son of a well-known artist, Witkiewicz adopted the name Witkacy—a conflation of his middle and last names—at least in part to distinguish his work from his father's. Witkacy was best known during his lifetime as a painter, making a living at one point from portrait commissions, but he was a prolific writer as well, authoring roughly fifty plays in addition to a series of novels and philosophical investigations.

Witkacy had no formal education and was relentlessly driven by his father to be unrestrained and passionate in his life's pursuits. "In intellectual concepts reach to infinity," his father wrote to his son. "Don't be encumbered by any cast limitations, by any professional prejudices, by any petty-minded forms of individual or class egotism . . . live in the future. Constantly stand on the heights from where you can see the farthest horizons and spread the wings of thought and action for flight beyond time."[33] This experiment in nonconformist childrearing worked in the sense that Witkacy grew into a young man with an unbridled interest in art, ideas, philosophy, nature—especially tropical forests—and non-Western cultures. But the list of passions didn't stop there; it also included various unattainable women, hallucinogenic drugs, and his own spectacular individuality which made Witkacy difficult to get along with.

Perhaps as a reaction to his overbearing anti-czarist father, Witkacy joined the Russian army as an officer at the outbreak of World War I, which, among other things, provided him with a ringside seat for the Communist revolution in 1917. This was a seminal event in his young life. Before this, Witkacy's creative energies were largely unconcerned with the real world, but the Russian Revolution served as a bridge between Witkacy's metaphysical musings and the social and political realities of his time. It would influence every aspect of his artistic vision for the remainder of his life.

After the war, Witkacy returned to a newly independent Poland and embarked on a staggeringly productive decade during which he wrote nearly all of his dramatic works (many of which were eventually lost.) Scanning the list of titles provides a glimpse into the ironic sense of humor of the author:

Metaphysics of a Two Headed Calf: A Tropical Australian Play in Three Acts,

The Madman and The Nun or There Is Nothing Bad Which Could Not Turn into Something Worse: A Short Play in Three Acts and Four Scenes,

Sluts and Butterflies: A Comedy with Corpses in Two Acts and Three Scenes,

The Crazy Locomotive: A Play without a Thesis in Two Acts and An Epilogue.

The other conclusion that emerges from the long list of curious titles is the author's affinity for mathematics. In 1921, Witkacy wrote *The Independence of Triangles: A Play in Four Acts*, and later that same year appeared two more plays with provocative subtitles: *Gyubal Wahazar: A Non-Euclidean Drama in Four Acts*, and *The Water Hen: A Spherical Tragedy in Three Acts*. It is an overstatement to call mathematics a focus of Witkacy's writing, but it was a regular color in his palette, and it turns out to be central to understanding his evolution as a playwright.

Of the scores of plays that Witkacy wrote during this period, his so-called non-Euclidean plays are examples of the artist at the pinnacle of his craft. Daniel Gerould is the scholar who introduced Witkacy to the English-speaking world in the 1960s by translating his plays and publishing multiple books of criticism. In Gerould's view, Witkacy's prolific output of scripts was a process of experimentation that eventually led to a small handful of skillfully wrought plays capturing Witkacy's artistic vision. On a very short list of what Gerould calls "theatrical masterpieces" appear *Gyubal Wahazar* and *The Water Hen*, plays that he declares "give drama a totally new shape with non-classical dimensions, almost abstract and geometrical, by analogy to modern mathematics and physics."[34]

What makes these plays non-Euclidean is not their content. In a way that he had to work out, Witkacy's intent was to create a theatrical form modeled on ideas that emerged in geometry from contemplating the veracity of Euclid's fifth postulate. This would require many iterations. An early play in this ambitious project was *Tumor Brainiowicz: A Drama in Three Acts with a Prologue*. More so than work to follow, this play actually features a robust amount of explicit mathematical source material, enough to give it a strong claim to being the first math play in modern drama. *Tumor Brainiowicz* was also the first of Witkacy's plays to be produced, and while not very well developed in terms of its inner structure, it provides an illuminating introduction to the author's eccentric personality and rogue sense of propriety with regards to the theatrical norms of his day.

Transfinite Theater

The eponymous lead character in *Tumor Brainiowicz* is described in the stage directions as "a famous mathematician of humble origins" who is not only built like a wild ox but, in his frequent gnashing and bellowing tantrums, has the disposition of one. As his name suggests, there is something dangerous and uncontrollable about him. Tumor has produced a gaggle of children by three marriages, and we get the sense early on that he is similarly prolific in terms of his mathematical output.

Although Tumor is completely fictitious, the primary inspiration for his research is the groundbreaking explorations into the infinite carried out by Georg Cantor about thirty years before. The major thrust of Cantor's work regarding the infinite was a rigorous means of distinguishing between the sizes of infinite sets. In particular, he developed a theory which yields a meaningful way to say that one infinite set is larger than another. One of his earliest results is a demonstration that the infinity of the set of all real numbers—numbers expressible as potentially nonterminating decimals—is of a distinct and larger order than the infinity of the set of positive integers. Cantor went on to show that this hierarchy of infinities extends ever upward beyond the infinity of the real numbers, and one of the upshots is the creation of a system of arithmetic that applies to these new "transfinite" numbers.

The impact of Cantor's work was staggering, but Cantor apparently pales in comparison to Witkacy's mathematical protagonist.

> BRAINIOWICZ: I have lived alone, immersed in my calculations. I begat a new world. Cantor, Georg Cantor, is a mere infant compared to my definitions of infinity, and Frege and Russell are the paltry decanting of the Greek void into the void of our own times with their definition of number, compared to what I thought up this morning. (80)[35]

Naming his new class of transfinite numbers after himself, Brainiowicz's "tumors" are so potentially sinister that the Mathematical Central and General Office sends special agent Alfred Green to investigate. His arrival coincides with the revelation that Tumor is in love with his stepdaughter Iza, and results in Iza being abducted by Green's agents while Tumor is wrestled and eventually hog-tied like the beast he is.

As should be apparent, there is a wild, unrestrained feel to the flow of the action. The mathematics is a catalyst for the chaos—the paradoxical nature of infinity being a good match for the overall turbulent tone of the play. Mixed in with the mathematics are a host of interpersonal dynamics that randomly crisscross to create sequences of passionate but incongruous emotional explosions. The effect is melodramatic, comic, and confusing. Tumor's genius puts him at odds with the authorities, his unexplained desire to be a poet is at odds with his mathematics, and his slovenly roots put him at odds with his current wife of noble blood and her daughter Iza who is, at least in act II, the object of Tumor's uncontrollable passions.

BRAINIOWICZ: (*fists clenched, to Iza*) Oh! If I could first calculate your differential, analyze each infinitesimal particle of your cursed, russet-colored blood, each element of your hot, parched whiteness, and then take it, mash it, integrate it and finally comprehend wherein lies the infernal strength of your unattainability that burns and consumes me down to the very last tissue of my lowborn, slobbish flesh. (66)

"My infinity is no symbol," is Iza's retort.

Regardless of its degree of coherence, the dense vocabulary of mathematical terms makes for an effective linguistic landscape, similar to the way Witkacy uses strong swaths of colors in his elaborate and detailed set directions. The sterilized white children's room of act I is replaced by the purple flowers and red cliffs of the tropical island of Timor, where we find Iza and Brainiowicz at the opening of act II. Events on Timor include an angry volcanic eruption, an interracial kiss, and the wild Brainiowicz finding he is no more suited to the primitive island culture than he is to the hypocritical European world that Green represents. Tumor murders the island king and installs himself as leader, but to no avail. In fact, he is "joyful" when Green arrives at the end of act II to bind him in ropes a second time and return him to the so-called civilized world.

Although this may sound unlikely, the third act follows an even more bizarre narrative arc. Green—who has married Iza during the act break—begs Brainiowicz to teach him about his latest research into the transfinite number "tumor–one," which Green explains is "the number of a certain plurality higher than all the pluralities known to him and to Satan alone." For what it is worth, there are more nods to Cantor here who employed the Hebrew letter aleph in his notation. In Cantor's theory, "aleph-one" refers to a particular class of infinite sets, although in no sense was it the largest. Cantor, and most likely Satan, knew of infinities larger than aleph-one.

Tumor is alternately betrayed and defended by his family members, and along the way his wife hurls their infant son from the playroom window. Despite the fact that he is now married to Iza, Green makes a play for Tumor's wife while simultaneously smearing Tumor's reputation in the public papers. In a final effort to set himself free, Tumor declares his intention to escape to Australia, but falls dead when he sees the ghost of the tropical island king he killed back on Timor. Not content to simply kill off his mathematical hero, Witkacy continues to run roughshod over the rules of playwriting and introduces an entirely new character in the last moment of the play. Lord Arthur Persville, the "sharpest geometrician on the planet" and "leader of fashion" strolls in wearing a striped tailcoat to announce that his cunning has enabled him to steal Tumor's latest research, as well as his latest girlfriend, and the two depart for Australia to create "true transcendental dynamics."

Witkacy's Debut

Written late in 1919, *Tumor Brainiowicz* was performed in Crakow the following year, but with echoes of Jarry's premier, it clashed so mightily with the tastes of the time that critics set out to destroy not only the play but the playwright as well. The tenor of the reviews was so intense as to border on disturbing:

> It seems that we are watching and hearing the ravings of a syphilitic in the last stages of creeping paralysis. . . . Despite the relativity of our ideas about art, despite the experiences of history, which teaches restraint and cautious appraisal of transitional phenomenon which at the beginning are embryonic and not fully formed—Witkiewicz's play is a total uncertainty from which nothing can ever arise. For it does not have a place in any line of development. It is an unnatural clinical abortion. It should be put in alcohol and studied by psychopaths.[36]

Amid this rocky beginning for the math play, the Cracow Mathematical Society had to act quickly to present Witkacy with a bouquet of lilies—which reportedly occurred during the intermission on one of the two nights that the play actually ran.[37] A revival of *Tumor Brainiowicz* five years later fared even worse. In that case, the actors simply refused to perform, and the run was canceled before it had a chance to start.

Witkacy would eventually manage to get a handful of his other plays performed, in part by founding his own amateur theater company, but his dramatic work was not embraced on any level during his lifetime—nor in the first few decades that followed. "In the old days, a Polish artist could at least count on his death to bring him recognition," Witkacy quipped in an interview from the late 1920s.[38] And Witkacy's day would finally come, first in Poland in the late 1950s and then across Europe in the '60s and '70s, where Witkacy came to be widely recognized as a pioneer of modern avant-garde theater. His posthumous fame rose to the point that the year 1985 was declared the "Year of Witkacy" in Poland, and his self-portrait appeared on a postage stamp. In 1988 his remains were exhumed from a nondescript country grave and returned to his hometown where he was given an extravagant state funeral accompanied by a weeklong arts festival.

Tumor Brainiowicz reveals Witkacy's uninhibited versatility along with his dark and self-effacing sense of humor. These characteristics are staples of his writing, but this exuberance alone would not have sufficed to earn Witkacy the respect as an innovator that he eventually received. Underneath the chaotic action of his plays was a philosophical idea driving his creative instincts. Witkacy was prolific as a playwright, and each new play he wrote represented a refinement of a conception that the highly skilled painter had for transferring his ideas about abstract form from the canvas to the stage. Witkacy sketched out his ideas in *An Introduction to the Theory of Pure Form in the Theater*, a treatise whose publication coincided with the self-described non-Euclidean plays *Gyubal Wahazar* and *The Water Hen*. While Witkacy riffed on Einstein's theory of relativity and Cantor's analysis of the infinite in *Tumor Brainiowicz*, it was the nineteenth-century revolution in geometry that provided the theoretical roadmap for his new kind of theater. To create his vision of "pure form" in a world dominated by naturalistic drama, Witkacy modeled his artistic guerrilla war on the overthrow of Euclid's *Elements*, an event in the history of mathematics whose impact is often compared to the Copernican revolution for the way that it changed some of the most fundamentally held assumptions about the nature of the universe.

Non-Euclidean Geometry

I have created a new universe from nothing.

—JÁNOS BOLYAI, FROM A LETTER TO HIS FATHER[39]

From the beginning, Euclid's fifth postulate was a thorn in the side of geometers:

Postulate 5: That, if a straight line falling on two straight lines make the interior angles on the same side less than two right angles, the two straight lines, if produced indefinitely, meet on that side on which are the angles less than the two right angles.

Although it was universally accepted as a true statement about the behavior of straight lines, it was just as universally agreed that it ought to be proved as a theorem rather than simply accepted as a self-evident postulate. One of the more notable eighteenth-century mathematicians to work on this problem was Johann Lambert, who gives a very clear summary of the issue in his *Theory of Parallels.* "This work deals with the difficulty encountered in the very beginnings of geometry and which . . . confronts every reader of Euclid's *Elements,* for it is concealed not in his propositions but in the axioms with which he prefaced the first book." Lambert is of course referring specifically to Postulate 5. "Undoubtedly, this basic assertion is far less clear and obvious than the others," he goes on to say. "Not only does it naturally give the impression that it should be proved, but to some extent it makes the reader feel that he is capable of giving a proof, or that he should give it. However, to the extent to which I understand this matter, this is just a *first* impression."[40]

In fact, Lambert never found a proof that gave him enough confidence to publish *Theory of Parallels* during his lifetime. Someone who did go public with a proof was Girolamo Saccheri, who in 1733 published *Euclid Freed of All Blemish or A Geometric Endeavor in Which Are Established the Foundation Principles of the Universal Geometry.* Despite the boldness of the title, the author makes the confession that he was not able to complete his proof of Euclid's fifth postulate "without previously proving that the line, all of whose points are equidistant from an assumed straight line lying in the same plane with it, is equal to [a] straight line."[41] Remembering the Euclidean lexicon (line = curve, whereas straight line = line), Saccheri is saying that if we start with a straight line l and take the collection of points a fixed distance away from l on the same side, we indeed get another straight line (Figure 2.7).

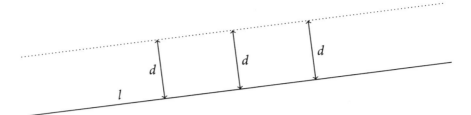

FIGURE 2.7. Does the locus of points a fixed distance from a given straight line *l* form another straight line?

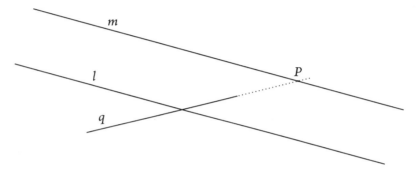

FIGURE 2.8. If the straight line *q* crosses *l* must it also intersect *m*, assuming *m* and *l* are parallel?

Euclid's definition of parallel says nothing about the lines being everywhere equidistant, requiring only that the straight lines do not intersect, and so the very intuitive-sounding idea that Saccheri describes represents a substantial mathematical leap. In fact, it turns out to be logically equivalent to Euclid's fifth postulate. Because Saccheri's argument regarding this equidistant property of straight lines is ultimately flawed, his entire proof of Postulate 5 becomes circular—and thus invalid.

And so it was with all the attempted proofs of Postulate 5. Proclus produced a proof in 450 CE that effectively assumed that if a straight line crossed one of two parallels, it would necessarily cross the other also (Figure 2.8). In the seventeenth century, John Wallis found what appeared to be a compelling argument for the parallel postulate, but one of the steps required the construction of a triangle similar to a given triangle. "Similar" means that the new triangle has precisely the same angle measures and represents a rescaled copy

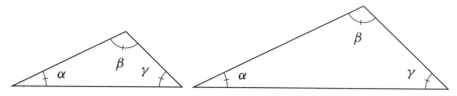

FIGURE 2.9. Similar triangles have equal angles, and thus the same shape.

of the original (Figure 2.9). In both of these cases, what the authors had really done was not prove Postulate 5, but simply *replace* it with a new assumption that turned out to have the same logical strength.

Now one could reasonably argue that these new assumptions were more primitive than Euclid's formulation of Postulate 5, and that they therefore had the benefit of sounding more like a self-evident axiom. Although there is some merit in this point of view, it failed to advance the more fundamental quest, which was to remove Postulate 5 from the list of postulates altogether. As close as mathematicians seemed to get to this holy grail, it somehow managed to stay just out of reach. Proof after proof was proposed, but in every case a flaw was discovered, and most often this flaw was a variation on the same theme. Every purported proof, although ostensibly based on only the first four Euclidean postulates, inevitably invoked some innocuous-sounding property that required the fifth as well. Logical circularity was the death knell, and it rang steadily throughout the eighteenth century and into the nineteenth.

The author of a series of particularly noble attempts was the distinguished mathematician Adrien-Marie Legendre. Lacking a proper justification for why an innocent-sounding property in one of his more promising proofs was valid, Legendre resorted to asserting that it would be "repugnant to the nature of the straight line" for it to be otherwise, a turn of phrase that Saccheri had notably invoked for a similar purpose almost a century earlier.[42] Legendre was among the last of the true believers that a proof would be found—indeed, he convinced himself on several occasions that he had done it, although he never satisfactorily convinced the larger mathematical community of his achievement. Other veterans of the struggle were losing confidence. One of the most telling quotations from this era comes from the Hungarian mathematician Wolfgang Bolyai, who wrote the following advice to his mathematically talented son János:

> You must not attempt this approach to parallels. I know this way to its very end. I have traversed this bottomless night, which extinguished all light and

joy of my life. I entreat you, leave the science of parallels alone. . . . I thought I would sacrifice myself for the sake of the truth. I was ready to become a martyr who would remove the flaw from geometry and return it purified to mankind . . . yet I have not achieved complete satisfaction. . . . I turned back when I saw that no man can reach the bottom of the night. I turned back unconsoled, pitying myself and all mankind.[43]

János, however, belonged to a new generation, and he came at the problem with a new idea. Staring out at the morass of mathematical wreckage that had accumulated around Euclid's fifth postulate, there was one assumption that every one of these attempts had in common: They were all carried out with the unshakable belief that Postulate 5 was an inviolable truth. The parallel postulate had achieved the status of natural law, and regardless of whether it was taken as a postulate or proved as a proposition, no one doubted that it was included in the universal geometry that governed space. But why did this have to be? Because Euclid said so? And Newton agreed? Challenging the most fundamental assumptions in mathematics and metaphysics at the time, János Bolyai asked the nonrhetorical question: *What happens if the parallel postulate isn't true?*

Strange New Universe

In fact, this question had been asked before but always in an attempt to *prove* the parallel postulate via an indirect argument. The logic goes like this: One starts by assuming the negation of Postulate 5 (or something logically equivalent) and then heads off in search of a contradiction, which would then imply that Postulate 5 had to be true. But the contradiction never seems to come. What does come is a series of strange and counterintuitive conclusions.

One of the first things we can do is construct a triangle whose angles add to less than 180°. Isn't this a contradiction? Actually, no. Because we *used* Euclid's fifth postulate in our earlier proof of Proposition 32 (which says the angle sum of any triangle equals 180°), this bedrock theorem is no longer an established truth of our geometric universe. Using only the first four Euclidean postulates, it is possible to prove that the angles of a triangle cannot sum to more than 180°, but proving angle sums equal 180° requires the parallel postulate, which is no longer among our axioms.

The spotting of this first mythical beast—a triangle with angle sum less than two right angles—is just the beginning. One quickly shows that *every*

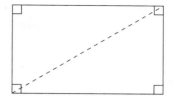

FIGURE 2.10. Because the four right angles of the rectangle add to 360°, one of the two triangles must have an angle sum of at least 180°, which is not allowed. Thus, there are no rectangles.

triangle has angle sums that are less than two right angles, from which it follows that *rectangles don't exist*. Why not? Because if we start with a four-sided figure whose four right angles add up to 360°, then inserting a diagonal produces two triangles, at least one of which must have an angle sum of 180° or more, and such a triangle cannot exist (Figure 2.10).

Even more perplexing than the disappearance of rectangles is the fact that it is no longer possible to rescale figures without deforming them. To see precisely what this means, consider the similar triangles depicted in figure 2.9. In a Euclidean frame of mind, it is natural to imagine the two triangles as having the same shape with the triangle on the right possessing proportionately longer sides. As utterly natural as this seems, it represents another scenario that can never happen if Postulate 5 is replaced with its negation. If the corresponding angles of two triangles are equal, then so are the corresponding sides. Said another way, if two triangles are similar, they must be congruent.

Determining Area with Angle Sums

So what does happen if we take, say, an equilateral triangle and double the length of each side? It turns out that the enlarged triangle has *smaller* angles, which add to even less than they did before. The sum of the three angles varies from triangle to triangle, and the amount by which the sum falls short of the old 180° mark is proportional to the triangle's area. Because the angle sum of an equilateral triangle determines its size, we now have a way of specifying a *length* by specifying an *angle*. For instance, a civilization living in a world where Postulate 5 is false could declare one "meter" to be the side of an equilateral triangle where each angle is precisely 45°.

To appreciate just how bizarre this is, let's compare it to the Euclidean world where Postulate 5 is among the axioms. If Postulate 5 holds, every equilateral

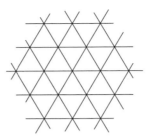

FIGURE 2.11. A standard tiling of the Euclidean plane using equilateral triangles.

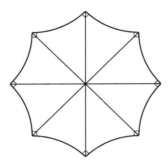

FIGURE 2.12. Equilateral triangles with 45° angles form a right-angled octagon.

triangle has three angles of 60° each. Consequently, six of these triangles fit perfectly together to form a regular hexagon (6 × 60 = 360°), and this pattern can be extended to tile a large floor. The size of the tiles is of absolutely no consequence. Practically speaking, if the floor were exceptionally large, we could use larger triangular tiles, and they would fit together just as well (Figure 2.11).

None of this is true for our imagined civilization living in a world governed by the negation of Postulate 5. Every equilateral triangle has three equal angles that are strictly less than 60°, so putting six together at one vertex always leaves a gap. They no longer fit! But not to worry. By selecting different size tiles, we can adjust the angles of our equilateral triangle. Thus, for some *fixed* length we get tiles with three 45° angles, and these *do* fit together, provided we use eight at a time (Figure 2.12).

The resulting regular octagons have eight right angles—one at each corner formed by the two 45° triangular corners—and these right-angled octagons therefore fit perfectly together in a tiling pattern, for the same reason that square tiles work so nicely on Euclidean floors. A visual image of this tiling appears in figure 2.13, using a model for this new geometry due to

FIGURE 2.13. A Poincaré model illustrating a tiling of the non-Euclidean
plane using equilateral triangles with 45° angles.

French mathematician and philosopher Henri Poincaré.[44] (Not coincidentally, Poincaré is listed among Witkacy's sources in the preface to *Tumor
Brainiowicz*.[45])

When Lambert realized that negating Postulate 5 led to an absolute unit
of length, he temporarily let himself muse that "there is something exquisite
about this consequence, something that makes one wish that [it] were true."[46]
But he quickly reined himself in from this type of heretical thinking. For those
with Euclid's geometry encoded into their DNA, the notion of a right-angled
stop sign or parallel lines that grow farther apart was ultimately "repugnant to
the nature of the straight line" and had to be rejected as logically contradictory.

But these results weren't actually contradictory. As disorienting as it
sounded, they all resided, contradiction free, inside an *alternative* system of
geometry. This was the younger Bolyai's insight. In a letter to his father in 1823,
János wrote:

> I have now resolved to publish a work on parallels. . . . I have not yet com-
> pleted the work, but the road that I have followed has made it almost
> certain that the goal will be attained, if that is at all possible: the goal is not
> yet reached, but I have made such wonderful discoveries that I have been

almost overwhelmed by them, and it would be the cause of continual regret if they were lost. When you see them, you too will recognize them. In the meantime I can say only this: *I have created a new universe from nothing.*[47]

Others were also starting to discover the possibility of a consistent geometry other than Euclid's. In the far reaches of Siberia, Nikolai Lobachevsky was charting a similar path, and he would start to publish his ideas in 1829. The third name attached to the discovery of this new geometry—in fact the one who coined the term "non-Euclidean"—was the incomparable Carl Friedrich Gauss. Gauss was the preeminent mathematician of his age, and it was to Gauss that Bolyai's father sent János's manuscript on parallels. Would Gauss endorse these revolutionary ideas, or would he reject them as nonsensical? This was likely the question on János's mind, but Gauss's reply represented a third possibility that János did not anticipate:

> If I begin with a statement that I dare not praise such a work, you will of course be startled for a moment: but I cannot do otherwise; to praise it would amount to praising myself; for the entire content of the work, the path which your son has taken, the results to which he is led, coincide almost exactly with my own meditations which have occupied my mind for some 30 to 35 years. On this account I find myself surprised to the extreme.[48]

There is ample evidence from Gauss's private correspondences that he had indeed not only found his way to this new geometry but worked out the details to his complete satisfaction. "The theorems of this geometry appear to be paradoxical and, to the uninitiated, absurd," he wrote to a colleague in 1824. "But calm, steady reflection reveals that they contained nothing at all impossible."[49]

Why did Gauss hesitate to go public with his ideas? His prestige was unmatched, even in his own time, so it is inconceivable to think he was concerned about damaging his reputation. In an 1829 letter to a colleague, he wrote that he feared "the howl from the Boeotians," and he later told another he had "a great antipathy against being drawn into any sort of polemic."[50] For Gauss, these were ideas ahead of their time, and he was content to hand the revolution over to the next generation. In 1920, the thirty-five-year-old artist Stanislaw Ignacy Witkiewicz must have also sensed he was ahead of his time, but unlike Gauss, Witkiewicz set about to push his fellow artists into the future.

Gyubal Wahazar

Oh, Reality—what are you really?!

—FATHER UNGUENTY, FROM *GYUBAL WAHAZAR*

A starting point for understanding Witkacy's fascination with modern geometry is his interest in pushing against the rising tide of realism in the theater. What Ibsen, Strindberg, and others had started was being carried forth in Witkacy's day by writers like George Bernard Shaw and Anton Chekhov. These great playwrights probed the social and psychological realities of their day, creating lifelike portraits of what they found. Like Alfred Jarry before him, Witkacy had no interest in writing plays in this style.[51] In Jarry's frontal assault on conventional theater, Euclid was a natural target. Euclidean geometry was mathematical realism—the geometry of empirical space that provided the framework for Newtonian mechanics. Using Euclid's geometry and applying his style of logic, Newton made rational sense out of everything we see. Jarry's strategy entailed replacing established science with a concoction from his own imagination, but Witkacy took a different path. Witkacy was sympathetic to Jarry's desire to undermine the certainty of the Euclidean universe, but what the Polish playwright understood better than his French predecessor was that contemporary mathematics had already done the hard work of supplanting Euclid for him. His 1921 play *Gyubal Wahazar*, which has the alternate title *Along the Cliffs of the Absurd*, includes the moniker "a non-Euclidean Drama in Four Acts" on the title page. What kind of artistic translation did the playwright have for these mathematical ideas?

Gyubal Wahazar gets its name from its central character, who is a maniacal dictator of a state where "everyone knows Einstein's theories," where "they're teaching differential calculus in high school," and where daily life is carried out in a "six-dimensional continuum," whatever that might mean. Along with his crimson jacket, wide, light green pants, and violet shoes, Wahazar's most distinctive feature is an uncontrollable personality that oscillates regularly between wild, roaring tantrums and meditative, almost comatose dream states. The frequency of these oscillations is about two complete cycles per act. At his frenetic peak he becomes a ranting lunatic, frothing at the mouth (the stage directions recommend soda tablets to achieve this effect) and usually ordering someone executed or tortured. When his mood starts to contract, a gaggle of other characters from Wahazar's various attendants sweep in to fill the void and challenge his power. The overall action of the play is like a series of foaming ocean waves that crash violently on the shore and then recede for a short respite before the next swell.

The play opens with a crowd of agitated petitioners waiting in a chamber of Wahazar's palace. Some have been there for hours, some for months, when Wahazar bursts in, spraying white foam over his crimson jacket. "Haaaaaaaaaaaaa!! Throw that old carcass out," a frothy Wahazar screams at a lady who has started to go mad from the long passage of time. "Oh! How atrocious!" a gentleman laments as the lady is carried away. "She over-waited." The comic edge makes for an incongruous accent to the dark, dystopian world Witkacy creates, in part by incorporating a distorted and sinister form of biological science. Among the other petitioners are several women who we learn are to be systematically tested to see whether they are female enough to become "mechanical mothers," or if instead their fate is to become "masculettes." In the latter case, the women will be "turned into men by means of the appropriate transplant of certain glands." Despite the totalitarian circumstances, there is no simple way to summarize the relationship between the people and their oppressor in this play. Throughout his surging mood swings, Wahazar is variously worshipped, mocked, threatened, feared, and loved.

"Even Einstein's theory is incorporated in my system, as a small detail." *Gyubal Wahazar*; actors unknown (Father Unguenty and his disciples); Teatr Narodowy, Warsaw, 1968 (photo: Lubak).

Amid the twisted assimilations of modern science, Witkacy regularly draws attention to his play's geometric subtitle by having Wahazar's subjects refer to him as "your psychic non-Euclideaness," presiding over a "non-Euclidean state." One challenge to Wahazar's rule comes from his old mathematics professor, Father Unguenty, who is now a high priest of a sect of "Perpendicularists." In an effort to woo more followers, Father Unguenty preaches his own strange philosophical cocktail of science and religion:

> FATHER UNGUENTY: Even Einstein's theory is incorporated in my system, as a small detail. For physicists, the world is finite and non-Euclidean, for me it's infinite and amorphous. Real space has no structure—that is the Absolute Truth, which includes Physical Truth as a mathematical convenience, good for a certain method of grasping phenomena. Understand? (138)[52]

Well, no, we don't really understand, but then this is heady high priest material.

Dream Reveries

As with *Tumor Brainiowicz*, the standard rules of narrative consistency are suspended in *Gyubal Wahazar* as the playwright addresses a higher calling. In *An Introduction to the Theory of Pure Form in the Theater*, Witkacy describes his dramaturgical philosophy this way:

> On leaving the theater, the spectator ought to have the feeling that he has just awakened from some strange dream, in which even the most ordinary things had a strange, unfathomable charm, characteristic of dream reveries, and unlike anything else in the world.[53]

It is in pursuit of this goal—creating a world where ordinary things have a strange, unfathomable charm—that the confluence of non-Euclidean geometry and theater begins to pay dividends. More or less simultaneously, a host of nineteenth-century mathematicians realized that if imitating the empirical world was no longer the starting point for geometry, then Euclid's fifth postulate could be rejected in favor of something else. Putting a different axiom in its place resulted in a new geometry that described a mathematical world full of ordinary things behaving with a disorienting allure—parallel lines that grow farther apart, octagons with eight right angles, triangles whose area is determined by their angles.

Witkacy creates his dream reveries by rejecting a subset of the familiar psychological axioms describing human behavior, replacing them with something unexpected. The result is a strangely rational kind of nonsense. The world of *Gyubal Wahazar* feels unnaturally distorted. It is full of familiar objects, but like the view we get looking through Poincaré's window, these familiar objects appear out of proportion and in unnatural relationship to each other. Consider, for instance, this typical outburst:

> WAHAZAR: (*roars*) Haaaaaaaaaa!!! (*Foam gushes from his ugly mug.*) Those two sluts go tomorrow to the Commission on Supernatural Selection! Shoot all the old bags in the Fourth district! Cancel all authorizations for exceptional marriages! Summon all third class educators for a special meeting today! Haaaaaaaaaaaa!!! (*He froths at the mouth.*) (118)

He's obscene, fascinating, even a bit humorous if we can keep from being offended, but not wholly incoherent. In Wahazar's world, the masses being tormented accept that Wahazar is "leading them where no one else could" because they understand that their dictator is suffering for them. Wahazar's suffering is connected to an intense form of loneliness which increases with his greatness and power. His power, meanwhile, is most unpredictable and terrifying in moments of weakness, which Wahazar has to somehow fabricate. "You covet your own weakness," Morbidetto the chief executioner tells Wahazar. "You lacked material in yourself to torture yourself with, so out of an excess of strength you artificially created your own weakness." This twisted loop is at the center of Wahazar's mood swings, creating a cycle that will be broken at the end of the play when Wahazar succeeds in becoming a martyr for the people he is eternally punishing. Before this happens, Morbidetto is the instigator of a conspiracy to overthrow Wahazar. The mutiny is hatched during one of Wahazar's catatonic states, but the dictator eventually roars back to life to regain control. Wahazar's lead guard stayed loyal during the coup attempt, and Wahazar thanks him with a bullet to the head. "As a reward you get to die with your boots on, Captain," Wahazar declares before pulling the trigger himself. And what becomes of the traitor Morbidetto?

> WAHAZAR: The more dangerous a creature you are, my dear Morbidetto, the better I like you. (*Morbidetto kisses Wahazar's hand.*) But we'll give up torture absolutely. From now on you won't torture anyone but me. Agreed?

MORBIDETTO: (*still bending over*) I can't stand it any longer! I'm going to burst!! (*Suddenly jumps for joy.*) What a mad comedy all this is!

Gyubal Wahazar is not a comedy, or not just a comedy, precisely because it is not entirely mad. Theater scholars like Gerould employ the term "non-Euclidean" in reference to Witkacy's plays to describe the way they distort reality into an unfamiliar world that exists independently of normal life, typically with some heightened sense of psychological intensity. The specific type of non-Euclidean geometry obtained from negating Euclid's fifth postulate is called *hyperbolic geometry*, and there is a natural way to apply the term "hyperbolic" to a play like *Gyubal Wahazar* where everything feels exaggerated to the point of absurdity. More recent scholarly analysis, particularly by Nicolas Salazar-Sutil, has sharpened up the analogy between Witkacy's theater and the geometry of Bolyai and Gauss. Rather than nonsense, Salazar-Sutil notes, "the non-Euclidean theatre Witkiewicz had in mind concerned itself with a reordering of sense. . . . The reason why Witkiewicz refused the term 'absurd' with which his theatre would later be equated is because his theatre was as consistent as non-Euclidean geometry is."[54]

Probing Salazar-Sutil's more refined assessment of the parallels between Witkacy's theater and geometry yields some unexpected consequences. The typical way to prove that a system of axioms is consistent in mathematics is to find an interpretation, or a model, where the meaning of all the axioms comes out true. If the axioms can all truthfully coexist in some concrete model then we can be sure there are no contradictions lurking somewhere. "Concrete" is a relative term here. The models that were used in the consistency proofs for alternatives to Euclidean geometry were constructed in a purely mathematical environment, but their successful execution generated a thorny question about the physical environment—a question with direct bearing on Witkacy's artistic agenda. Specifically, which geometry is the correct one for describing reality?

Debating the Real Geometry

Gauss and Bolyai and Lobachevsky all recognized the *mathematical* legitimacy of an alternative geometry to Euclid's, but what about *physical* legitimacy? A natural first instinct is to stick with what we know. Although non-Euclidean geometry makes for a fascinating intellectual exercise, the universe we inhabit is still Euclidian, isn't it? Maybe, but this turns out to be complicated for much the same reason that Euclid's fifth postulate was so enigmatic to begin with.

Because it involves extending lines indefinitely, there doesn't appear to be any obvious means for testing whether this postulate represents an empirical truth.

It might be tempting to cut the debate short in favor of Euclid by identifying a large triangle and measuring the angles to show that they add to 180°. The first problem with such an endeavor is to agree on the physical manifestation of a straight line, but even then, experimental error would foil any possibility of proving we live in a Euclidean world.[55] Despite what Immanuel Kant said about Euclidean geometry being an inherent component of our conception of space, any confidence we might profess in the "real" behavior of parallel lines is just bias built up over thousands of years of living by Euclid's rules. And whereas Newton constructed his laws of motion within a Euclidean framework, Einstein, as it turned out, relied heavily on non-Euclidean geometry in the development of his general theory of relativity.

This is a stunning turn of events, to say the least. What started out as an intellectual exercise to create new forms of geometry became useful; the new geometries, which originated as purely abstract constructions, were actually *more* appropriate for describing the nature of space. If the distorted view could become the more practical one in mathematics, then might Witkacy's apparent lunacy turn out to be a new kind of realism in disguise?

Witkacy created his protagonist in *Gyubal Wahazar* by negating a handful of universally assumed postulates of human psychology central to naturalistic playwriting. The result was a totalitarian dictator governed by mood swings and an unbridled individualism surrounded by subjects with an unexplained bond to their oppressor. To an audience in 1921, the dystopian state of *Gyubal Wahazar* would have appeared outlandish and far-fetched, repugnant to the nature of any conceivable collection of governing principles. Modern audiences see things very differently. Adolf Hitler would not become a known personality in Poland for many years after *Gyubal Wahazar* was written, but in the spirit of life imitating art, there is an uncanny set of parallels between the future Nazi leader and Witkacy's despot. This starts with their mutually volatile moods and extends to the deep sense of loneliness and isolation that each experiences in relation to a grandiose feeling of personal sacrifice.[56] By some twisted logic they each see themselves as instruments of a larger plan.

What is internally consistent can be imagined, and what is imagined can come to pass. It is this unforeseen and inadvertent form of realism that distinguishes Witkacy's assaults on traditional theater from Alfred Jarry's and most clearly separates Wahazar from the intellectually vacuous Père Ubu. By the end of the play, Wahazar recognizes that his own powerful ego is at odds with

"Couldn't you be a little quicker about it?" *The Water Hen*; James Curran (Edgar), Betty LaRoe (Water Hen); Theatre Off Park, New York City, 1983 (photo: Brad Mays).

his utopian vision of the future, and he has himself executed. Before his body cools, his vital glands are removed and injected into Father Unguenty, who declares himself Wahazar II, the new dictator of what he calls "this infernal six-dimensional continuum of Absolute Nonsense."

The Water Hen

A tiger devoured him in the Janjapara Jungle. . . . I assure you he died beautifully. His belly was torn to pieces and he suffered terribly. But up to the last moment he was reading Russell and Whitehead's *Principia Mathematica*. You know—all those symbols.

—LADY ALICE OF NEVERMORE, FROM *THE WATER HEN*

Despite what the name might suggest, the title of Witkacy's play *The Water Hen* refers to a person—a twenty-six-year-old woman, "pretty, but not at all seductive"—who we meet in the opening scene. She is standing on a small mound in an open field waiting impatiently for Edgar to finish loading his double-barreled shotgun. Witkacy employs his painter's eye to create a hypnotic opening montage. A red sunset and yellow poppies frame Edgar, who is dressed in an eighteenth-century three-cornered hat and boots. The Water Hen stands under an octagonal lantern mounted on a crimson pole and fitted with tinted green glass that casts an eerie glow. This oddly placed streetlight is just as incongruous as the conversation taking place:

> WATER HEN: (*gently reproachful*) Couldn't you be a little quicker about it?
> EDGAR: (*finishing loading*) All right—I'm ready, I'm ready. (*Shoulders the gun and aims at her—a pause*) I can't. Damn it. (*He lowers the gun.*) (45)[57]

In a manner that suggests they are old friends, the two then begin haggling over the meaning of existence, with Edgar exhibiting his habitual indecisiveness:

> EDGAR: I waged a futile battle against myself for ten years, and after all that, you wonder why I can't make up my mind about such a trivial matter as killing you. Ha! Ha! (*He knocks the shotgun against the ground.*)
> WATER HEN: How stupid he is! Greatness is always irrevocable.
> EDGAR: There are limits to my endurance. Let's not have any contrived scenes. Even in the most idiotic plays it's definitely against the rules.
> WATER HEN: All right, but even you'll agree it's a vicious circle. Everything irrevocable is great, and that explains the greatness of death, first love, the loss of virginity, and so forth. Whatever one can do several times is by its very nature trivial. You want to be great, and yet you don't want to do anything that can't be undone. (46)

The Water Hen makes a compelling argument and, sure enough, Edgar eventually concedes and fires both barrels of his gun in her direction. "One miss," she remarks with no inflection in her voice, "the other straight through the heart."

This surreal beginning is the start of a long, strange journey for Edgar. As the Water Hen lies dying, a young boy named Tadzio appears claiming to be Edgar's son. "Who knows?" Edgar says, unfazed. "Maybe I'm a father, too." As a general rule, family relationships are essentially indeterminate throughout the play. We've already heard the Water Hen say to Edgar, "You were my child and my father," and before the dust settles she will become a candidate for Tadzio's mother as well as his lover.

So, what in the world is going on here? As we have been discussing, the answer depends on what we mean by "world." Returning to the scene of the Hen's execution, we get our first clue as to what "a spherical tragedy in three acts" might refer to. Watching the Water Hen expire, Tadzio becomes unnerved and uneasy.

TADZIO: I'm afraid. What happened to the lady?

EDGAR: I'll tell you the truth. She's dead.

TADZIO: Dead? I don't know what dead means.

EDGAR: (*surprised*) You don't know! (*Somewhat impatiently*) It's exactly as though she went to sleep, but shall never wake up.

TADZIO: Never! (*In a different tone of voice*) Never. I said I'd never steal apples, but that was different. Never—I understand now, it's the same forever and ever.

EDGAR: (*impatiently*) Well, yes, that's the infinite, the eternal. (49)

Spherical Geometry

Once the radical idea was hatched that Euclid's axioms could be formally manipulated, multiple versions of non-Euclidean geometry ensued. The geometry of Bolyai and Lobachevsky, which is obtained by keeping Euclid's first four postulates intact and negating the fifth, eventually became known as hyperbolic geometry. The spherical geometry mentioned in the subtitle of *The Water Hen* refers to the system that arises when doing geometry on the surface of a sphere.

What sorts of axioms would characterize geometry in this setting? The most natural interpretation for the undefined term "line" on the surface of a sphere is a *great circle*. Great circles are the circles obtained by taking the intersection of the surface of the sphere with a plane that contains the center point of the sphere. The equator is a great circle on the earth, as are all the longitudinal lines, which stretch from the North Pole to the South Pole. Most people learn about great circles on their first international flight because great circles give the shortest distance between two cities—and this is very much the reason why they make ideal candidates for the lines in spherical geometry.

Already we can spot a host of non-Euclidean phenomena emerging. With this interpretation there are no parallel lines at all in this model because every two great circles cross—twice actually, at a pair of points on the sphere that are diametrically opposed. Before deciding which is more repugnant to the

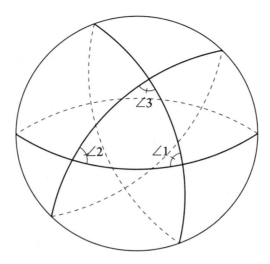

FIGURE 2.14. A triangle on the surface of a sphere satisfies
$\angle 1 + \angle 2 + \angle 3 > 180°$.

nature of the straight line—the banishing of parallelism altogether or a pair
of lines intersecting in two points—notice that lines have also lost the infinite
character they possess in Euclidean geometry. Although we won't go into it
here, it is possible to develop a consistent set of axioms that capture the geometry suggested by this model, and a look at figure 2.14 should make it plausible
that the triangles of this strange new universe have angle sums of strictly *more*
than 180°.

No Boundary $\neq$ Infinite

Returning our attention to Edgar and Tadzio's father-son conversation about
the eternal, the provocative question on the table is what might infinity look
like if one adopted a spherical point of view—not so much about the nature of
physical space, but a spherical analog to psychological space. What would the
implications be for a non-Euclidean existence where all lines intersect and the
universe of possibilities is without boundary but yet finite and cyclical? What
would "never" mean in such a place? Would anything be irrevocable?[58]

The spherical imagery in *The Water Hen* increases when Edgar's father
arrives to host a party for Lady Alice of Nevermore. Lady Alice's late husband, we learn, was also named Edgar and was purportedly in love with the
Water Hen. If having two characters named Edgar isn't confusing enough,

"How's the greatness problem coming?" *The Water Hen*; James Curran (Edgar),
Betty LaRoe (Water Hen); Theatre Off Park, New York City, 1983
(photo: Brad Mays).

Lady Alice's new suitor is described in the stage directions as looking very much like Edgar—the Edgar we have gotten to know, that is. By the end of the first act, our Edgar has soundly defeated his doppelgänger in a fistfight and unwittingly gotten himself engaged to Lady Alice.

Amid all the repeated names and themes, the most jarring turn of events occurs in the second act when the previously deceased Water Hen enters in the middle of a dinner party to check on the progress of her old friend. "Will you be angry if I ask you something?" she says to Edgar, who is miserable in his new marriage and being pressured by his father to become an artist. "How's the greatness problem coming?" (64)

The play is dark and comic, illuminating and maddening. Art, capitalism, love, torture, death—and as somewhat of a last resort, mathematics—are scoured for their potential to give meaning to life, but none of them holds any lasting legitimacy (the example of Alice's first Edgar who died reading Russell and Whitehead notwithstanding). By act III, Tadzio is twenty years old and is as aimless as his father when the Water Hen turns up at the door yet again to find Tadzio alone with his books. "Me? I'm not doing anything," Tadzio tells her. "I study. Mathematics. They torture me with mathematics, even though I don't have any talent for it." The Water Hen, who for some non-Euclidean reason has not aged in the ten years that have passed since act I, at least pretends to be sympathetic, and in so doing successfully seduces the young man.

In a spherical tragedy there is nowhere to go except where you have already been, and a fit of desperation finally overwhelms the aging Edgar when he discovers the Water Hen luring Tadzio into her spell. Edgar redons his outfit from act I, reloads his shotgun, and executes the Hen a second time. In keeping with the self-deprecating jokes that have been ubiquitous throughout, Edgar's father proposes one final time that his son pursue the arts—"You could even become an actor; after all, actors are now creative artists too, ever since Pure Form became the rage." (77)

Edgar responds by pulling a revolver from his pocket and shooting himself in the head.

Pure Form and Pure Mathematics

Pure Form was, of course, not the rage in Polish theater at the time—but it was very much at the center of Witkacy's own writing. Pure Form was the name Witkacy gave to the movement he was trying to bring about in theater, the tenets of which he set out in *An Introduction to the Theory of Pure Form in the Theater*, written in 1920. *The Water Hen* is arguably his most successful attempt at producing an example of a play in this style.[59]

When "pure" is used to modify mathematics—as in "pure mathematics"—the intent is to distinguish this enterprise from applied mathematics. The theorems of pure mathematics are arrived at through abstraction and logical reasoning, and they are judged for their artful beauty. Empirical methods are scorned, and the practical usefulness of the results is never a consideration. Witkacy's Pure Form was also very much in favor of more abstraction. A guiding principle of Pure Form was to move theater away from the burden of being evaluated in terms of its constituent parts—plot, character, sound, setting—and to liberate it to explore what Witkacy called the "mystery of existence."

> In the theater we want to be in an entirely new world in which the fantastic psychology of characters who are completely implausible in real life, not only in their positive actions but also in their errors, and who are perhaps unlike people in real life, produces events which by their bizarre

interrelationships create a performance in time not limited by any logic except the logic of the form itself of that performance.[60]

With all the allusions to Cantor and Russell and Gauss in his scripts, it is at this conceptual level that Witkacy's plays achieve their deepest affinity with mathematics. At its core, the non-Euclidean movement in mathematics was created by the recognition that the axioms of geometry need not be chosen for their ability to mimic the empirical world. For two thousand years, mathematics was constrained by the misconception that there was a single geometry for the space we inhabit, and the shattering of that myth in the nineteenth century set the mathematical imagination free to explore new worlds of its own creation. Witkacy wanted the same thing for himself and his fellow artists. What happens when theater is freed from the usual restrictions of narrative consistency prescribed by the so-called real world? "The idea is to make it possible to deform either life or the world of fantasy with complete freedom," Witkacy explains, sounding very much like he is describing János Bolyai's new geometric world, "so as to create a whole whose meaning would be defined only by its purely scenic internal construction."[61]

The analogy here is very rich because Witkacy and the pioneers of non-Euclidean geometry were both fighting against the demon of deep-seated bias. The raw material of geometry is "points" and "lines," and as much as one tries to make them undefined terms, we've seen that their age-old Euclidean interpretations are a hard habit to break. When we read the word "line," our mind's eye conjures up an image of a Euclidean straight line and all the familiar properties of parallelism that come with it. In drama, the basic elements of a theatrical production are actors and language—"beings who act" is how Witkacy phrases it[62]—and they bring with them the suggestion of realism not present in other art forms like abstract painting. When an audience sees a person on stage, they inevitably draw parallels between themselves and the actor, and these parallels unconsciously get extended to the respective world each inhabits. Witkacy was well aware of this phenomenon, pointing out that it was "difficult to imagine Pure Form on the stage, essentially independent, in its final result, of the content of human action."[63]

In a very real sense, however, these natural prejudices can be converted into an asset, because it is the disjointedness between what we expect and what actually happens that provides the intellectual electricity—in both the geometry and on Witkacy's stage. We are all raised to believe that triangles have angles that sum to 180°, and when someone is shot dead in act I, we don't expect her to arrive at a dinner party in act II. When these inviolable rules are violated, we can either reject the entire enterprise as "repugnant" or we can go back to the fundamental assumptions we're making—the axioms of the system in question—and ask whether these really do deserve the mantle of "self-evident truths." If mimicking our long-held perceptions of the world is no longer the central goal, then the mathematician and the playwright are free to adopt a different set of axioms that suits their purpose.

There are some restrictions, however. A fundamental feature of any geometry, Euclidean or otherwise, is the internal consistency required of the axioms. They don't have to jive with our empirical preconceptions of the real world, but they are required to have their own internal logic that must be free of contradictions. Here again, Witkacy's conception of Pure Form follows the mathematics:

> We can imagine such a play as having complete freedom with respect to absolutely everything from the point of view of real life, and yet being extraordinarily closely knit and highly wrought in the way the action is tied together. The task would be to fill several hours on the stage with a performance possessing its own internal, formal logic, independent of anything in "real life."[64]

To suggest that the creative act of writing plays is somehow the same as deriving theorems would be to push the notion of "non-Euclidean drama" further than Witkacy intended, but the ordained and assured quality of a properly proved theorem was very much a feature that the Polish playwright wanted his plays to emulate. "We cannot *prove* the necessity for [each part of the play], but it should appear inevitable in so far as each element is a necessary part of the work of art once it has been created," Witkacy writes, sounding very much like Aristotle in his *Poetics*.[65] "While we are watching the play unfold, we ought not to be able to think of any other possible internal interrelationships."[66]

The Final Act

The connection between Pure Form and pure mathematics is striking—not just in the way that Witkacy presents his theory but also in the way these two self-identified formal endeavors exhibit prophetic capabilities. In geometry and elsewhere, abstract mathematics has routinely been a roadmap for physics, laying out theoretical possibilities that, with time, have led to revelations about the natural world. The so-called "unreasonable effectiveness" of

"Pass." *The Water Hen*; Richard Ladd (Evader), Nat Warren-White (Typowicz), Stanley Keyes (Spector), James Brown (Father); Theatre Off Park, New York City, 1983 (photo: Brad Mays).

pure mathematics as a tool for science has been borne out routinely through history, and this phrase applies to Witkacy's formal theater as well.[67] "The historian speaks of what has happened," Aristotle writes, "the poet of the kind of thing that *can* happen." Aristotle is describing the prescient power of playwrights in this quotation, but it can be adapted to mathematicians, and it points to how abstraction has the potential to reveal truths that transcend time, place, and culture. To his psychological detriment, Witkacy understood this in a visceral way. By exploring what is possible in a formal world he glimpsed the future of the real one, with foreboding results.

At the conclusion of *The Water Hen*, as is often the case in Witkacy's plays, a revolution is starting in the streets. Amid reports of her financial ruin, Lady Alice manages to escape with her old Edgar look-alike lover, and Tadzio also flees into the night. The corpses of Edgar and the Water Hen are dragged away. Left alone with a few stray house guests, Edgar's father orders up a game of bridge and some women to accompany them later at dinner. In a statement hauntingly apropos of the real-life insanity that is to be visited upon Poland and the rest of Europe in the years ahead, the play concludes with the incongruous sounds of bids and bombs:

FATHER: There's no need to worry gentlemen, perhaps we can still get jobs in the new government.

TYPOWICZ: One club.

EVADER: Two clubs.

FATHER: (*sitting down*) Two diamonds. (*A red glare floods the stage, and the monstrous boom of a grenade exploding can be heard.*) Banging away in fine style. Your bid, Mr. Spector.

SPECTOR: (*in a quivering, somewhat plaintive voice*) Two hearts. The world is collapsing. (*Fainter red flashes and immediately afterward two shells exploding a little farther off*)

TYPOWICZ: Pass.

(*End of play.*) (79)

Witkacy remained active and productive as an artist, eventually turning his attention to fiction and philosophy in the 1930s but all the while growing more despondent at the political realities closing in around him. When the Nazis invaded Poland in September 1939, Witkacy fled on foot with other refugees, but his health was not good and his heart was not in it. Aware that the Russians were invading as well, Witkacy methodically committed suicide in a country field by first thinning his blood with ephedrine tablets and then opening several veins in his wrists and throat. He died in the company of his traveling companion and lover at the time, to whom he reportedly said, "I will not live as less than myself."[68]

Witkacy's body was buried nearby in the small village of Jeziory, where it was marked with a simple pine cross and essentially neglected—until it was triumphantly returned, in April 1988, to a flag-waving funeral in his hometown of Zakopane. It is difficult to believe that Witkacy would have approved of the postage stamps and parades in his honor fifty years later, which is why there was a second celebration among his most ardent admirers when it was determined—from an X-ray of the sealed coffin—that the reburied bones were in fact the skeleton of an anonymous young woman.[69] True to the life he lived, the irreverent playwright had orchestrated one final comedy with corpses.

3

Beckett: The Language
of Incompleteness

Thus, let's proceed like that crazy (?) mathematician who at every stage of the
calculation applied a new principle of measurement.

—SAMUEL BECKETT, FROM A LETTER TO AXEL KAUN

QUAD IS a piece for television created and directed by Samuel Beckett. It was
first broadcast in Germany in 1981, when Beckett was seventy-four years old.
By this point in his career, the Irish novelist, poet, and playwright was firmly
established as one of the most important artists of the twentieth century.
This was not just the opinion of critics and scholars; Beckett's work—most
notably his stage plays—had earned him a devoted following among the wider
public. Beckett's strange and desolate landscapes have an alien quality, but
audiences found something personal and identifiable in them. His characters typically exist in fundamentally hopeless situations, but through stoicism,
self-delusion, or a naive persistence they never fully succumb to hopeless-
ness. While scholars pored over his writing, aspects of Beckett's influence
seeped into the folklore and lexicon of popular culture. The two tramps passing time on a country road in *Waiting for Godot* have become a universal
symbol of unrequited aspirations. The last phrase in *The Unnamable*, "I can't
go on, I'll go on," is so familiar as to sound almost cliché. The citation for his
Nobel Prize from 1969 said that Beckett was being recognized "for his writing
which—in new forms for the novel and drama—in the destitution of modern man acquires its elevation." His work has a depth that continues to attract
academics and literary theorists, but it also resonates with audiences and

Quad; German television, 1981. © SWR/Simone Christ.

readers across cultures, languages, and now generations. It seems everyone is interested in what Samuel Beckett has to say.

Quad

This is why *Quad* is so vexing—it is not clear that it says anything at all. There are no words, no characters, and no conflicts to speak of. Even by Beckett's minimalist standards, *Quad* comes off as uniquely austere and inaccessible.

The first thing one notices about it is its mathematical construction. The set consists of a single square roughly six meters along each side. At the outset, we see a single walker, completely covered in a long, shapeless white gown with a hood and cowl obscuring its (her? his?) face. Starting from the upper left corner, the white walker proceeds to trace out a carefully choreographed path through the square. The pacing rules are simple and rigid: at each corner, the walker makes a sharp 135° left turn and heads off toward a new corner. Eight segments later, white ends up back at the corner where it started and, with no cessation in movement, heads off again to start another cycle.

At the moment white arrives at its original corner, it is joined by a blue-cloaked walker who embarks on the same eight-legged journey through the square, only starting from the lower left corner. Without interruption, the two

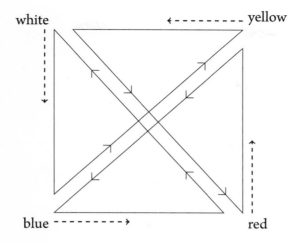

FIGURE 3.1. Each walker in *Quad* follows a rotated version of the same path through the square, turning 135° left at each corner until it makes a cycle.

walkers move in unison along their respective paths. When they complete their courses, a red walker joins the choreographed march starting from the lower right corner, and when this trio has finished a complete cycle, a yellow walker joins from the upper right to fill out the square (Figure 3.1).

At least this was Beckett's original intent. What happens in practice is that when multiple walkers try to navigate the stage, they arrive at the center point of the square at the same time! When this was discovered during rehearsals, Beckett modified the diagonal parts of the path, having the walkers all avoid the square's center by taking an angled step to the left just before they reach it. The effect is engaging. The faceless robed figures all move wordlessly in rhythm along mirrored copies of the same path through the square. The pace is quick—a modest allegro—so there is a hint of suspense every few seconds when it appears as though a collision is imminent, only to be avoided at the last moment as each of the walkers deviates slightly left of the center.[1]

To expunge the human element, Beckett stipulates that the players be indistinguishable in size and shape. By contrast, he goes to extra lengths to highlight the patterns and combinations in the piece, giving each colored walker its own accompaniment—drum, gong, triangle, and wood block—that plays when they are on the stage. The growing complexity of the movement within the square is accompanied by an increasingly cacophonous soundtrack, eventually becoming a four-voice fugue of movement, color, and sound when all the walkers are simultaneously navigating the square stage.

After the four walkers complete the eight-bar phrase, the white walker exits into the darkness around its home corner, leaving behind the trio of blue, red, yellow. The diminishing continues when blue departs at the end of the next cycle. Red is next to go, leaving yellow on its own to march out the pattern. After yellow completes a solo cycle, the walkers began to reappear, although in a different order than before. This process of increasing and diminishing continues until every combination is achieved.[2]

Quad fades out where it began—with the single white walker navigating through the square.

Sounds and Symbols, Signifying Nothing

Quad is not a piece about mathematics, but each of its building blocks is conspicuously mathematical. The square play-space comes from geometry; the path each walker repeatedly traces is from graph theory; the succession of all possible arrangements of the four colored players comes from combinatorics; the perfect symmetry of the performance (it would look the same from any of the four sides) points to abstract algebra; the incremental increase and decrease of walkers suggests a connection to number theory. By making his players faceless, shapeless, and mute, it is as though Beckett is deliberately trying to remove everything he possibly can from the performance that isn't explicitly mathematical. What is left is rhythmic, patterned, and impersonal.

The up-tempo pace, together with the colored robes and raucous percussion section, provides the energy and, at the outset at least, a hint of levity. Four colorful objects are navigating with great purpose in tight quarters, narrowly avoiding each other so that they can press on quickly with the task at hand—which is to avoid each other so that they can press on quickly with the task at hand. But the levity does not last long. With no human qualities to access, we begin to sense in *Quad* a more foreboding portrayal of the human experience—one consisting of a mindless conformity.[3]

One way to get a foothold into the significance of *Quad* is to lash it to some established literary guideposts. "Dante and Virgil in Hell always go to the left, the damned direction," Beckett wrote in reference to a different work, but perhaps it applies here as well.[4] One needs a longer rope to get from *Quad* to Shakespeare, but Macbeth's iconic passage where he describes human life as a "poor player that struts and frets his hour upon the stage," has been suggested as a text that might illuminate the action of *Quad*.[5] A drawback to this interpretation is that, while the imagery may be shared, Macbeth's famous speech—even out of context—is richly textured and immediately gripping. *Quad*, meanwhile, is sterile and aloof. Without characters or conflict, *Quad* has barred all the usual points of emotional entry, achieving a high level of abstraction by forfeiting any obvious psychological accessibility.

Waiting for Gödel

The one confident observation to make about *Quad* is its mathematical composition. This in itself is surprising, and it raises the question of what purpose mathematics is serving in Beckett's creative agenda. Beckett is revered for using various forms of writing—fiction, poetry, and theater—to explore deep questions about human identity, an enterprise that does not traditionally have much use for the tools of mathematics.

So, is *Quad* an anomaly among Beckett's works? No, actually. Mathematical ideas are ubiquitous throughout his long career, and the best way to get beneath the surface of *Quad* is to follow the mathematical stepping stones from the beginning. What emerges from this journey is a distinctive portrait of mathematics as a regular Beckettian device, employed initially as a weapon for dismantling the way literary art is constructed and then as scaffolding to reassemble the pieces into new forms.

As *Quad* foretells, combinatorics and symmetry are points of regular interest for Beckett, as are number theory and elementary algebra. There is no evidence that Beckett was interested in non-Euclidean geometry, but the original Euclidean version makes a number of appearances. Although it is a good first step, focusing too intently on these regular instances of mathematical content actually obscures the main story. Where Beckett's relationship to mathematics is most potent is around developments in twentieth-century logic, most notably the role of paradox and its significance in developing the languages of axiomatic mathematics.

In the aftermath of the discovery of non-Euclidean geometry, mathematics was left to wrestle with a philosophical crisis. Axiomatic systems emerged as the unambiguous winner in explaining how mathematics should be conducted, but it was no longer clear that the axioms in these systems actually described something real. How could they, for example, if there was more than one valid system for geometry? There was also a growing sense of unease around the notion of proof. The Euclidean model of deductive reasoning was compelling, but its dependence on natural language cast an aura of uncertainty over whether some overlooked assumptions might be corrupting the arguments. The response to these kinds of questions brought about a period of intense introspection in mathematics, and one of the outcomes was the evolution of highly specialized notational systems—or languages—designed with the audacious goal of providing a perfected and reliable foundation for mathematical truth. As these languages were created, the languages themselves became the subject of intense scrutiny. Did they describe something real or did they bring some new reality into existence? Were they free of contradictions? Were they as robust as advertised? Did provable statements in these new systems always correspond to true statements? Did true statements correspond to provable ones, or did mathematical truth elude axiomatic codification?

This is where Beckett's relationship to mathematics takes its most interesting turn. Beckett's chosen subject was not mathematical introspection but human introspection, and the primary means by which he took up this topic was by challenging the established linguistic conventions that governed the fiction and theater of his day. Does language point to something real? In what way does language define or limit the reality we perceive? What, if anything, lies beyond its expressible domain? The task that aligns the Irish novelist and playwright most directly with the metamathematicians from the early part of the century is the charge to investigate the integrity of their respective languages, and in each case what transpires involves a negotiation with the power and paradoxes of self-reference. This is the result of conflating the object to be studied with the means for studying it. In literature and in mathematics, language defined the discipline while that very same language was the tool for investigating the discipline's legitimacy.

Key landmarks in the development of modern logic include Gottlob Frege's construction of the whole numbers in 1884, Russell and Whitehead's *Principia Mathematica* from 1910–1913, Hilbert's program from the 1920s, and Kurt Gödel's expectation-shattering incompleteness theorems published in 1931. Samuel Beckett's first significant forays into fiction began in the early 1930s and, after a slow start, led to a long and distinguished career that lasted until his death in 1989. Scholars of Beckett's work have made note of the strong resonances between Beckett's creative agenda and the revolutionary developments in logic that took place during his youth, a few making the case that Beckett explicitly alluded to Gödel in his correspondences. This is a fascinating claim, one that we will presently take up, but it is not clear that the answer matters as much as it might seem. What is more important is recognizing how theater once again becomes a vehicle for extracting personal insights from mathematical ones. Intentionally or not, Beckett found common cause between his desire to repurpose theater as a tool for exploring the human condition and the rise of formalism that characterizes the journey from Frege to Gödel.

Murphy and *Watt*

Drowned in a puddle, for having divulged the incommensurability of side and diagonal.

—NEARY, FROM *MURPHY*

A search for mathematically inspired moments in the early education of Samuel Beckett comes up empty. Born in 1906 in a relatively affluent suburb of Dublin, Beckett was subjected to an Irish Protestant education that eventually took him to Portora, a boarding school located in what is now Northern Ireland. Known as much for his athletic prowess as for his scholarship, Beckett had an adversarial relationship with the mathematics teacher at Portora, who was also Beckett's cricket and swimming coach. The following lines from "For Future Reference," a poem Beckett wrote some seven years after he left Portora, offer a sense of the residual animosity he carried with him:

Well of all the...............!
that little bullet-headed bristle-cropped
red-faced rat of a pure mathematician
that I thought was experimenting with barbed wire in the Punjab
up he comes surging to the landing steps
and tells me I'm putting no guts in my kick.
Like this he says like this.

In 1923, Beckett enrolled at Trinity College in Dublin. He studied fine arts and languages, and when he graduated he was awarded the honor of being a lecturer in Paris for two years at the École Normale Supérieure. Among all that Paris had to offer Beckett, perhaps the most significant was his friendship with James Joyce. Already a great admirer of Joyce's work, which at this point included *Ulysses* as well as *Dubliners* and *Portrait of the Artist as a Young Man*, the young Beckett found in his fellow countryman both a kindred spirit and mentor. Joyce, like Beckett, had degrees in French and Italian and a broad love of language more generally. Both were intimately familiar with the Bible and all its potent imagery, but neither had any personal stake in religious practice. They shared a common love of Schubert, Cézanne, and Charlie Chaplin.[6]

When they met, Joyce was at work on what would become *Finnegans Wake*, a daunting project that involved abstracting language into a form that is nearly impossible to read without some kind of assistance or training. Beckett served as a part-time assistant to Joyce and, at the author's invitation, wrote a critical essay on the work in progress called "Dante . . . Bruno . Vico . . Joyce." (Each period is meant to stand for a century.)[7] "Here is direct expression—pages and pages of it," wrote the twenty-two-year-old Beckett. "And if you don't understand it, Ladies and Gentlemen, it is because you are too decadent to receive it. . . . Here form *is* content, content *is* form. You complain that this stuff is not written in English. It is not written at all. It is not to be read—or rather it is not only to be read. It is to be looked at and listened to. His writing is not *about* something; *it is that something itself.*"[8] (Italics included in the original.)

When, in a few years, Beckett sets his mind on becoming a creative writer, his success will hinge on emerging from Joyce's shadow and heading in a distinct, almost antipodal direction. But even as he does, this notion that the subject matter of his writing is the writing itself will be a relevant comment about Beckett's work as well.

When his two-year position in Paris ended, Beckett returned to Dublin to teach modern languages at Trinity and complete his master's degree. Even at the outset of this appointment, Beckett had a strong sense that academic life was not a good fit, and in 1932, two years after returning to Trinity, he submitted his resignation in order to devote his full attention to writing. Beckett was unable to secure a publisher for his first novel, *Dreams of Fair to Middling Women*, but parts of this project did appear in a collection of short stories published in 1934 under the title *More Pricks Than Kicks*. In the meantime, Beckett's personal life entered a difficult period. Beckett had greatly disappointed his strict mother by forfeiting a promising academic career, and in June 1933 his father, with whom he was very close, died of a heart attack at age sixty-one. This event came on the heels of other bad news. Beckett's first cousin, a young woman he had fallen in love with a few years earlier, died from tuberculosis only a month before Beckett's father unexpectedly passed away. Gripped by a combination of depression and anxiety, Beckett moved to London, in part so that he could undergo an extended course of psychotherapy, a practice that was not legal in Ireland. A year later, while still in treatment, Beckett gathered up a handful of the autobiographical details of his life and started writing what would become his first published novel, *Murphy*.

Matrix of Surds

Appreciating the significance of mathematics in Beckett's dramas, which he would not start writing for another decade, requires a brief survey of the way mathematics seeps into the architecture of his early novels. Scouring *Murphy*, *Watt*, *Molloy*, *Malone Dies*, and *The Unnamable* for mathematical inspiration is a treasure hunt worthy of its own extended investigation, but we will content ourselves with a few observations meant to provide some necessary context for undertaking Beckett's plays.

In *Murphy*, the eponymous protagonist spends the entire novel engaged in a Cartesian mind-body negotiation that mirrors Beckett's own struggles during his time in London. To soften the sharp edges, Beckett pushes Murphy's predicament up to the level of farce so that Murphy's journey comes off as comedy despite its unmistakably tragic undertones. Full of curious geometric allusions to squared circles and circumcenters of love triangles, *Murphy's* most instrumental mathematical device for elucidating the psychological core of the novel is the concept of irrational numbers.[9]

Rational numbers are the numbers that can be expressed as the quotient of two integers—fractions like $1/3$, $22/7$, and $-5/9$. Although it is not obvious, there is no rational number whose square is equal to 2. This simple truth turns out to be a very big deal—$\sqrt{2}$ is irrational.

Attributed to the school of Pythagoras from around 500 BCE, the discovery of the irrationality of $\sqrt{2}$ had a profound impact on the mathematics and the minds of the ancient Greeks.[10] It is natural to suppose—and the early Greek geometers made this assumption by default—that the rational numbers *are* the numbers. The Pythagorean Theorem applied to a triangle with legs of length 1 makes it clear that $\sqrt{2}$ exists as a length, and certainly every geometric length ought to correspond to an arithmetic number (Figure 3.2). Because the integers grow without bound, intuition suggests that we can endlessly adjust both the numerator and the denominator of a fraction, fine-tuning it so as to obtain any possible length we desire. In the presence of this kind of infinite precision, how could there be any gaps?

Alas, it turns out that the universe of numbers has a vast dark zone that extends beyond the well-charted reach of the rational numbers. The trauma this inflicted in 500 BCE was extensive. Pythagorean philosophy was a fascinating mix of mathematics, music, and mysticism in which the integers carried a spiritual significance. What fate then would befall the person who unearthed the truth that the integers were insufficient for

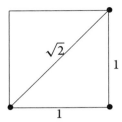

FIGURE 3.2. "Drowned in a puddle," said Neary, "for having divulged the incommensurability of side and diagonal."

expressing something as primitive as the hypotenuse of an isosceles right triangle?

Here is how the apocryphal tale is recounted in an early passage from *Murphy*:

> "But betray me," said Neary, "and you go the way of Hippasos."
> "The Akousmatic, I presume," said Wylie. "His retribution slips my mind."
> "Drowned in a puddle," said Neary, "for having divulged the incommensurability of side and diagonal."
> "So perish all babblers," said Wylie. (47)[11]

Where this mathematical metaphor pays dividends, here and in Beckett's future work, is as a means for imagining the extra-lingual part of the human psyche. In the case of Murphy, Beckett describes the mind of his protagonist as being composed of three zones—the light, half-light, and dark—each situated further from the physical world and less accessible to Murphy's mental manipulations. By the time he gets to the third and all-important dark zone, Beckett is so far removed from anything in empirical experience that it becomes nearly impossible to describe within the confines of the written word:

> The third, the dark, was a flux of forms, a perpetual coming together and falling asunder of forms ... nothing but forms becoming and crumbling into the fragments of a new becoming, without love or hate or any intelligible principle of change. Here there was nothing but commotion and the pure forms of commotion. Here he was not free, but a mote in the dark of absolute freedom. He did not move, he was a point in the ceaseless unconditioned generation and passing away of line.
> Matrix of surds. (112)

A "surd" is an irrational number, and a "matrix of surds" is an early attempt by Beckett to conjure up a component of our mental existence that lies beyond our descriptive power. The rational numbers are the tangible, empirical numbers of everyday life; their reach and precision are sufficient to handle all the tasks required of engineering and commerce. They permeate the real number line in the sense that they can be found in infinite supply in any chosen segment of it, no matter how small a segment we consider. Most of us, like the early Greek mathematicians, live out our days under the impression that being rational is simply what it means to be a number. Could there really be something else?

Yes, as it turns out, and in unimaginably vast supply. With the discovery of irrational numbers like $\sqrt{2}$ (and other discoveries to come) it became clear that to consider the number line and see only the rational numbers is akin to looking at the ocean and seeing only the salt. The truth of the matter is that the irrational numbers account for most of the number line, the water in this analogy, with the salty rational numbers making up such a modest proportion of the totality of the numbers as to be almost undetectable. For Beckett, this provided an irresistible portrait of reality, with the rational numbers standing in for the portion of the universe accessible through linguistically dependent observation while the unfathomably more numerous irrationals represented what lay beyond, undetected and unexplored.[12]

The Ingenuity of Despair

A second use of mathematics in *Murphy* which informs Beckett's career as a playwright is the inclusion of a chess game, in its entirety, toward the end of the novel. By this point in the story Murphy has stumbled into employment as a nurse at a mental institution, and during his evening rounds he settles in for a game with an elegantly dressed schizophrenic patient named Mr. Endon. Interrupting the narrative prose, Beckett proceeds to list all forty-three moves played by White (Murphy) and Black (Mr. Endon), and supplies an entertaining commentary of footnotes next to significant moments in the match. In response to Murphy's opening move, P-K4, we have "The primary cause of White's subsequent difficulties." At Murphy's fifth move, Kt-Q5, "Apparently nothing better, bad as this is." As events continue to go poorly for Murphy we get "Ill-judged," "The flag of distress," and "The ingenuity of despair." Comments that accompany Mr. Endon's play include "An ingenious and beautiful debut," "Exquisitely played," and "Black has now an irresistible game." Following Mr. Endon's forty-third move, Murphy concedes.

The strong sense one gets from these footnotes is that Murphy has been methodically outmaneuvered by his mental patient, but Beckett is actually engaged in some sleight of hand that is half prank and half test of his readers' acumen. Murphy has indeed been bested, but his defeat has nothing whatsoever to do with chess strategy. To understand what has transpired in this scene, *which is absolutely crucial to the core of the story*, Beckett makes a bold, implicit request of his reader. Specifically, it is necessary to decipher the chess notation and follow the flow of the match.

Although it starts in a typical way, it is evident after just a few moves that something curious is afoot. Mr. Endon is moving his knights around Black's end of the board in a totally legal fashion but showing no inclination to attack, or even engage, his opponent. Murphy, meanwhile, attempts to copy every one of Mr. Endon's moves. This strategy runs into trouble because, playing white, Murphy had to move first ("The primary cause of White's subsequent difficulties") and his pawn to king-4 opening eventually forces him to improvise a substitute on the fifth move ("Apparently nothing better, bad as this is.") After eight moves, Mr. Endon succeeds in returning his pieces to their original position ("An ingenious and beautiful debut"), a feat Murphy cannot duplicate because he has advanced a pawn.

As the game continues, Mr. Endon manipulates the eight pieces on his back row so that they all return to the back row but occupy a different position ("Black has now an irresistible game"). He then proceeds to move his permuted pieces back to their original locations. As this is happening, Murphy grows restless and begins moving his pieces forward ("The flag of distress"), exposing his knight and then even offering his queen ("The ingenuity of despair"). But no matter what Murphy tries, Mr. Endon does not acknowledge him. Murphy tries one final time to sacrifice a knight ("No words can express the torment of mind that goaded White to this abject offensive"), and then, finally, turns his king on its side ("Further solicitation would be frivolous and vexatious, and Murphy, with fool's mate in his soul, retires").

In addition to the logical roots of chess, what makes the inclusion of this match in *Murphy* notable from a mathematical perspective is Beckett's willingness to temporarily entrust his narrative to the formal notation of the game. For what is arguably the culminating event of the story, the novel sheds its reliance on old-fashioned words and communicates in a purely symbolic, logographic language. It's not exactly mathematics, but it is a close cousin, and it points to the mathematically inspired stress tests Beckett is about to inflict on his native language in his follow-up novel written during the war.

War Stories

Beckett finished *Murphy* in 1936 and, after a number of rejections, succeeded in finding a publisher the following year. The ensuing years included time back in Ireland and an extended tour of Germany, but ultimately Beckett returned to live in Paris in October 1937. In 1938, Beckett formed a romantic friendship with Suzanne Deschevaux-Dumesnil, a strong, savvy woman six years his elder with a talent for music and a fondness for the arts. Although Beckett was cautious in his initial assessment for how long the relationship would last, the two would become lifelong partners, officially marrying in 1961. On September 4, 1939, France declared war on Germany.

Although generally humble about it, Beckett's conduct during the war years deserves recognition. Seeing friends arrested and interned by the Nazis, Beckett joined an undercover French resistance group in 1941. His main task was to serve as a central collection point for pieces of information obtained by other group members—usually troop locations or maps of infrastructure. One of Beckett's major suppliers of information was the artist Alfred Péron, who had the perfect cover for frequent visits with Beckett because he was simultaneously helping with the French translation of *Murphy*. Eventually the group was exposed, and Beckett and Suzanne had to flee, literally at a moment's notice. Péron was not so fortunate. He was arrested and imprisoned, and he died just after being released from a concentration camp in May 1945.

After several months of hiding in and around Paris, Beckett and Suzanne successfully escaped from German-occupied territory. They settled in the small town of Roussillon in October 1942. Living in primitive quarters but out of imminent danger, Beckett took a job on a local farm where he was paid with food and friendship. Although this arrangement lasted several years, Beckett grew restless and frustrated as the stories of Jewish persecution continued to mount. With the war drawing to a close, Beckett once again sought out the local resistance group and in this case was even issued a gun. To his relief, he never faced a situation where he was required to use it.

Beckett started writing *Watt* in 1941 while he and Suzanne were in Paris. He returned to it during their time in Roussillon, finishing it in 1945. His first major creative work since *Murphy*, *Watt* is a curious novel that shows Beckett heading in an uncharted new direction. *Murphy* poses a few hurdles for its reader, not the least of which is its required erudition. The writing is dense with recondite allusions (e.g., how many readers know that the early members of the Pythagorean school of philosophy were known as Akousmatics?), but even so, the basic ingredients of the novel—character, plot, conflict, resolution—are present in some recognizable form. In *Watt*, Beckett begins the process of dismantling these familiar narrative structures in pursuit of a new agenda. The result is an enigmatic book where his interest in using mathematics to alter the traditional way literature functions is arguably at its most transparent.[13]

Like *Murphy*, *Watt* is grounded in the mental perambulations of its protagonist but has its focus much more tightly restricted to an internal, cognitive space. There is very little narrative scaffolding to speak of. In part I of the novel, Watt travels to the house of the mysterious Mr. Knott. Parts II and III recount Watt's employment on the various floors of the house, and in part IV he departs. The book feels more like an allegorical portrayal of successive stages of consciousness than it does a story about an actual person. The character of Watt takes a pedantic, hyperrational approach to understanding his surroundings, constantly observing, recording data, playing out hypotheticals in his mind but never progressing toward any kind of enlightenment. In fact, there is a palpable sense of mental deterioration as the novel progresses through its four stages, with Watt becoming less and less able to extract meaningful conclusions from even the simplest events that surround him.

And Sometimes He Went Barefoot

Throughout *Watt*, Beckett appeals to the refined austerity of mathematics for the raw material he needs to create the strained psychological portrait of his central character. When Watt hears three frogs croaking at distinct but regular intervals, Beckett includes a multiple-page chart illustrating the 120 beats required for the pattern to repeat (Figure 3.3). Watt hears songs full of numerical calculations, he puzzles over a simple picture of a circle with its center moved outside the circumference. There are anagrams, poems created from permutations of a single sentence, and passages where Watt speaks in code by reversing various combinations of letters, words, and sentences (Figure 3.4). In early drafts of *Watt* there are paragraphs that are transcriptions of algebraic formulas that Beckett converted into English by assigning words to the different variables. Most notoriously, there are multiple instances where the empirically minded Watt attempts to glean some insight into Mr. Knott

Krak!	—	—	—	—	—	—	—
Krek!	—	—	—	—	*Krek!*	—	—
Krik!	—	—	*Krik!*	—	—	*Krik!*	—
Krak!	—	—	—	—	—	—	—
—	—	*Krek!*	—	—	—	—	*Krek!*
—	*Krik!*	—	—	*Krik!*	—	—	*Krik!*
Krak!	—	—	—	—	—	—	—
—	—	—	—	*Krek!*	—	—	—
—	—	*Krik!*	—	—	*Krik!*	—	—

FIGURE 3.3. "And the three frogs croaking Krak! Krek! And Krik!, At one, nine, seventeen, twenty-five, etc., and at one, six, eleven, sixteen, etc., and at one, four, seven, ten, etc., respectively." (112)[14]

> We shall be here all night,
> Be here all night shall we,
> All night we shall be here,
> Here all night we shall be. (38)

FIGURE 3.4. A poem of permutations from *Watt*.

(whom he never encounters) by exhaustively cataloging the space of possibilities regarding Mr. Knott's appearance or what he wears on his feet. The problem is that Mr. Knott, as his name suggests, eludes description and in a sense becomes all things at once:

> As for his feet, sometimes he wore on each a sock, or on the one a sock and on the other a stocking, or a boot, or a shoe, or a slipper, or a sock and a boot, or a sock and a shoe, or a sock and slipper, or a stocking and boot, or a stocking and shoe, or a stocking and slipper, or nothing at all. And sometimes he wore on each a stocking, or on the one a stocking and on the other a boot, or a shoe, or a slipper, or a sock and boot, . . . (164)

The previous quotation has been mercifully truncated here, but Beckett includes the entire list of possibilities. Three choices (sock, stocking, nothing) times four choices (boot, shoe, slipper, nothing) yields twelve possibilities. With two feet that makes $12^2 = 144$ options, although Watt apparently does not distinguish between left and right feet. Thus, the list is reduced to a mere seventy-two entries, concluding *two pages later* with "and sometimes he went barefoot." As exhaustive and *even longer* lists like this appear more frequently in the late stages of the novel, Watt's disturbing slide into madness becomes literally unbearable to read.

That Crazy Mathematician

The breadth and scale of these mathematically inspired ambushes to traditional writing beg for a more thorough explanation. Yes, *Watt* comes across as a pointed attack on the limits of rational thinking. One might also see in them a coded reflection of the horrors of the war taking place as the book was being written. But there is something else at work in the text of *Watt*, of a more technical nature, centered on the function of the words themselves. A clue to what Beckett was doing with these experiments melding mathematics and prose can be found in his correspondences. In 1937—so post-*Murphy*, but pre-*Watt*—Beckett penned a fascinating letter to a German friend and colleague named Axel Kaun. After some pleasantries and an opinionated rant about the shortcomings of a particular poet, Beckett opens up about his frustrations with the state of affairs among contemporary writers. "Writing in English is getting more and more difficult and senseless," Beckett confesses midway through the letter, which is written in German:

> More and more my language seems like a veil that must be ripped apart to reach the thing that lies behind it (or the nothingness) that lies behind it. Grammar and style. To me they seem to have become as irrelevant as a Biedermeier bathing suit or the unflappability of a gentleman. A mask. Hopefully the time will come—fortunately in some circles it already has— where language is used best where it is most abused. Because we can't just suddenly turn language off, we should at least make sure we don't miss anything that might contribute to it acquiring a bad reputation. To drill one hole after another into it, until whatever is hiding behind it, be it something or nothing, leaks through—I cannot imagine a higher goal for today's writers.[15]

Beckett has defined the enemy—or the challenge at least—and language is in the crosshairs. In Beckett's view, music and painting have found ways to evolve into new and unforeseen modes of expression, but literature is hamstrung by its inability to repurpose words for anything other than their originally intended function. When Beckett wonders aloud in the letter,

> why this terrible arbitrary materialism in the wordscape can't be dissolved . . . so that we cannot perceive entire pages in any other way than as a dizzying path of sounds that connect bottomless abysses of silence,

the mindless pages of combinatorial lists in *Watt* immediately come to mind. In that same vein, all the linguistic antics at work in this strange novel might be interpreted as pinpricks in the so-called wordscape, designed to weaken it in an effort to reveal what lies beneath. But they are by no means the solution to Beckett's dilemma:

> Of course, for now, one has to be satisfied with little. The first important step should be to somehow find a method to express with words this scornful mocking position of words. In this dissonance of means and use, one should perhaps already feel a whisper of the endmusic of the universal silence.

The comment about "means and use" gets to the crux of the issue. Words are the fundamental obstacle to be negotiated, but they are also the instrument to be used in the negotiation. Herein lies the predicament. What is telling is that twice in the closing paragraph of his letter, Beckett turns to mathematicians for inspiration on how to proceed:

> Perhaps the logography of Gertrude Stein is closer to what I have in mind. . . . The poor lady (is she still alive?) is without doubt still in love with her vehicle, if only in the way the mathematician is in love with his numbers—the solution of his problem is of purely secondary interest and, furthermore, it must seem positively terrible because it is the death of the numbers. To connect Stein's method with that of Joyce, as is the fashion, appears just as meaningless to me as the yet unknown attempt to compare nominalism (in the sense of the Scholastics) with realism. But on this path toward the literature of the unword it is possible that some form of nominalistic irony is a necessary stage. But it is not enough if the game loses something of its holy seriousness. It should just stop. Thus, let's proceed like that crazy (?) mathematician who at every stage of the calculation applied a new principle of measurement. A storm of words in the name of beauty.

Beckett experimented with all manner of mathematical devices in his early novels, but to more effectively proceed like a crazy mathematician would require a larger degree of precision over the components of his art. To achieve this precision, he started writing for the stage.

Endgame

Moment upon moment, pattering down, like the millet grains of . . . that old Greek.

—HAMM, FROM *ENDGAME*

"When I was working on *Watt*, I felt the need to create for a smaller space, one in which I had some control of where people stood or moved, above all, of a certain light. I wrote *Waiting for Godot*."[16] *Waiting for Godot*, or *En attendez Godot*, was originally written in French during a period of intense creativity after the war. The switch to a new language (French) and a new medium (theater) can be viewed as two distinct mechanisms aimed at the common goal of tightening up the experimental environment in which language was being employed. Of the decision to start writing in French, Beckett has offered a number of reasons, including that it "had the right weakening effect" and that it was easier to write "sans style." The problem with English, he said, is that "you couldn't help writing poetry in it."[17]

Beckett's comments about writing *Godot* indicate his desire to extend that same precision and formality to the sound and space around the words. One view of theater is that it is a richly collaborative endeavor between playwright, director, actors, and designers. There are scores of wildly varied productions of *Hamlet*, for instance, each expressing a different artistic vision inspired by Shakespeare's original script. This malleability is distinctly lacking in Beckett's plays. Each one is constructed to look and sound a certain way. His stage directions are detailed and include instructions for gestures and tone of voice. Pauses in the dialogue are written into the script. Silence is a word in Beckett's vocabulary, and the shape, the sound, and the structure of the performance matter as much as the lines of dialogue. While he was alive, Beckett often directed his own plays, and when he was not directing he still tried to maintain some control over the performance. On one occasion, Beckett was made aware of a production of *Endgame* that took some liberties with the script, setting it in an abandoned New York City subway car. Threatening legal action, Beckett wrote to the director saying, "Any production which ignores my stage direction is completely unacceptable to me."[18]

Waiting for Godot premiered in Paris in 1952, and its popularity transformed Beckett's relationship to the public. While he maintained that critical success never mattered much, the proliferation of performances of *Godot* that followed earned the relatively unknown Irish novelist a devoted audience. It also had a profound and widely acknowledged impact on twentieth-century theater. Tom Stoppard explained it this way: "I was immobilized for weeks after I saw *Godot*. Historically, people had assumed that in order to have a valid theatrical event you had to have x. Beckett did it with x minus 5."[19] The most famous summary of this quixotic play is from scholar and critic Vivian Mercier. Noting the parallel structure of the two acts during which Vladimir and Estragon wait in vain for the arrival of Mr. Godot, Mercier wrote that Beckett had "achieved a theoretical impossibility . . . he has written a play in which nothing happens, twice." This movement toward minimalism—toward a restricted, more austere focus—starts to align Beckett with twentieth-century mathematics, but in *Godot* this connection is still far below the surface. Other than Vladimir being a trademark Beckettian logician like Watt, there is not much mathematics to be found in the script. This changes with Beckett's next play.

Zeno's Impossible Heap

Beckett's follow-up to *Waiting for Godot* was *Endgame*, or *Fin de partie* as it was called in the original French. The title recalls Beckett's fondness for chess, and the opening speech recalls Beckett's fondness for mathematical paradoxes. Tom Stoppard imports Zeno's paradoxes of the Arrow and of the Tortoise and Achilles into *Jumpers*—a play that acknowledges its author's debt to Beckett in its last line: "Wham, bam, thank you Sam." In *Endgame*, Beckett creates a central image for his play using Zeno's lesser-known paradox of the Millet Seed.

The opening tableau of *Endgame* features a bare interior with two small windows to either side. In the center of the room is Hamm, initially covered with a sheet and sitting in a large armchair where he will reside for the entirety of the play. Off to the side is Clov, Hamm's servant. Clov laboriously limps about the stage, folding up sheets and looking out the windows before delivering the play's opening line:

CLOV: (*fixed gaze, tonelessly.*) Finished, it's finished, nearly finished, it must be nearly finished. (*Pause.*) Grain upon grain, one by one, and one day, suddenly, there's a heap, a little heap, the impossible heap. (*Pause.*) I can't be punished anymore. (*Pause.*) I'll go now to my kitchen, ten feet by ten feet by ten feet, and wait for him to whistle me. (*Pause.*) Nice dimensions, nice proportions, I'll lean on the table, and look at the wall, and wait for him to whistle me. (8)[20]

If a single millet seed makes no sound as it falls, Zeno argued, how then does a heap of seeds make an audible thump? Generalizing, how can a collection of nothings—the word "zero" is uttered upward of ten times at different points in *Endgame*—accumulate into something nontrivial? Later in the play, Hamm makes the allusion more explicit:

> HAMM: Moment upon moment, pattering down, like the millet grains of
> ... (*he hesitates*) ... that old Greek, and all life long you wait for that to
> mount up to a life. (79)

Beckett's penchant for mathematical imagery is also evident in Clov's perfectly cubical kitchen. We never see this space. The entirety of the play takes place in Hamm's room, which is described as a circle with Hamm located at its center. On several occasions, Hamm commands Clov to move him about but always he wants to be returned to his proper location:

> HAMM: Back to my place! (*Clov pushes chair back to center.*) Is that my
> place?
> CLOV: Yes, that's your place.
> HAMM: Am I right in the center?
> CLOV: I'll measure it.
> HAMM: More or less! More or less!
> CLOV: (*moving chair slightly*): There! (34)

The sparse interior world of *Endgame* is more claustrophobic than the country road in *Godot*. There is a picture hanging, but its face is to the wall. As the play progresses, we learn that out one of the windows is the "earth" and out the other is the "sea," but looking out either one requires a step ladder and the use of a telescope. Like outer space, the world beyond the windows is alien and inaccessible. Hamm and Clov's relationship is complicated, equal parts master and servant, patient and nurse, father and stepson. Both are showing the physical effects of a burdensome life. Hamm is blind and lame; Clov's eyesight is shoddy, and his stiff legs—a familiar ailment among Beckett characters—make it impossible for him to sit. Hamm is the more obvious king of the chessboard, full of bluster and bitterness but just short of abusive. Clov does what he is told but speaks his mind and is not afraid to challenge Hamm. In fact, Clov's responsibility for Hamm's care gives him a certain

"Gone from me you'd be dead." *Endgame*; Mark Rylance (Hamm), Simon McBurney (Clov); Duchess Theatre, 2009. © Donald Cooper/photostage.co.uk.

kingly authority of his own, and thus the game is on. As Clov indicates in his opening speech, the negligible millet seeds have somehow accrued into an impossible heap, and this day is going to be different. "It's finished, nearly finished," he says. The two adversaries have entered the endgame of their lifelong match:

> HAMM: Why do you stay with me?
> CLOV: Why do you keep me?
> HAMM: There's no one else.
> CLOV: There's nowhere else. (*Pause.*)
> HAMM: You're leaving me all the same.
> CLOV: I'm trying. (13)

Clov's intention to leave Hamm forms the central pillar of the play, a prospect that becomes less and less feasible the more we realize that Clov's existence is tied to Hamm's in some axiomatic way. There are faint whiffs of tenderness that add bits of color to an otherwise gray canvas. On the edge of being overtaken by loneliness or their physical ailments, Clov and Hamm persevere in a moment-by-moment way, in part by simply having their existence acknowledged by the other person. This is a major motif that stretches from

Beckett's fiction to his plays to his television work. In *Godot*, each time the boy arrives to announce that Mr. Godot is not coming, Vladimir always asks the same favor: "Tell him you saw us. You did see us, didn't you?" Sticking out obtusely in the middle of an otherwise jumbled soliloquy earlier in *Godot* is the name Bishop Berkeley.[21] Berkeley was an Irish philosopher from the early eighteenth century who took the position that there is no such thing as matter; that all there is in the universe are ideas and minds to perceive them. *Esse est percipi* is the Latin phrase associated with Berkeley's philosophy of immaterialism—to be is to be perceived.

When only the two kings remain in a game of chess it is impossible for one to put the other in check without simultaneously moving into check. Thus, all they can do is stalk each other around the board. In a world governed by Berkeley's maxim, a similar predicament faces Clov and Hamm.

> HAMM: What? Neither gone nor dead?
> CLOV: In spirit only.
> HAMM: Which?
> CLOV: Both.
> HAMM: Gone from me you'd be dead.
> CLOV: And vice versa. (79)

Nicely Put, That

Amid these enacted paradoxes of Zeno and Berkeley, there is a third significant assault on our sense of logic that bubbles up through a performance of *Endgame*. Intermittently throughout the evening, Hamm composes aloud what he refers to as his chronicle, an ongoing tale told in the first person that falls somewhere between fiction and autobiography.

> HAMM: (*Narrative tone.*) The man came crawling towards me, on his belly. Pale, wonderfully pale and thin, he seemed on the point of— (*Pause. Normal tone.*) No, I've done that bit.

Hamm is artist and editor. He uses his narrative tone to compose his tale, switching to his normal tone when he needs to reflect on his creative work. In accordance with Berkeley, he always requests an audience to listen when he is orating but, even so, he serves as his own critic.

> HAMM: (*Narrative tone.*) . . . It was a glorious bright day, I remember, fifty by the heliometer, but already the sun was sinking down into the . . . Down among the dead. (*Normal tone.*) Nicely put, that. (*Narrative*

tone.) Come on now, come on, present your petition and let me resume my labours. (*Pause. Normal tone.*) There's English for you. Ah well . . . (*Narrative tone.*) It was then he took the plunge. It's my little one, he said. Tsstss a little one, that's bad. My little boy, he said, as if the sex mattered. (60)

Watching the fiction writer at work, we start to feel the entire structure of the play wobble, and looking down we notice cracks in the foundation that had eluded us on first glance. The cause of these cracks is a familiar culprit to mathematicians. It was a demon that had brought the foundations of mathematics tumbling down at the end of the previous century and, when Beckett was writing, it was still wandering the mathematical hallways making mischief in the newly erected structures. We shall presently take up the mathematical story of self-referencing paradoxes, but Beckett's incorporation of them is equally potent.

Hamm is weaving images out of words, manipulating them to manipulate us, and then explicitly confessing that he is doing so. Like a magician who performs illusions while simultaneously telling his audience the trick, Hamm creates a logical quagmire around the distinction between real experiences and invented ones. It sounds like he might be recounting the actual past except that he keeps finding ways to make it clear he is improvising. To what end? By his own declaration, Beckett was trying to get beyond the veil of language but employed language as his primary means for doing so. By having Hamm reveal the duplicity of the fabric of reality that words can point to, the hope is that the fabric will become translucent, and as it does, something else gets revealed. Whatever it is cannot be described obviously, but its presence can be acknowledged in this negative way—as the thing that's left when the thing we can see loses its integrity.

In the particular case of Hamm's chronicle, we are constantly at sea trying to parse fact from fiction. Hamm's last line, "Tsstss a little one, that's bad. My little boy, he said, as if the sex mattered," is especially devious. In it, he invents a detail which adds some flesh to the characters he is conjuring and then makes the details irrelevant. The sex doesn't matter to Hamm, but it matters enormously to us because upon hearing it, we can't help but feel that the little boy might be Clov. Hamm's story is about a man who comes desperately seeking employment and then asks if he can also bring his son:

HAMM: (*Narrative tone.*) . . . It was the moment I was waiting for. (*Pause.*) Would I consent to take in the child. (*Pause.*) I can see him still, down on

"As if he were asking me to take him for a walk." *Endgame*; Mark Rylance (Hamm), Simon McBurney (Clov); Duchess Theatre, 2009. © Donald Cooper/photostage.co.uk.

his knees, his hands flat on the ground, glaring at me with mad eyes, in defiance of my wishes. (*Pause. Normal tone.*) I'll soon have finished with this story. (*Pause.*) Unless I bring in other characters. (*Pause.*) But where would I find them? (*Pause.*) Where would I look for them? (*Pause. He whistles. Enter Clov.*) Let us pray to God. (62)

Hamm and Clov pray, as Hamm instructs, but it doesn't work. "The bastard!" Hamm laments. "He doesn't exist!"

The existence of God was never of much concern to Beckett, but the existence of the extra-lingual self most certainly was. Another way to investigate what might be lurking behind the parts of ourselves that we can verbally articulate is to peel away the delusions we unconsciously rely on to keep ourselves unaware. "We always find something," Estragon says in *Godot*, "to give us the impression that we exist." One question *Endgame* implicitly asks is: Does it matter if the somethings we find are real? God? Memories? A well composed history? At his blind master's request, Clov makes a stuffed dog that Hamm has requested be white and stand at attention "as if he were asking me to take him for a walk." (49) But the unfinished toy Clov passes off is black and has only three legs.

The toys and jokes in *Endgame* are a modest balm on the overwhelmingly mournful arc of the play. The most compelling part of the quest for meaning in *Endgame* is focused on relationships—on seeing and being seen, or hearing and being heard. In the closing moments of the play, Clov attempts his final move and exits the stage. Alone, Hamm tries to navigate about the room in his chair, but it is useless. Resigned, he tips his hat and tries once more to compose a little poetry where, again, the improvised edits toggle the piece in and out of the realm of autobiography:

> HAMM: You prayed—(*Pause. He corrects himself.*) You CRIED for night; it comes—(*Pause. He corrects himself.*) It FALLS: now cry in darkness. (*He repeats, chanting.*) You cried for night; it falls: now cry in darkness. (*Pause.*) Nicely put, that. (91)

And now? Hamm returns to his chronicle one last time:

> HAMM: (*narrative tone.*) If he could have his child with him. . . . (*Pause.*) It was the moment I was waiting for.

Accentuating the logical paradoxes at play, Clov has silently reappeared, observantly standing in the doorway wearing his traveling hat and carrying his coat over his arm. Is Clov the child? If Hamm is perceived but does not know it, does he still exist? Hamm casts away his whistle and his toy dog, then sniffs the air. Does he sense Clov is there? The blind, chair-bound Hamm calls out, "Clov! (*Long pause*) No? Good." Hamm's stoicism is like a restraining strap pulled taut around a bursting chest of conflicting emotions. Devastated? Yes, it would seem. Heartbroken, betrayed? That too. Resigned, relieved that his adopted son has set himself free—or distraught that he has not? It's unclear, and in fact it could be that the game has not ended. "Since that's the way we're playing it," is the blind king's last desperate gambit, "let's play it that way. . . and speak no more about it." (93)

Charming Evening We're Having

The first play Beckett wrote was not *Waiting for Godot* but an unruly play called *Eleutheria*. In accordance with Beckett's wishes, it has never been performed. It requires seventeen actors and draws heavily on features of traditional drama with the goal of parodying them. In the third act, a character called the Spectator emerges from the audience and walks on stage to critique the play he is watching. This is one way to add a metatheatrical voice to a play, but it is a bit heavy-handed and jarring to the dramatic flow. In *Endgame*, Hamm's self-referencing chronicle is a more organic way to make the self-commentating voice a purely internal component of the play. Beckett creates a similar effect in *Godot*. Over the course of this full-length play, the only real action is the inaction of waiting. This makes passing the time the primary task of both Vladimir and Estragon as well as the audience, and it results in the rise of a comic, self-referential voice in the two tramps' music-hall banter:

VLADIMIR: Charming evening we're having.
ESTRAGON: Unforgettable.
VLADIMIR: And it's not over.
ESTRAGON: Apparently not.
VLADIMIR: It's only beginning.
ESTRAGON: It's awful. (23)

How, then, does a play that consistently draws attention to its potential monotony become so profoundly moving? The answer is complicated. The first production of *Godot* in the United States was mounted at the Coconut Grove Playhouse in Miami in 1956. Advertised as "the laugh riot of two continents," the show was a flop. Reviews were harsh, audiences walked out, and the larger tour was prematurely canceled.[22] Later productions fared much better, one in particular at the San Quentin penitentiary in San Francisco the following year. Apparently, the inmates detected a voice more sophisticated theatergoers could not hear.[23]

Godot acquires its agency when a sense of empathy is built up between the audience passing the time watching the characters on stage doing their best to pass the time. Amusing at first, this deep identification between audience and actor has a chilling effect when the self-deprecating, vaudevillian moments are gradually replaced in the course of the evening by something darker. As the play moves inexorably toward its nonexistent denouement, a revelation lingers just out of reach—an uneasy truth about who we are and what our lives consist of between the boredom and the busyness. Very late in the play, the typically flummoxed Vladimir is given a glimpse of introspective enlightenment:

VLADIMIR: Was I sleeping, while the others suffered? Am I sleeping now? Tomorrow, when I wake, or think I do, what shall I say of today?

"We always find something, eh Didi, to give us the impression we exist?"
Waiting for Godot; Ian McKellen (Estragon), Patrick Stewart (Vladimir);
Theatre Royal Haymarket, London, 2009. © Donald Cooper/photostage.co.uk.

That with Estragon my friend, at this place, until the fall of night, I waited
for Godot? . . . Astride of a grave and a difficult birth. Down in the hole,
lingeringly, the grave-digger puts on the forceps. We have time to grow
old. The air is full of our cries. (*He listens.*) But habit is a great deadener. (*He looks again at Estragon.*) At me too someone is looking, of me
too someone is saying, He is sleeping, he knows nothing, let him sleep
on. (58)

It is the quiet but audible metatheatrical voice in *Godot* that creates the
visceral sense of desperation. A similar comment applies to *Endgame*, where
Hamm's chronicle allows him to be both a character and the play's narrator.
The desire to experiment with reflexive structures is a motif that Beckett pursued in his playwriting in progressively tighter, shorter, and more abstract
pieces. More than his regular use of combinatorics or geometric imagery, it was
his instinct for exploiting the fragility of logical structures that enabled Beckett to proceed like the crazy mathematicians who reconfigured the landscape
of mathematical truth a few decades before.

Russell's Paradox

Mathematics may be defined as the subject in which we never know what we are talking about, nor whether what we are saying is true.

—BERTRAND RUSSELL, FROM "RECENT WORK ON THE PRINCIPLES
OF MATHEMATICS," 1901

In June 1902, Gottlob Frege was working on volume 2 of *The Basic Laws of Arithmetic*. It is not an exaggeration to say that Frege's goal was to provide a foundation for the entirety of mathematics, and that in his own estimation the finish line was in sight. It was at this time that a letter arrived from Bertrand Russell. "I find myself in full accord with you on all main points," Russell wrote. "I find discussions, distinctions, and definitions in your writings for which one looks in vain in other logicians." That was the good news. Then Russell continued, "I have encountered a difficulty only on one point."[24]

Constructing Mathematics from Logic

The overarching theme of Frege's work was to resolve, in the affirmative, the metaphysical question of whether mathematical objects really exist. They did, Frege believed, and he was going to explicitly construct them.

In the conclusion of his letter to Axel Kaun, Beckett brings up the philosophical positions of realism and nominalism, which are very much relevant to the predicament faced by mathematics after alternatives to Euclid's geometry emerged. If different, equally valid, systems of geometry exist, and none of them has primacy in describing physical space, then what precisely do these different axiomatic systems describe? What, in other words, is mathematics really about?

Realism—the philosophical kind, not the artistic movement associated with Ibsen and others—takes its cue from Plato, who argued that the subject matter of mathematics is permanent and preexisting, albeit in some idealized mathematical world separated from the physical one in which we live. For the realist, points and lines and numbers and functions are objective entities that exist independently of the mathematician, and well-formed statements about these objects possess truth values that do not depend on any particular mental processes or notational convention. "[E]ven if all rational beings were to take to hibernating and fall asleep simultaneously our proposition would not be cancelled for the duration," Frege said. "For a proposition to be true is just not the same thing as for it to be thought."[25]

Nominalism takes a different view in holding that language does not point to an actual object in some far-off Platonic world. To the nominalist, existence requires a concrete, sensory-based justification. This relegates the abstract objects of mathematics—along with other types of universal statements—to being linguistic constructions. This is a fairly radical position, one not likely to find much support among working mathematicians, but developments in geometry and arithmetic forced these issues into the foreground at the turn of the century, and a satisfying resolution was hard to come by.

Frege's elegant solution was to make a compelling case for realism by showing how mathematics could be viewed as a subset of logic. For Frege, the rules of logic that govern deductive thinking were entities that did not rely on empirical experience or some kind of Kantian intuition for their existence. This made them ideal ingredients from which to build mathematical objects. *The Foundations of Arithmetic*, published by Frege in 1884, begins with the question, "What is the number 1?" Frege's answer is to quite literally construct the number 1, and the rest of the whole numbers, from the raw material provided by the laws of logic. In Frege's lexicon, this raw material consists of *concepts*, *objects*, and *extensions*. In the introduction to *Foundations*, Frege instructs his reader to "never lose sight of the distinction between concept and object,"[26] an admonition that Russell showed was harder to heed than Frege realized.

For Frege, objects are the nouns of the logical world. Proper names like Hamm and Samuel Beckett point to objects, as do sufficiently specific phrases like "the blind character in *Endgame*." Concepts, meanwhile, are not objects. Concepts are best understood as properties of objects, and Frege typically references them as predicates such as

is a character in *Endgame*,

or

is one of Goldilocks's bears.

Concepts properly describe some objects but not others. To decide whether a concept is a property of some particular object, we combine the object with the concept and ask whether the now completed statement comes out true or false. The concept "is a character in *Endgame*" applies to Hamm and Clov. It does not apply to Vladimir, the book *Murphy*, Julius Caesar, or the Golden Gate Bridge. In Frege's terminology, the objects to which the concept applies are said to *fall under* that concept. The collection of all the objects that fall under a concept is called its *extension*. The extension of "is a character in *Endgame*" includes Hamm, Clov, and a few others. It does not include Watt, Julius Caesar, or the Golden Gate Bridge. The extension of the concept "is one of Goldilocks's bears" consists precisely of Papa Bear, Mama Bear, and Baby Bear.

A Difficulty Only on One Point

Although concepts are not objects, extensions are objects. This is important, in part because it opens up the possibility that an extension of a concept might fall under that very same concept. Thinking of extensions as sets, or collections, the question is whether there exist concepts with extensions that contain themselves. This is not the case for any of the concepts we have encountered thus far. The concept "is one of Goldilocks's bears" has an extension containing Papa Bear, Mama Bear, and Baby Bear. Notice that there are three objects in the extension and none of them is, on its own, the whole collection. A concept like "wrote *Endgame*," which applies to a single object, doesn't work either, although this requires a bit more concentration. The object Samuel Beckett is not the same thing as the extension whose only member happens to be Samuel Beckett. The former is a writer, the latter an extension, and extensions don't write plays.[27]

Although at first it is hard to think of a concept with an extension that contains itself, there is nothing in Frege's formulation to prevent such a beast. Consider, for instance, the concept "is an object." Remembering that extensions are objects, it follows that the extension of "is an object" is an object, and thus contains itself.

Concepts with this exotic property are, well, exotic. The concept "is the extension of some concept" is another exotic concept whose extension contains itself. By contrast, "wrote *Endgame*" and "is one of Goldilocks's bears" are regular, not-at-all-exotic concepts in that their extensions do not fall under the concept in question. Having made this distinction between regular concepts and exotic ones, the table is set for Russell's proposal. Let R be the concept

is the extension of a regular concept.

This sounds harmless enough. Given some concept, let's call it F, we know it has an extension, and we can ask the unambiguous yes-or-no question of whether the extension falls under F. If it does, then F is exotic and does not fall under R; if it does not then F is regular and does fall under R. This is all to say that the concept R seems perfectly reasonable.

And then Russell asks: Is R regular or exotic?

The paradoxical answer is neither, or more accurately, if it is one then it is also the other. If R is exotic, then the extension of R falls under R. Because objects that fall under R must be extensions of regular concepts, we conclude that R is regular, a blatant contradiction. Starting from the assumption that R is regular fares no better. The definition of regular means that the extension of R does not contain itself, which is to say that R is not regular after all—it is exotic. The full force of the logical contradiction is upon us. Neither option is acceptable, which is, in the world of logic, unacceptable.

With the new volume of his book already at the printer, Frege went hastily to work on an appendix to patch up the problem, but in the end, Russell's observation would prove fatal.[28]

The Epimenides Paradox

Russell's paradox, as it has come to be known, has its roots in a family of logic puzzles that all share a common theme. Russell himself referenced a version involving the town barber. Assume there is a town with a barber who shaves every man in the town who does not shave himself. Who shaves the barber? If we assume the barber shaves himself then the barber does not shave himself. The other option—that the barber does not shave himself—leads to the conclusion that the barber shaves himself. The question leads to a contradiction either way and can't be answered. The consensus seems to be that the earliest recorded riddle of this nature stems from around 600 BCE and is attributed to the philosopher Epimenides, who is supposed to have asserted, "All Cretans are liars." Keeping in mind that Epimenides was a Cretan, notice that we encounter a similar paradox when we ask whether Epimenides's statement is true or false.

Perhaps the tightest version of the Epimenides paradox is formed by considering the sentence:

This sentence is false

and asking whether the sentence is true or false. Generally speaking, there is an innocuous mix of charm and mental trauma that results from the logical purgatory created by these kinds of self-referential puzzles. The consequences of Russell's version of this paradox, however, were dire. Logical systems were being proposed as the foundation of mathematics. Self-referential statements were at the root of the contradictions corrupting these systems. Therefore, an attempt had to be undertaken to create a logical foundation for mathematics that was impervious to the perils of self-referential constructions—and thus logical contradictions—but still powerful enough to support the full capabilities of arithmetic. This, in a nutshell, was the goal of Russell's historic tome *Principia Mathematica*. *Principia Mathematica* was a Herculean effort. Russell spent the best part of a decade working on it with his colleague and former professor Alfred North Whitehead. In three large volumes published between 1910 and 1913, Russell and Whitehead laid out a formalized system of notation and deterministic rules for deriving mathematical propositions. To distinguish the logical system from the three dense books where it is meticulously constructed, the books are collectively abbreviated as *PM* and the logical system as PM (without the italics).

To guard against the possibility of PM being susceptible to Russell's paradox, its authors implemented a hierarchy of stratified levels for the different objects within the system. When a new object was created as a collection of other objects, the new object was considered to be on a distinct, higher level, making it impossible for a collection of objects to contain itself. "Whatever involves all of a collection must not be one of the collection," was how Russell summarized it.[29] Without getting any further into the weeds, the upshot was that self-referencing concepts like "is an object" or "is the extension of a regular concept" would be forbidden in Russell and Whitehead's system, thus inoculating PM from the scourge of Epimenides and his descendants.

At least that was the intention.

Film and *Not I*

what? . . . who? . . . no! . . . she!

—MOUTH, FROM *NOT I*

Self-reference brought about the demise of Frege's attempt to anchor mathematics on solid footing, and banning it from the pages of *Principia Mathematica* was at the forefront of Russell and Whitehead's agenda. Self-reference, meanwhile, was central to the artistic power of *Godot* and *Endgame*. In 1963, Beckett wrote a script, or screenplay rather, for a black-and-white silent film to more directly explore this phenomenon. To emphasize that self-reference is at the core of the thirty-minute comic drama, Beckett titled the piece *Film*. As a header to the script, he included the Berkeley dictum *Esse est percipi*.

.

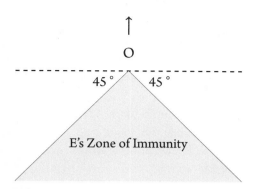

FIGURE 3.5. With O facing away, E must stay within the shaded region
in order to remain undetected.

Film

The two main characters in *Film* are called O and E in the script—O stands
for Object and E is shorthand for Eye. To play O, Beckett originally hoped
to get Charlie Chaplin but was ultimately content to secure the services
of sixty-eight-year-old actor and comedian Buster Keaton. The part of E is
essentially played by the camera.

The premise of *Film* is that E is in pursuit of O while O is desperately seeking refuge from being perceived. The first action we encounter is a view of O
hurrying along a sidewalk with his face obscured and his head pointed away
from the camera. Having spotted O, E takes up the chase and maintains a
position about ten meters behind, slightly off O's right shoulder. As the chase
continues, it becomes evident that there is a very specific region in which
E must remain so that O is unaware that E is watching him. Beckett speci-
fies the critical angle to be 45° from the horizontal of O's periphery (Figure
3.5). When E inadvertently moves to a position outside the specified "zone of
immunity," O winces in anguish and tries to rectify the situation by finding a
way to elude the sense of being watched. If "to be is to be perceived," as Berkeley would have it, then O's agenda is to become unconscious of his suffering
by remaining unobserved.

The last half of *Film* takes place in a sparsely furnished one room apartment
that does not belong to O. This is evident because O must, piece by piece,
disable the various possibilities of being perceived that exist in the room. He
locks the door, closes the window, puts a blanket over the mirror, puts out
the cat, then puts out the dog, then puts out the cat who sneaked in when he
was putting out the dog, then puts out the dog who sneaked in when he was

putting out the cat. He covers up the fishbowl and the birdcage and even rips up the poster of a watchful Jesus tacked to the wall. Satisfied that he is unobserved, O settles into a rocking chair to look at a series of photographs that chronicle a boy's life from infancy to early adulthood. Although momentarily at ease, the sight of the boy at age thirty is too close to O's current state of mind, and he proceeds to rip up the entire stack. Watching O's attempt to expunge from his environment anything that might lead to the possibility of O being perceived provides a poetic illustration of Russell and Whitehead's efforts to ban the possibility of self-referential constructions inside PM. And O fails for the same reason that Russell and Whitehead ultimately did.

By the conclusion of *Film*, director Alan Schneider has succeeded in creating the sense that the camera is an autonomous character in the film it is recording. Eventually, the anxious O, whose face we have not seen to this point, dozes off in his rocking chair. Seizing the moment, E audaciously moves out of its zone of immunity. It circles the room, imitating a path we saw O traverse earlier, and comes face-to-face with O who wakes in horror to discover E=O. Provocatively wearing an eye patch, suggesting a juxtaposition of inward and outward vision, an agonized Buster Keaton stares into his own face and crumbles under the revelation of the inescapability of self-perception.

The critical response to *Film* was underwhelming. Its comic intentions do not come through clearly and, unlike *Godot* and *Endgame*, it does not resonate in any emotional way. Beckett's attraction to *Film* seems to be very much a structural one, and that is the level on which the piece is most interesting to pursue.[30]

By conflating E and O—perceiver and perceived—Beckett is playing a game similar to the one Russell played with Frege's system of concepts and extensions. Russell saw the possibility for a concept to describe extensions in general (since extensions are objects) and describe its own extension in particular. By crafting the concept "is an extension that does not fall under itself," Russell effectively concentrates this reflexive capability much like what happens to the characters at the end of *Film*. The gaze of the perceiver in question—*Film*'s E, Russell's concept—is focused squarely on the object doing the perceiving, and the result is a sense of instability, or subversion, of the system in question.

In his letter to Axel Kaun, Beckett challenged his contemporary writers "to drill one hole after another into [language], until whatever is hiding behind it, be it something or nothing, leaks through." Beckett's suggested line of attack in that letter was to create a "dissonance of means and use," which is a

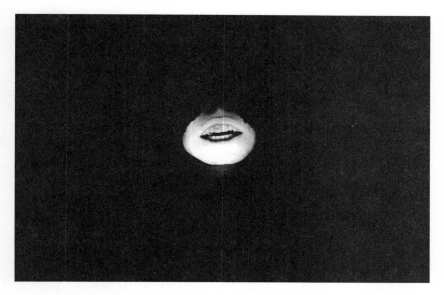

"gradually she felt . . . her lips moving . . . imagine!" *Not I*; Billie Whitelaw (Mouth); Royal Court Theatre, London, 1973. © Donald Cooper/photostage.co.uk.

reasonable description of what the mathematics and *Film* have in common in this instance. Frege's system of concepts and extensions is the means by which he is constructing his foundation for arithmetic, but extensions are also the objects being used in Russell's example. By the same token, O's sense of perception is the means by which he detects whether he is being perceived. To create the dissonance, there needs to be some kind of negation within the self-referential loops—Russell's concept applies to extensions that do *not* contain themselves, O seeks *not* to be perceived. From these twisted, self-referencing negations arise the potential for paradox and the unsettling consequences that ensue in each case.

Not I

Although *Film* offers a structural incarnation of an Epimenides type paradox, a more psychologically effective example of this same phenomena in Beckett's theater is the short play *Not I*, written in 1972. *Not I* is a monologue delivered by a disembodied mouth floating roughly eight feet above stage level. The actress playing the part—always a female—has her face blacked out and her body occluded in some way so that no other part of her is visible. The stage is in complete darkness except for a single narrow spot illuminating her lips, and so it is imperative that her mouth does not move about. In the script, Beckett refers to the character as Mouth. Physically restrained and separated from its body, Mouth spends the length of the short play exercising its one potential function by spewing forth a torrent of words.

When the curtain rises, Mouth is already talking at a low, unintelligible volume, but as soon as it has the audience's attention it starts in with an emphatic tone:

> . . . out . . . into this world . . . this world . . . tiny little thing . . . before its time . . . in a godfor— . . . what? . . . girl? . . . yes . . . tiny little girl . . . into this . . . out into this . . . before her time . . . godforsaken hole called . . . called . . . no matter . . . parents unknown . . . unheard of . . . he having vanished . . . thin air . . . no sooner buttoned up his breeches . . . (215)[31]

Beckett was insistent that the pace be extremely fast, bordering on incomprehensibility. In the first performance of *Not I*, at Lincoln Center in New York City, actress Jessica Tandy completed the monologue in twenty-four minutes. Although he didn't see it, and reviews were generally positive, Beckett was unsatisfied.[32] The London premiere a few months later featured Billie Whitelaw as Mouth. With Beckett involved in rehearsals and Whitelaw engaged in speed-speaking drills, the London performance of the play had a running time of about fifteen minutes. More recent productions have reduced that number to under ten in some cases.

At this speed, the words sound as though they are being generated in a profuse and uncontrollable manner. There is a mechanical aspect to Mouth, but that is not to suggest that the words are random or without sentient intent. Quite the opposite, actually. From the short clip that opens the play it should be apparent that Mouth has the framework of a story to tell, although gleaning anything too specific is unrealistic. Mouth speaks as though it is responding to an overabundance of impulses, with no filter or parsing devices at its disposal. Like a musical score intended to be played at a hundred and twenty beats per minute, the rhythm of the words is as important as their meaning. "Words at the speed of thought," is how Beckett described what he wanted.[33]

Roughly speaking, Mouth's monologue is about an Irish woman who, while walking across a grassy field, experiences a cognitive event corresponding loosely to a loss of consciousness:

found herself in the dark . . . and if not exactly . . . insentient . . . insentient
. . . for she could still hear the buzzing . . . so-called . . . in the ears . . . and
a ray of light came and went . . . came and went . . .

With sight and sound reduced to flashes and buzzing, the woman then starts
to lose a sense of her physical body:

feeling so dulled . . . she did not know . . . what position she was in . . . imagine! . . . what position she was in! . . . whether standing . . . or sitting . . . but
the brain— (216)

The images contained within the rushing phrases create a strobe light effect,
briefly illuminating moments from the woman' s life. We learn a small scattering of details: she was abandoned by her parents, unloved, has recently turned
seventy, and still clings to an old-fashioned Irish piety. All the while, the mental space Mouth describes descending around her in the field gets smaller and
smaller. The ringing in her ears is "not in the ears at all . . . in the skull . . . dull
roar in the skull," and she is skeptical that the sensation of closing her eyelids
that she feels is authentic.

And what is left when so much of reality dissolves and all becomes "dead
still . . . sweet silent as the grave"? The answer is words:

words were coming . . . imagine! . . . words were coming . . . a voice she did
not recognize . . . at first . . . so long since it had sounded . . . then finally had
to admit . . . could be none other . . . then her own . . . (218)

The single pinpointed spotlight on the ceaselessly blabbering lips is an
effective theatrical illustration of the psychological experience that Mouth is
describing. Walking in a field, this elderly woman is unexpectedly caught in a
powerful loop of self-perception with an ever-increasing focus. What do we
see when we sharpen the introspective lens and look past our experiences and
our bodies and our ears and eyelids? Is there a part of the self left to see when
we zoom in past all those things? In *Not I* there are words.

The ending of *Film*, when Object=Eye, has a curious intellectual impact
on the viewer, but the ceaseless yammering of Mouth makes its way under
the skin to work on the nerves.[34] With no body attached, Mouth is pure
consciousness, and the effect is unsettling in part because Mouth's story, like
Hamm's, begins to sound autobiographical. We learn that the old Irish woman
had trouble speaking during her life, with the exception of a few traumatic
bouts of uncontrollable logorrhea that would send her back into a state of

humiliated silence. In self-defense, the woman would delude herself that the voice she heard speaking during these times was not hers:

> not her voice at all . . . and no doubt would have . . . vital she should . . . was on the point . . . after long efforts . . . when suddenly she felt . . . gradually she felt . . . her lips moving . . . imagine! . . . her lips moving! . . . (219)

And so Mouth tells us that she—the Irish woman, that is—faces the truth that this insuppressible and intractable voice was hers alone. What Mouth cannot do, however, is relinquish telling her tale in the third person. Four times during the monologue Mouth responds to some unheard internal questioner suggesting that Mouth is perhaps talking about herself:

> what? . . . who? . . . no! . . . she!

These violent refusals stop the piece in its tracks, momentarily halting the flow of words.[35] They crack like an arc of electricity in a logic machine experiencing a short circuit trying to parse Russell's paradox. Mouth is telling us the lonely tale of a loveless old woman who could not speak but for a maddening voice that she recognizes as her own but cannot control—and that voice is the one telling the story. Mouth recounts the woman's attempts to make sense of the voice, or make it cease—"something begging in the brain . . . begging the mouth to stop." Comprehending the odd word here or there sends the woman—via the voice of Mouth—back through her life in a frantic search for some elusive truth that might reveal what is animating the voice in the first place. But the search is doomed. Mouth is unwittingly the subject of its own scrutiny and there is no way for it to understand its inability to understand itself.

Incompleteness

The wide range of self-referential devices at work in Beckett's writing has been a regular source of interest for scholars. Although storytellers becoming paradoxically engulfed in the stories they tell is a broad tenant of postmodern literature, Beckett distinguishes himself in this pursuit with the depth of his emulation of the mathematicians' techniques. From *Godot* (1949) to *Endgame* (1956) to *Film* (1963) to *Not I* (1972), Beckett parallels the increasingly introspective arc of twentieth-century logic by writing under progressively tighter self-imposed restrictions. This is true of the psychological space explored within each drama; it is also an apt description of the physical components that go into the dramas themselves. The plays get shorter and more sparse, the characters more physically restrained, the content more abstract and inwardly directed.

In his German letter from 1937, Beckett was restless with the traditional ways that language was being employed by writers, but he does not offer much in the way of specifics for how to proceed. Over time, Beckett made other attempts to articulate his core philosophy, but every time he did so a version of the Epimenides paradox was lurking about complicating matters. In 1949, Beckett attempted to explain himself to Georges Duthuit in a published conversation ostensibly about painting. Beckett complains that the Italian masters, with all their imagination, "never stirred from the field of the possible, however much they may have enlarged it."

Duthuit: What other [field] can there be for the maker?

Beckett: Logically none. Yet I speak of an art turning from it in disgust, weary of puny exploits, weary of pretending to be able, of being able, of doing a little better the same old thing, of going a little farther along a dreary road.

Duthuit: And preferring what?

Beckett: The expression that there is nothing to express, nothing with which to express, nothing from which to express, no power to express, no desire to express, together with the obligation to express.[36]

Duthuit adopts a highly skeptical attitude, sarcastically wondering how "the form of expression known as painting ... [could] be rid of the misapprehension ... that its function was to express, by means of paint." But Beckett holds fast to his untenable position, and he does so with a favorite mathematical metaphor:

Beckett: The history of painting, here we go again, is the history of its attempts to escape from the sense of failure, by means of more authentic, more ample, less exclusive relations between representer and representee, in a kind of tropism towards a light as to the nature of which the best opinions continue to vary, with a kind of Pythagorean terror, as though the irrationality of pi were an offense against the deity, not to mention his creature.[37]

Beckett falls back on the image of irrational numbers residing in some dark, inaccessible domain. To the Greeks they were the nameless numbers existing in the invisible gaps of the accepted—i.e., rational—part of the number line. Their discoverer was put to death, but the real heresy, Beckett is saying, is ignoring their potential to alter the evolution of arithmetic.

This is an analogy with which Beckett was very comfortable, but there is another one begging to be addressed. Noting his propensity for mathematics and logic, scholars of Beckett have speculated about a potential connection between Beckett's pursuit of the inexpressible with the celebrated incompleteness theorems of German logician Kurt Gödel. "Thus let's proceed like that crazy mathematician," Beckett wrote to his German friend, "who at every stage of the calculation applied a new principle of measurement." To say definitely if Beckett was referring to someone in particular, and if so who, is too much to ask, but a serious argument has been advanced that the "crazy mathematician" is Gödel himself.[38]

This is a remarkable conjecture.

In 1931, the twenty-five-year-old Gödel published a short and highly technical paper in a German scientific periodical with the translated title "On Formally Undecidable Propositions of *Principia Mathematica* and Related Systems." The impact of this paper was profound. Within the realm of mathematical logic, it ranks as one of the most important discoveries in history, and its subsequent aftershocks vibrated throughout the foundations of mathematics and philosophy for decades to follow. They are still being felt today. In the search for certainty, Gödel permanently altered the mathematical landscape.

To assess Beckett's relationship to Gödel requires a thoughtful look inside Russell and Whitehead's *Principia Mathematica* mentioned in Gödel's title. This distinctly formal notational system of mathematics was at the center of an impassioned debate about the power and potential of axiomatic systems. The central question was whether the notion of mathematical provability could be refined in such a way that it would be fully commensurate with the notion of truth, at least within the limited jurisdiction of PM or some other "related system." As the title of Gödel's 1931 paper suggests, the answer to this question was a humbling no. Mathematical truth, even in a domain as well-defined and concrete as the arithmetic of whole numbers, could not be completely codified by the formalized languages being developed by Russell and others.

Starting in the late 1930s, Beckett's artistic interest was similarly focused on the relationship between language and truth. Beckett instinctively believed that large swaths of the deepest truths about the human experience lay beyond the reach of language, even as he committed himself to the obligation to use language as the vehicle to express this conviction. As unreasonable as this task sounds, Kurt Gödel succeeded in demonstrating this very relationship within the domain of mathematics. Whether or not Beckett was indeed familiar with Gödel, Gödel's groundbreaking discoveries about incompleteness provide a compelling metaphor for Beckett's agenda, as well as a template for how he might proceed.

Principia Mathematica

In pursuing our inferences we must never be led to assert both p and not-p; i.e., $(p \wedge \sim p)$ cannot legitimately appear.

—BERTRAND RUSSELL AND ALFRED NORTH WHITEHEAD,
FROM *PRINCIPIA MATHEMATICA*

Although daunting in size, scope, and the density of its notation, it is possible to get a sense of the overall project of *Principia Mathematica* by looking at the opening few sections (with the aid of some modern polish). Section A of Part I is titled "The Theory of Deduction," which, the authors explain, is "concerned with deduction itself; i.e., with the principles by which conclusions are inferred from premises."[39]

Rational thinking comes with an a priori sense of what constitutes a valid argument. Given the two premises:

(1) If Hamm whistles Clov comes, and (2) Clov has not come,

we don't need a course in logic to reach the proper conclusion that

(3) Hamm has not whistled.

This does not mean that all three statements must be true; it does mean that *if* the first two are accepted as true, *then* the third must be as well. Logical thinking is rule-bound, and Russell and Whitehead's agenda in Section A, Part I is to write down the rules.

Perfecting the *Elements*

The influence of the *Elements* over the evolution of mathematics is impossible to overstate. In this sense, *Principia Mathematica* is a part of the natural progression to perfect the axiomatic method that Euclid championed. Like Euclid, Russell and Whitehead start from a set of accepted axioms—five in fact—which they refer to as "primitive propositions" and list with the accompanying parenthetical names:

$((p \vee p) \supset p)$ (Tautology)
$(q \supset (p \vee q))$ (Addition)

$((p \vee q) \supset (q \vee p))$ (Permutation)

$((p \vee (q \vee r)) \supset (q \vee (p \vee r)))$ (Associative)

$((q \supset r) \supset ((p \vee q) \supset (p \vee r)))$ (Sum)[40]

The first thing that is evident from this list of axioms—and this may be the most important observation of all—is the purely formal, symbolic nature of this new language. These formulas appear to be constructed according to some unknown rules of grammar, but without any background or training it is not obvious that they have an external meaning. This is the essence of a purely *formal* system. Working with the system developed in *PM* can be compared to playing a game of chess. There are specialized symbols in place of the pieces and a rigid set of mechanical rules for how to manipulate these symbols. The theorems of PM, which take the form of strings of symbols, are then like the legal positions of the pieces on the board. The difference between PM and chess is that the symbols and rules of PM are carefully crafted to mirror the processes of mathematics, and in so doing, the generated formulas acquire an external, outer meaning that is absent in the arrangements of chess pieces.

Most of the symbolic alphabet for Section A appears in the stated axioms. The symbols p, q, and r are variables which are intended to represent general declarative statements such as "5 is prime," "God exists," or "Hamm whistles." It doesn't matter whether the statements are true or false; they just need to be statements to which "true" or "false" can be applied. The parentheses— (and)—are grouping devices employed much as they are in algebra, although subjected to very strict rules of usage.[41] The last four symbols are so-called connectives meant to represent the following fundamental concepts of logic:

$\sim$ stands for "not"

$\vee$ stands for "or"

$\wedge$ stands for "and"

$\supset$ stands for "implies" or "if . . . then . . . "

Thus, the string of symbols $(p \vee q)$ gets interpreted as the compound expression "p or q," which is understood to be true if at least one of p or q is true (or possibly both). By contrast, $(p \wedge q)$ is read as "p and q" and is true if and only if both p and q are true. The string $(p \supset q)$ is interpreted as "p implies q," or more verbosely, "if p is true then q is also true." When is this compound statement true? Let's let *PM* speak for itself:

The essential property we require of implication is this: "What is implied by a true proposition is true." It is in virtue of this property that implication yields proofs. But this property by no means determines whether anything, and if so what, is implied by a false proposition. What it does determine is that, if p implies q, then it cannot be the case that p is true and q is false; i.e., it must be the case that either p is false or q is true. . . . Hence "p implies q" is defined to mean: "Either p is false or q is true."[42]

The last line of the previous quotation, cast in the symbolic notation of PM, says that $(p \supset q)$ and $(\sim p \vee q)$ are, logically speaking, synonyms. The implication is that they can be interchanged in this language without changing the truth value of what is being expressed.

Formulas, Axioms, and Theorems

To earn the title of *formula* in the language of PM, a string of symbols has to be created according to strict rules of grammar, which the five axioms very much are. What makes these particular five formulas so distinctive is that, in the opinion of the authors, they correspond to truths about the nature of logical reasoning that are so fundamental as to be accepted without further proof.

For instance, the second formula says that "if q is true, then p or q is true." This is compellingly self-evident. It encompasses something fundamental about how the logical connective "or" functions within deductive reasoning. Russell and Whitehead call it "addition" because, as they point out, "if a proposition is true, any alternative may be added without making it false."[43] To repeat the previous point, this formula, along with the other four on the list, are accepted as valid without any other justification. They represent five primitive truths about reasoning from which all the other valid formulas can then, in theory, be derived.

This brings us to the pivotal question: How, precisely, are these derivations carried out?

The other component of a formal system is the collection of *transformation rules*. The transformation rules provide a rigid procedure, based purely on formal syntax, for producing new formulas from old ones. In our lexicon, we shall refer to any formula that can be generated from the axioms and transformation rules as a *theorem* (using a small "t" to distinguish them from proper objects like the Pythagorean Theorem). Thus, the five axioms stated above are theorems, and any new formula that can be created by starting from these axioms and applying the transformation rules is also a theorem.

For example, there is a transformation rule called the Rule of Substitution which says that given a theorem containing a variable like p, a new theorem can be created by replacing every instance of that variable with some other formula. Thus, starting with the first axiom on the list:

$$((p \vee p) \supset p)$$

we can swap out the variable p, replacing it with the formula $\sim p$ to get the new formula

$$((\sim p \vee \sim p) \supset \sim p). \tag{1}$$

Because we arrived at it using the Rule of Substitution, this formula is a theorem. Looking back at the earlier discussion about how $(p \supset q)$ is a symbolic synonym for $(\sim p \vee q)$, it should not be too surprising that another transformation rule related to the definition of the symbol $\supset$ justifies replacing $(\sim p \vee \sim p)$ with $(p \supset \sim p)$ in the formula from line (1) to get the theorem

$$((p \supset \sim p) \supset \sim p). \tag{2}$$

This theorem, in fact, is the first theorem highlighted in *PM* after the list of axioms.

Returning to our chess analogy, there are two very distinct ways to think about the string of symbols in line (2). On the one hand, it is just that—a string of symbols generated by taking one of the designated starting axioms and following the prescribed symbol-manipulating rules; i.e.,

$((p \vee p) \supset p)$	Axiom 1
$((\sim p \vee \sim p) \supset \sim p)$	Rule of Substitution
$((p \supset \sim p) \supset \sim p)$	Definition of $\supset$

PM refers to this chain of formulas as a "demonstration" of the "asserted proposition" in (2). We will call it a *proof* of the *theorem* in (2). Although we use the terms "proof," and "theorem," it should be clear that what is happening here is a purely mechanical algorithm based solely on syntax that could be carried out by a computer. Similar to a sequence of valid chess moves, a proof in PM is no more and no less than a list of formulas where each entry is either an axiom or is justified by applying one of the transformation rules to formulas appearing earlier on the list.

But having arrived at the theorem in (2), we can emerge from our mechanical mindset and unpack what this new formula says. The theorem in (2) asserts that "if p implies its own falsehood, then p is false." Although not earthshaking, this new statement agrees with our sense of logic; it's true, in other words, and that is the crucial point. Following the rules at the symbolic level leads to true statements at the interpreted level. To make this point a different way, consider the following two formulas:

$$((p \vee q) \supset q) \qquad \text{and} \qquad ((p \wedge q) \supset q).$$

Understanding how the concepts of "and" and "or" function in propositional logic, one of these formulas corresponds to valid reasoning and the other does not. (Before reading on, take a moment to decide which is which.) The second formula, which translates as

If the compound statement "p and q" is true, then q is true,

forms a true statement no matter how the variables p and q are interpreted. This is what it means for a formula to represent valid reasoning. Russell and Whitehead's goal was to craft the language of PM so that following its rules would lead to the creation of the formula $((p \wedge q) \supset q)$ while avoiding $((p \vee q) \supset q)$.[44] "In ordinary written language a sentence contained between full stops denotes an asserted proposition," PM's authors explain, "and if it is false the book is in error." Making this analogy with "ordinary written language" is helpful but a bit misleading because Russell and Whitehead are setting a higher bar for themselves. They literally intend for PM to be incapable of generating any statements that are false. If, for example, the formula $(p \supset p)$ is arrived at in PM they note, "it is to be taken as a complete assertion convicting the authors of error unless the proposition $(p \supset p)$ is true (as it is)."[45]

Having set out their symbolic language for propositional logic in the opening section (which the authors credit largely to Frege), Russell and Whitehead then set off to create a more articulate formal system that can enunciate the truths of mathematics. While the formulas we've seen thus far assert general truths about reasoning such as "if p and q is true then p is true," the goal of PM is to create a system of formulas whose interpretations include statements like "5 is prime," "$\sqrt{2}$ is irrational," and "there are no integer solutions to $a^n + b^n = c^n$ when n is greater than 2." The implications of the existence of such a system should start to be apparent. If indeed we could convert Fermat's Last Theorem into a formula of PM, then the resolution of a three-hundred-year inquiry by history's most creative mathematical minds could, in theory at least, be carried out by a mindless and mechanical symbol-shifting algorithm that generates one theorem after another until, potentially, the desired formula is reached.

Consistency and Completeness

All this, of course, comes back to the question of whether Russell and White-head succeeded in their overall goal, which they beautifully summarize in the introduction this way:

> The proof of a logical system is its adequacy and its coherence. That is: (i) the system must embrace among its deductions all those propositions which we believe to be true and capable of deduction from logical premises alone, though possibly they may require some slight limitation in the form of an increased stringency of enunciation; and (ii) the system must lead to no contradictions, namely in pursuing our inferences we must never be led to assert both p and not-p; i.e., $(p \wedge \sim p)$ cannot legitimately appear.[46]

Item (i) in the above quotation is about the strength of PM. It says that if a formula asserts a proposition that is true, then this formula should be producible from the axioms and transformation rules. While the authors use "adequacy," the modern term for this is *completeness*. A system like PM (where the initial axioms have an intended and truthful interpretation) is said to be *complete* if whenever a formula represents a true statement it can be derived as a theorem.

Item (ii) refers to the property of *consistency*, which in propositional reasoning takes a very concrete form. As Russell and Whitehead explain, consistency means "$(p \wedge \sim p)$ cannot legitimately appear." A compelling way to experience the viral potency of a contradiction like this is by considering the formula

$$((p \wedge \sim p) \supset q). \tag{4}$$

Although we have not introduced enough of the transformation rules to show it, this formula is a theorem. What truth does it assert? The formula in (4) says, "if p is both true and false, then q is true." In other words, from the contradiction "p and not p" we may conclude q. Notice that q is arbitrary here. Using the Rule of Substitution, we can put in *any* formula we like for q. Thus, if it ever turns out that $(p \wedge \sim p)$ is a theorem, then *every formula is a theorem* and the correspondence between truth and derivability dissolves. The problem is not with the formula in (4). This is a proper theorem of propositional reasoning. The problem comes if ever it turns out that $(p \wedge \sim p)$ is a theorem, because then a transformation rule known as the Rule of Detachment can be applied to conclude that the unspecified variable q is an asserted truth. Substituting whatever formula we like for q makes this substituted formula into a theorem regardless of whether or not it corresponds to a valid statement. This renders the entire system incoherent and pointless.

Any system like PM that incorporates the rules of propositional reasoning into its architecture is vulnerable to the same fate. One contradiction among its derived theorems dooms the entire system to inarticulate gibberish.

Hilbert's Program

Taken together, the concepts of completeness and consistency represent the ideal for an axiomatic system. In tandem, they declare that a system like the one in *PM* has achieved a level of perfection. To get a better sense of this let's consider a concrete proposition from arithmetic. Examples like "5 is prime," or "$\sqrt{2}$ is irrational" would suffice, but to raise the stakes let's take a statement whose validity is currently unknown. Consider the claim:

Every even number larger than 2 is the sum of two prime numbers.[47]

All evidence points to this being a true statement; e.g.,

$$12 = 5 + 7, \quad 20 = 7 + 13, \quad 56 = 19 + 37, \quad 200 = 97 + 103,$$

but the possibility that there exists some extremely large even number that is not the sum of two primes has not been definitely ruled out. Does PM, at least in theory, have the resolution to this riddle encoded into its architecture? It is a modest but manageable challenge to translate this statement into a formula in PM which we denote as T.[48] Adding the negation sign $\sim$ to the front of this formula gives the new formula $\sim T$ which translates as "Some even number larger than two is not the sum of two prime numbers." The question is whether T or $\sim T$ represents the true statement. If PM is complete then at least one of these two formulas is among its collection of theorems, and if PM is consistent then the other one is not. If PM is complete and consistent, then T is *decidable* within the system of PM.

The significance here is theoretical much more than it is practical. Formal systems like PM were not proposed to *find* proofs for open questions like the one in T—proofs in these systems are too pedantic and mechanical to incorporate the necessary creativity and intuition. The issue was a philosophical one: Could the language of the foundations of mathematics be crafted so that, in principle, there would be large domains of mathematical inquiry where truth was synonymous with provability?

The name most closely associated with this quest for a perfected codification of mathematical certainty is David Hilbert. In 1900, Hilbert delivered an address at the International Congress of Mathematicians in Paris in which he listed twenty-three unsolved problems that, in his estimation, should guide mathematical inquiry into the new century.[49] This was an audacious talk, but Hilbert's strength and breadth as a mathematician uniquely qualified him for the job. The second problem on his list was to establish a proof for the consistency of the axioms of arithmetic.[50] With the publication of *Principia Mathematica* and the advent of other related axiomatic systems, Hilbert's program, as it came to be called, took the specific form of proving the consistency and completeness of these systems. Foreshadowing Beckett's dilemma, Hilbert was faced with the problem that *PM* had essentially defined what proof meant, and so there was the circular issue of using PM to prove the efficacy of PM. Undeterred, Hilbert was steadfast in his conviction that a perfection of axiomatic mathematics was attainable, essentially by using an accepted fraction of PM's capabilities to establish its larger consistency.[51]

We Must Know, We Will Know

Upon his retirement from mathematics in 1930, Hilbert was asked to give a special lecture at a conference in Königsberg—the site of Euler's seven bridges problem and the town where Hilbert (and Immanuel Kant) was born. Hilbert spoke broadly on this occasion about the role of mathematics in science and the role of logic in mathematics. The lecture memorably concluded with six words that would appear on his tombstone thirteen years later:

Wir müssen wissen, Wir werden wissen. [We must know, We will know.]

Amazingly, exactly one day earlier at the same conference, a young mathematician named Kurt Gödel had participated in a roundtable discussion with the leading logicians in the field about the feasibility of Hilbert's program. A shocking thing had happened. In short, Gödel had explained to the distinguished group of mathematicians that Hilbert's dream could not be realized. If PM is indeed consistent, Gödel explained, then it would necessarily have to contain a formula—now ubiquitously referred to as G—such that neither G nor $\sim G$ was among its theorems. The formula G was *undecidable*, in other words, and represented a blind spot in the ability of PM to distinguish true from false. The formula G was complicated, to be sure, but like all formulas it articulated a claim about whole numbers—i.e., that some specific number possessed some particular property, not qualitatively different from a statement like "5 is prime." Moreover, by the way Gödel constructed it, G asserted a true statement, but PM would never derive it as a theorem:

Gödel's Theorem: Every sufficiently powerful consistent formal system for whole number arithmetic contains undecidable propositions.

The phrase "sufficiently powerful" needs some attention, but PM certainly meets this criterion. The conclusion is that any system of this nature is incomplete—and irreparably so—implying that provability in formal systems like these can never fully encapsulate the notion of truth.

So was Beckett familiar enough with these developments in logic to make a disguised allusion to Gödel in his personal correspondences? "I must admit I don't see any reference to Gödel in the Axel Kaun letter," says Chris Ackerley, the scholar most widely versed on Beckett's relationship to the mathematical sciences. "There seems to be an assumption . . . that Beckett had a mastery of complex maths, but I think his factual knowledge was quite limited whereas his capacity to use number theory as metaphor was almost infinite." Ackerley confirms that Beckett was very much attuned to the paradigm shifts in early twentieth-century mathematics and physics. Beckett was aware of the quantum revolution most certainly as well as the general effort by Russell, Wittgenstein, and others to ground mathematics in logic. "But, and this is a big but," Ackerley cautions, "his mode was essentially that of a spectator."

The speculation that Beckett knew something of the technical details of Hilbert's program is understandable, but it obscures the more interesting conclusion that the resonances between Beckett and modern logic are due to some deeper common core. In his published conversation with Georges Duthuit, Beckett admires the early Italian masters as artists but laments that they "never stirred from the field of the possible, however much they may have enlarged it." From Euclid to Bolyai to Frege to Russell to Hilbert, mathematics was driven by a search for certainty. These early masters each contributed to the refinement of axiomatic mathematics, sharpening the austerity of the language and expanding its domain but, Beckett might complain, never straying from the field of the possible. Gödel's discoveries are a watershed moment, much like the Greek discovery of the irrationality of $\sqrt{2}$, which Beckett knew well. Mathematical truth, it seems, has a dark zone that extends beyond the reach of Hilbert's beloved axiomatic methods in the same way that the familiar rational numbers were revealed to be engulfed in a number line full of irrational ones. Because of the primacy of axiomatic reasoning in early twentieth-century mathematics, Gödel's discoveries represent what at first may seem like a fatal blow, but that response is just a symptom of the long-standing assumption that the function of mathematics is to express by means of axiomatic reasoning. Could this assumption, pursued from Euclid to Russell and championed by Hilbert, really be a misapprehension? Although incompleteness appears to be a pernicious defect, perhaps it is more properly viewed as an unexpected but inevitable quality of the landscape of mathematical truth.

To pursue the logically impossible challenge of upending what is possible to express as a writer, Beckett took inspiration from the rogue Pythagorean who questioned whether the existence of irrational numbers was necessarily "an offense against the deity." Because they represent unpatchable holes in the facade of mathematical certainty, the undecidable propositions that Gödel unearthed serve Beckett's purpose even better.

Quad

Extraordinary how mathematics helps you to know yourself.

—MOLLOY, FROM *MOLLOY*

By the time he creates *Quad* in 1981, the diminishing trend in Beckett's writing that starts in earnest with *Watt* and continues through *Endgame* and *Not I* has reached something close to the end of the line. Speaking to biographer Deirdre Bair, Beckett said, "The best possible play is one in which there are no actors, only the text. I'm trying to find a way to write one."[52] With no actors in a traditional sense, and no text either, *Quad* goes a step further.

There is a way in which *Quad* is the logical end result of Beckett's earlier themes. In *Watt*, Beckett was drawn to permutations and combinations as a means to illustrate the ineffectiveness of a purely rational approach to enlightenment. *Watt*'s exhaustive attempts to glean some kind of knowledge about Mr. Knott by enumerating every possible arrangement of his furniture or physical appearance led to insanity rather than insight. As Beckett starts writing for the stage, especially his more abstract pieces, combinatorics and mathematical patterns are more centrally employed as the main scaffolding on which the textual elements are mounted. *Come and Go*, from 1965, features three older women in solid-colored dresses methodically permuting their positions

"Shall we hold hands in the old way?" *Come and Go*; Barbara Brennan (Ru),
Susan Fitzgerald (Vi), Bernadette McKenna (Flo); Beckett Centenary
Festival, Barbican, 2006. © Donald Cooper/photostage.co.uk.

on a shared bench as they whisper gossip to one another. The mathematical
choreography, much more than the sparse dialogue, forms the body of this
hypnotic piece. When the women take each other's hands "in the old way" in
the final moment, the knotted pattern of their intertwined arms is unexpectedly arresting.[53] In *Footfalls*, created ten years later, the one visible character
paces back and forth across a dimly lit stage following a rigidly specified nine-
step pattern. "Seven, eight, nine, wheel," the mother says in voice-over, making
the footsteps part of the script—like a metronome marking out the long
silences.[54] The text of *Footfalls* is spare and haunting, particularly in the way
that it is sprinkled with strange inconsistencies that subvert its coherence. It's
as though the dialogue is trying to undo itself and leave only the patterned
pacing.

When Beckett gets to *Quad*, there is no text left to subvert; he constructs the entire piece out of mathematical components. The patterns are
rigid and deterministic, the symmetry elegant. Harking back to the permutations of *Watt*, the four walkers appear in every possible arrangement, and
their colors and percussion accompaniments allow us to experience the combinations in multiple sensory ways. The difference is that *Watt* appeals to
mathematics as a potential survival mechanism. By the time Beckett creates

Quad, mathematics is all there is. Standing out among the many scholarly interpretations of *Quad* is Erik Tonning's observation that "*Quad* is not just 'abstract' in the sense of being quasi-geometrical. . . it has also been abstracted *from* previous Beckettian patterns. . . . It is a Beckettian *reductio ad absurdum* that blatantly advertises itself as such." For Tonning, *Quad* represents Beckett's artistic predilections pushed so far to the extreme that the result borders on a form of self-caricature, and in the process the playwright unwittingly ensnares himself into the essence of the drama. "Beckett the author should be pictured not simply as the ironic arranger," Tonning contends, "but also, even as he writes and produces *Quad*, as on stage with his pacers, circling, in his own way, the same void they are."[55]

Nothing, and Nothing Else

The two acts of *Waiting for Godot* are all Beckett needed to suggest the possibility of infinity, and *Quad* incorporates this same device, with walking in place of waiting. In the same way that Mouth in *Not I* can be heard unintelligibly before the play even begins, *Quad* fades in on the last few seconds of white and yellow completing a cycle before yellow exits and the piece begins in earnest. The effect is that the audience arrives to encounter an event already in progress. The ending, or lack thereof, is similarly ambiguous, as it is in so many of Beckett's plays. *Quad* fades out with the walkers still walking; Mouth continues to babble after the lights fade; Clov announces he is leaving but does not depart.

So often we find Beckett's characters engaged in habitual obsessions. The walkers in *Quad* are a more mechanical version of the pacing daughter in *Footfalls* who requests that her mother remove the carpet from her hallway so that she can hear the sound of her footsteps. Watt's absurd calculations about footwear options echo a moment from *Murphy* where the protagonist muses about the 120 different orderings in which he could eat his five biscuits.[56] A fascinating character we have not encountered is Molloy (from the novel of the same name), who takes comfort in sucking on sixteen smooth stones he keeps in his pockets. For upward of five pages, Molloy occupies himself with the challenge of how to arrange the stones in his four pockets in such a way that will ensure he sucks them each once, and in some established order, before sucking any of them a second time.

What is the common thread? "Habit is a great deadener," Vladimir says in *Godot*. At one level, the bantering, the pacing, the counting, and the stone

Mathematics, motion, nothing, and nothing else. *Quad*; Hammer
Museum/UCLA, 2020 (photo: Josh Concepcion).

sucking are devices we turn to so that, as Estragon says, "we won't think." They
keep our minds off the unimaginable truth of our predicament by overlaying it
with some invented agenda that we use to fabricate a sense of purpose.[57] The
narrator in *The Unnamable* explains it this way:

> I have my faults, but changing my tune is not one of them. I have only to
> go on, as if there was something to be done, something begun, somewhere
> to go. It all boils down to a question of words, I must not forget this, I have
> not forgotten it.[58]

And once again, we are back to the question of the words themselves. Beckett's relentless explorations of the sense of self are inextricable from his curious
and strained relationship to language. In Beckett's conception, our thoughts
form part but not all of our sense of consciousness, whatever consciousness
may be. Consciousness meanwhile forms a part but not all of our sense of
being, whatever being may be. "Consciousness [is] but the rational tip of the
psychic iceberg," is how Ackerley elegantly describes this aspect of Beckett's
philosophical orientation.[59] Although we can try to articulate our thoughts
by filtering them into language in some approximate way, this still leaves vast,
inaccessible regions of the human story unexplored.

The contrast between the poetry of a piece like *Endgame* and the austerity of *Quad* is stark. Language, and all its poetic potential, has been banned in *Quad*, and that seems to be a central feature of this highly structured piece. Are insights about the wordless part of the self best communicated without any words? Is there even anything left once we remove all of the linguistically dependent components of ourselves, or of reality more generally? For philosophical realists like Frege and Russell, there were primitive parts of mathematics, such as logic and the concept of whole numbers, that enjoyed a universal, a priori existence that did not require sensory experience to be accessed. Without getting mired too deeply in ontological questions, we can at least propose that when the "veil of language" is completely dissolved some kind of mathematical ether remains. To describe it any further is to violate the premise of the experiment even more than we already have. Recognizing that paradox is inevitable when trying to linguistically represent the linguistically unrepresentable, we can still acknowledge that the collage of mathematics and motion in *Quad* is an inventive stand-in for this wordless realm. It is universal in the literal sense that any sentient audience would be fluent in its constituent parts. It could be performed on any planet, in any solar system, at any time past or present. No text, no faces, no bodies to speak of. Just whole numbers and symmetry, fleshed out with primary components of color and rhythmic percussion. Anything else? Some geometry, including a primitive conception of space, where a pervasive sense of absence at least leaves behind the impression that something is missing.

When Beckett was working on the filming of *Quad*, one of the producers showed him a view of the piece on a black-and-white monitor. This was the inspiration for a second act, which was filmed more or less on the spot. Having constructed the world of *Quad* out of what seemed to be the most primary of mathematical blocks, Beckett repeats the experiment by dissolving *that* veil of mathematical structure to create an even more austere reality. In *Quad II*, the hooded players are all cloaked in the same gray as the play space, the pacing slows, and the cessation of the percussion instruments makes room for the soft sound of the walkers' footsteps, staggering in perfect rhythmic unison. Mathematics, motion, nothing, and nothing else.[60]

P and Not *P*

"Extraordinary how mathematics helps you to know yourself."[61] So says Beckett in the voice of Molloy. This is an effective summary statement for why

Beckett so frequently turns to mathematics as a source of inspiration, but it is a little disingenuous to take it out of context for this purpose. In the novel, Molloy makes this observation after counting his farts for a day and calculating his average rate of flatulence in farts per minute.[62]

Bodily humor aside, knowing yourself is at the heart of the matter with nearly everything Beckett wrote. In his lifelong struggle to understand himself, or help us understand ourselves, writing was Beckett's weapon of choice and also the target of his advances. It is in this collision of means and use that establishes Beckett's most robust affinity to mathematics. Yes, Molloy takes on the characteristics of an obsessive mathematician as he permutes his sucking stones between his various pockets in search of an algorithm that will satisfy his optimizing constraints. But what is more important here are the stones themselves. "A little pebble in your mouth," Molloy says, "round and smooth, appeases, soothes, makes you forget your hunger, forget your thirst."[63] Language for Beckett is like the stones to Molloy. It pacifies, it soothes, it takes the place of something and makes us forget we are not in possession of the real thing. It provides a sense of significance—until we start to push on it, at which point it loses its integrity. This is the experience of watching Hamm compose his history with Clov. The last lines of *Molloy* create a similar effect:

> Then I went back into the house and wrote, It is midnight. The rain is beating on the windows. It was not midnight. It was not raining.[64]

This juxtaposition of a statement p with its negation $\sim p$, a Beckettian trademark, connects the Irish author to the twentieth-century metamathematical search for certainty in a profound way. The symbolic theorem $((p \wedge \sim p) \supset q)$ stands as a warning sign perched over the foundations of mathematics. If it ever happens that $(p \wedge \sim p)$ becomes a theorem, then the game is up. The contradiction "p and not p" subverts the correlation between provability and truth, signaling the complete demise of the logical system.

This Is Not a Theorem

That Beckett was most likely unfamiliar with Gödel's mathematics actually makes the confluence between the Irish playwright and modern logic more interesting to consider. Beckett's deductive, chess-playing mind was keenly attuned to the destructive power of logical paradox, which like Russell and Hilbert and Gödel, he tried to harness for a constructive purpose. Gödel's discovery of incompleteness is not a model Beckett followed, but it is a useful metaphor and a robust endorsement of Beckett's agenda. The inevitable existence of undecidable propositions within the symbolic domain of systems like the one in *Principia Mathematica* is provocative evidence for Beckett's conviction that there are vast regions of the human experience that extend beyond the reach of the linguistically dependent part of our conscious minds.

This analogy between Beckett and Gödel runs deep. Gödel did not just argue that unprovable truths existed, he constructed one—and he did so by exploiting the metalanguage of the system in question. If this sounds strange and unexpected, that's because it is. Self-reference was the culprit that led to Russell's paradox and Frege's downfall; PM was designed specifically to guard against the paradoxes of self-referential constructions. And yet, Gödel recognized that, despite its careful architecture, PM nevertheless acquired a self-commentating ability that could be exploited to reveal its inherent limitations. Roughly speaking, Gödel created a formula G to have the following self-referential interpretation:

The formula G—i.e., this very formula—cannot be derived as a
theorem in PM.

Accepting for the moment that a formula of PM could in fact be interpreted in this way, what must we then conclude about G? In particular we ask:

Is G a theorem?

Rather than paradox, the answer this time is an unequivocal no. Assuming, as we are, that the axioms of PM are true and consistent, it follows that the theorems of PM must state truths. If G were a theorem then it would truthfully assert that it was not a theorem, an impossibility. Thus, G is not a theorem. This option is satisfactory. Unlike Russell's paradox, it does not lead to a logical impasse, but notice what it does give us. Because G asserts that it cannot be derived as a theorem, and because we have just convinced ourselves that G is in fact not a theorem, *we are forced to accept that G is true.* This is incompleteness—a formula G within PM asserting a truth that PM is unable to demonstrate.[65]

This brief sketch of the structure of Gödel's proof requires time to digest, in particular because it seems to violate the rules of the game. Formal systems in Gödel's day represented the definition of what is meant by proof. Gödel revealed the inherent limitations of formal mathematics to codify mathematical truth, but in the process *he unearthed a new, nonformal means by which truth could be established*. Gödel's formula G makes a number theoretic claim that cannot be proved or disproved by the formal system in which it resides. But, in its metamathematical voice, G simultaneously asserts its own unprovability. Thus, we come to recognize G's indisputable validity at the same moment we encounter the system's incompleteness.

Engaged in the elusive search for truth about the human condition, Beckett's assault on the linguistic facades we use to structure our world had a similar side effect. Amid the rubble wrought by the power of paradox, Beckett's intent was to bring about a new kind of insight into who we are. Trying to articulate this new insight into the extralinguistic self within the not-so-formal system of natural language is an obvious impossibility. The only point of access is to acknowledge its indisputable validity at the same moment that we experience our language's limitations.

It is midnight. The rain is beating on the windows. It was not midnight. It was not raining.

$(p \land \sim p)$

I can't go on, I'll go on.

4

Dürrenmatt, Frayn, and McBurney: The Shape of Content

A mathematician, like a painter or a poet, is a maker of patterns.

—G. H. HARDY, FROM *A MATHEMATICIAN'S APOLOGY*

ALFRED JARRY died in relative obscurity, and even as his impact on avant-garde theater began to take hold it was *Ubu Roi* rather than the more mathematical *Ubu Cocu* that received the attention. Stanislaw Witkiewicz also died long before his significance as a playwright was broadly acknowledged, and his name is still not widely recognized outside eastern Europe. Samuel Beckett did acquire a robust and devoted following during his lifetime, but only a small fraction of his audience made any association between Beckett and mathematics. The same is true of current Beckett audiences. The Irish playwright's fame is anchored to those pieces where his mathematical inclinations are hidden below the surface. Performances of *Waiting for Godot* and *Endgame* are ubiquitous in community theaters and on university campuses, whereas *Not I* and *Quad* arise as topics of conversation mainly among Beckett scholars.

An engagement of mathematics on stage, even a relatively modest one, was rarely associated with robust ticket sales before Rosencrantz and Guildenstern's probability-defying coin made its debut in the mid-1960s. That said, there are a few notable landmarks on the pre-Stoppardian landscape of twentieth-century European theater that deserve attention. Bertolt Brecht is regarded, along with Beckett, as one of the most influential playwrights of the previous century. A contemporary of Witkacy, Brecht lived in Germany until 1933, when the rise of the Nazis forced him into exile, first to Denmark, then Sweden, and eventually to the United States where he remained until the conclusion of the war. During this time Brecht wrote *Life of Galileo*. As the title suggests, the play chronicles the historical events surrounding Galileo from the advent of the telescope through his encounter with the Inquisition and subsequent house arrest. Brecht is inconsistent in his attention to biographical accuracy, but he does tell us in the opening scene that Galileo is a professor of mathematics from Padua and also that, for Galileo, mathematics is not where the action is going to be. "I regret that I cannot recommend it to the University," an administrator says in response to Galileo's request for a raise. "As you know, courses in mathematics do not attract new students. Mathematics, so to speak, is an unproductive art."[1]

There is some irony in these lines. Galileo is known for championing an empirical approach to the pursuit of truth, looking at nature with his own eyes rather than submitting to the established doctrine of an outside authority. Although this is fundamentally accurate, the historical Galileo's mathematical instincts—particularly his recognition that nature was following mathematical rules—were crucial to his propensity to make proper sense out of the observations he recorded through the telescope he misleadingly claimed to have invented. Brecht does not emphasize the role of mathematics in Galileo's thinking, but he does bring a bit of Copernican astronomy to life on stage. Early on, the script calls for a large wooden model of the earth-centric Ptolemaic system, which Galileo uses to enlighten his ten-year-old pupil Andrea Sarti. In the same tutoring session, Galileo puts young Andrea in a chair next to an iron washstand that is meant to represent the sun:

GALILEO: Where's the sun, right or left of you?
ANDREA: Left.
GALILEO: And how does it get to be on your right?
ANDREA: By you carrying it to my right, of course.
GALILEO: Isn't there any other way? (*He picks him up along with the chair and makes an about turn.*) Now where's the sun?
ANDREA: On my right.
GALILEO: Did it move?
ANDREA: Not really.
GALILEO: So what did move?
ANDREA: Me.
GALILEO (*bellows*): Wrong! You idiot! The chair!
ANDREA: But me with it!
GALILEO: Of course. The chair's the earth. You're sitting on it.[2]

"Isn't there any other way?" *Life of Galileo*; Simon Russell Beale (Galileo), Ryan Watson (Andrea); National Theatre, 2006. © Marilyn Kingwill / ArenaPAL.

Galileo is teaching science, but he is doing it as a mathematician would—helping his pupil experience the beauty of a simple explanation for what appears to be complex behavior.

Although thin on mathematics, *Life of Galileo* would become an early touchstone for the collaborative possibilities of theater with science more generally, even though Brecht's interest in Galileo had little to do with science and even less to do with dramatizing biography. In the late '30s and early '40s, politics was front and center for Brecht. Having been forced to flee Germany, the playwright saw something of his own predicament in Galileo's clash with the church, and it was the story of intellectual survival in the face of authoritarian oppression that originally attracted the German playwright to the Renaissance astronomer and mathematician.

Another important midcentury playwright whose interest in politics led him to an engagement with mathematics was Friedrich Dürrenmatt. Dürrenmatt was born in 1921 in the village of Konolfingen near Bern, Switzerland. Twenty-three years younger than Brecht, Dürrenmatt rose to international prominence in the 1950s, most notably because of his play *The Visit*. Of his dozens of plays and novels, *The Visit* stands out as Dürrenmatt's most widely acclaimed creative work, but in 1962 *The Physicists* debuted and was immediately recognized as an important play. Performances proliferated, making *The Physicists* the most widely produced play of the season in German-speaking Europe.[3] The following year, translations of the script went global, with *The Physicists* landing on Broadway in 1964.

Like so much of Dürrenmatt's writing, *The Physicists* is hard to pigeonhole in terms of its style. Comic and sharp-witted, this Cold War drama is equal parts satire and Greek tragedy, entertaining its audience with unexpected plot twists on its way to a foreboding conclusion. The entirety of the play takes place in the drawing room of a private sanatorium where three mental patients reside. The first purports to be Sir Isaac Newton and the second Albert Einstein. Given the specter of atomic devastation that loomed over the landscape in the 1960s, there was an obvious motivation for Dürrenmatt to co-opt the identities of these two pillars of modern physics for two of his patients. This is why the name of the third patient, Johann Wilhelm Möbius, arouses some curiosity. With so many famous options available, why did Dürrenmatt settle on the surname of the relatively obscure German mathematician Alfred Ferdinand Möbius as the moniker for the third lunatic scientist and protagonist of his play?

Although physics is the more evident motif, it turns out that *The Physicists* relies on mathematics more than science for its moral reckoning.

The Physicists

What was once thought can never be unthought.

—MÖBIUS, FROM *THE PHYSICISTS*

There are several ways in which Brecht's *Life of Galileo* provides context for understanding what Dürrenmatt intended—and did not intend—to accomplish in *The Physicists*. The first of these is simply acknowledging the reach and influence of Brecht's reputation. For the German-speaking Dürrenmatt, comparisons were unavoidable.

Throughout his career, Brecht was closely associated with a particular theory of drama known as "epic theater." The traditional theater of Brecht's day was geared to appeal on an emotional level, luring its audience into an illusory mindset where spectators could experience an entertaining and cathartic journey of conflict and resolution. The audience emerged satisfied and content but intellectually unaltered. Brecht thought theater should be more confrontational and political, that it should work on the mind more than on the senses. A committed Marxist, Brecht believed that not only are human beings capable of change but that art, and theater in particular, could be a vehicle for that transformation. For this to transpire, the audience had to be prevented from descending into an escapist, emotional state and instead prompted to a more heightened level of awareness. Implementing this theory in practice took many forms. Characters directly addressing the audience, bright lighting, placards announcing scene changes, and inserted songs were meant to remind the audience they were watching a play, emphasizing the didactic nature of theater over its potential to entertain.

Although he was regularly associated with Brecht, Dürrenmatt expressed little sympathy for Brecht's theory of drama. In fact, Dürrenmatt went so far as to say that Brecht was at his creative best when he wasn't being rigidly Brechtian.[4] Dürrenmatt's orientation was more pragmatic. In the published version of The Physicists, Dürrenmatt includes an addendum titled "21 Points to The Physicists." The first two of these read:

1. I don't start out with a thesis but with a story.
2. If you start out with a story you must think it to its conclusion.

Dürrenmatt was interested in the potential of art to shed light on human questions, but as a playwright he believed that a script should be allowed to evolve in whatever direction its story required, unfettered by theoretical restrictions. One consequence of this freedom was an acknowledgment that a play was not likely to effect the kind of social change that Brecht envisioned. For Dürrenmatt, a play could hold a mirror up to the world, revealing its tragic circumstances, but as he says in his 1954 essay Problems of the Theater, "Today art can only embrace the victim, if it can reach men at all."[5]

This difference of opinion between Brecht and Dürrenmatt is on vivid display in a comparison between the scientifically themed Life of Galileo and The Physicists. Through modern eyes, the historical Galileo is viewed very sympathetically—even heroically—but with each draft of his play, Brecht evolved toward a less kindly view of the founder of modern science. Implicit in Galileo's survival strategy was the maxim that scientific progress took precedence over social responsibility. Eventually Brecht came to the opinion that, by recanting, Galileo missed his chance to initiate a moral code that asserted science only be used for the collective benefit of the greater good. As this darker picture came into focus, two nuclear bombs exploded over Japan. "The atomic age made its debut at Hiroshima in the middle of our work," Brecht wrote in his diary. "Overnight the biography of the founder of the new system of physics read differently."[6]

In the postwar version of *Life of Galileo*, a grown-up Andrea Sarti makes the argument to his old teacher that a commitment to scientific truth justified Galileo's recantation, but the defeated Galileo will have none of it:

ANDREA (*loudly*): Science makes only one demand: contribution to science.

GALILEO: And I met it. Welcome to the gutter, brother in science and cousin in betrayal! Do you eat fish? I have fish. What stinks is not my fish but me. I sell out, you are a buyer.[7]

In performance and on the page, Galileo's self-condemnation is jarring to the ears. There is not much in the first thirteen scenes to prepare an audience for this about-face, but Brecht is insistent on turning Galileo into a modern villain. Even after extensive edits, a definitive interpretation of the character of Galileo in Brecht's play remains a point of contention between the playwright and his audience. What is unambiguous, however, is that Brecht laid down the ethical gauntlet for science and mathematics going forward.

It was in the context of the ensuing Cold War that Dürrenmatt wrote *The Physicists* in 1962. The title suggests that viewers should be prepared for some science on the stage, but in his appended "21 Points" Dürrenmatt makes it clear there will be no technical requirements made of his audience. Ethical requirements, on the other hand, will be in full force:

15. It [the play] cannot have as its goal the content of physics, but its effect.
16. The content of physics is the concern of physicists, its effect the concern of all men.
17. What concerns everyone can only be resolved by everyone.

It is interesting then, that in a play full of scientists—or mental patients pretending to be scientists—no actual science is discussed. What is invoked, however, is a mathematical mindset. Elsewhere in his list appear the following observations:

12. Playwrights, no less than logicians, are unable to avoid the paradoxical.
13. Physicists, no less than logicians, are unable to avoid the paradoxical.
14. A drama about physicists must be paradoxical.

Although mathematics provided the grammar of quantum mechanics required to create the atom bomb, Dürrenmatt looks to mathematics for an entirely different reason. He needs the expertise of mathematics to help him probe the paradoxical because that is where his play about physicists is inevitably heading:

19. Within the paradoxical appears reality.

Dürrenmatt did not delude himself into thinking his play could make a measurable difference in the world, but he held out the possibility that it might "dupe the spectator into exposing himself to reality."[8] To this end, Dürrenmatt cleverly leaned on mathematics to help him construct a play that, although it could not bring about a different reality, might at least enrich our understanding of the one in which we find ourselves.

Harmless, Lovable Lunatics

Although the preamble describes the three physicists as "harmless, lovable lunatics," the play itself opens with a dead nurse on the floor of the drawing room and a police inspector asking questions. Visible across the back of the set are a trio of doors numbered 1, 2, and 3—all closed for the moment, with the sound of a piano and violin duet coming from room number 2.

The paradoxical nature of Dürrenmatt's rendition of reality is apparent from the outset:

INSPECTOR: The murderer?
SISTOR BOLL: Please, Inspector—the poor man's ill, you know.
INSPECTOR: Well, the assailant?
SISTOR BOLL: Ernest Heinrich Ernesti. We call him Einstein.
INSPECTOR: Why?
SISTOR BOLL: Because he thinks he is Einstein . . .
INSPECTOR: (*turns to the [doctor]*) Strangled, doctor?
POLICE DOCTOR: Quite definitely. With the flex of the standard lamp. These mad men often have gigantic reserves of strength. It's phenomenal.
INSPECTOR: Oh. Is that so? In that case I consider it most irresponsible to leave these mad men in the care of female nurses. This is the second murder—
SISTOR BOLL: Please, Inspector.

"But that was something quite different Inspector. I'm not mad you know."
The Physicists; George Voskovec (Newton), Martyn Green (Inspector), 1964.
Friedman-Abeles © New York Public Library for the
Performing Arts.

INSPECTOR: —the second accident within three months in the medical
establishment known as Les Cerisiers. (14)[9]

Murders are accidents, criminals are patients, and Les Cerisiers—which translates as The Cherry Trees—is either a sanatorium or a madhouse. Either way,
two nurses have now been killed, the earlier one at the hands of Newton, who
confesses as much when he emerges from door number 3 to straighten up the
place (he cannot stand disorder) and grab a nip from the bottle hidden in the
fireplace:

NEWTON: That poor Ernesti. I'm really upset. How on earth can anyone
bring himself to strangle a nurse? (*He sits down on the sofa and pours out
a glass of brandy.*)
INSPECTOR: I believe you strangled one yourself.
NEWTON: Did I?

INSPECTOR: Nurse Dorothea Moser.

NEWTON: The lady wrestler?

INSPECTOR: On the twelfth of August. With the curtain cord.

NEWTON: But that was something quite different, Inspector. I'm not mad, you know. Your health. (19)

Newton justifies his actions as necessary for the overall progress of science, and ups the ante significantly when the Inspector gets addled with Newton's various rationalizations for taking a human life. "Is it because I strangled the nurse that you want to arrest me," he asks, "or because it was I who paved the way for the atomic bomb?" (22) The sharp, comic edge of the writing makes the play work—the satire masks the tragedy, luring the audience in before the play's haunting undertones become more audible later in the evening.

The last of the three physicists is Johann Wilhelm Möbius, who emerges from room 1 midway through the first act. Rather than imagining himself to be someone else, his mental instability manifests itself in the form of visions of King Solomon, the Old Testament figure whose name is synonymous with wisdom. The question of whether Möbius is mad, pretending to be mad, or just misdiagnosed is cleverly batted about, but what is clear is that the institutionalized Möbius is precisely where he thinks he belongs. When nurse Monika Stettler confides to Möbius that she has secured his release—and also that she loves him—we do not need the slowly dimming stage lights to tell us that poor sweet Monika is in trouble. Her silhouetted suffocation under the ripped-down curtains at the hands of Möbius is the grotesque image that closes the first act.

One-Sided or Two?

Focusing on its content, *The Physicists* is about physics, or at least the effect of physics. Focusing on its form, *The Physicists* becomes an inventive example of a math play. Dürrenmatt's declaration that logicians and playwrights are prone to the paradoxical opens the door to a mathematical device that shapes the narrative arc of the play as well as its psychological repercussions. To help his audience, Dürrenmatt hides a clue for this metaphor in plain sight.

Although most people are not familiar with nineteenth-century mathematician August Ferdinand Möbius, they have at some point very likely encountered his eponymous discovery of a non-orientable two-dimensional manifold, otherwise known as a Möbius strip. The phrase "two-dimensional manifold" is mathematical jargon for a surface such as a plane, a sphere (the two-dimensional outer surface, not the three-dimensional inside), or a cylinder (same comment). These three examples are all *orientable* surfaces. If a person walking around on the surface of a sphere returns to her original starting point she will always be oriented the same way—i.e., she will not be upside down relative to how she was when she started. She may be facing a different direction, of course, but the orientation of her head to her feet will be precisely as it was. The same is true on a piece of paper or a length of tubing. Imagine a bug with an antenna that points straight up out of its body, perpendicular to the surface on which it is walking. Assuming the bug does not cross any edges, every time the bug returns to a point it has already visited, it does so with the antenna pointing in the same direction it was before.

A physical model of a Möbius strip can be formed by taking a long, thin, rectangular piece of paper, giving the strip a half twist, and then taping the two short ends together. The image in figure 4.1 gives a good sense of what the result looks like, but there is no substitute for the intuition provided from examining the real thing. What should be both apparent and also counterintuitive is that on this Möbius strip it is possible for our bug to go for a walk—again not crossing any edges—and return to the place it started but with its antenna pointing in the diametrically opposite direction. There is a temptation to say that the bug is not at the same place it was before, that it is on the "other side" of the strip. But this misses the point. There is no other side—*there is only one side.* The simplicity of a Möbius strip, and to some extent its familiarity, can shroud its subtle magic and its beauty. A Möbius strip is not, strictly speaking, paradoxical, because of course it exists—a mathematical surface with one side and, as can be empirically verified, one edge. From the point of view of our very small bug or, as mathematicians like to say, "locally," a Möbius strip gives the impression of being two-sided in the same way that we think of a piece of paper as having a front and back or a sphere as having an inside and an outside. But that distinction on a Möbius strip is completely illusory. If we take a pencil and put two Xs anywhere we like on our paper model, it is always possible to draw a line connecting one X to the other without crossing any edges or picking up the pencil.

One way to heighten the sense of strangeness is to get rid of the edge on the Möbius strip. To achieve this we can take two Möbius strips, imagine they are made out of a stretchable fabric, and mentally stitch their edges together. Again, each individual Möbius strip has a single edge, so conceptually we can

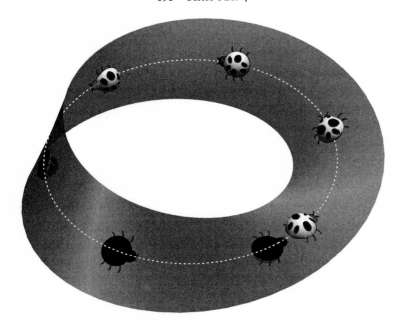

FIGURE 4.1. The bug returns to where it started but arrives oriented in the
opposite direction.

align the corresponding edges of each strip and attach them together. The
result is a finite surface with no edges. This in itself is not so remarkable. After
all, a sphere is a finite surface with no edges. But a sphere is orientable—it has
two sides. The edgeless surface that results from combining two Möbius strips
together in this way has only one side. It is called a Klein bottle, in honor of
Felix Klein, who discovered it in 1882. August Ferdinand Möbius identified
the Möbius strip in 1858.

Trying to carry out the construction of a Klein bottle by physically sewing
two Möbius strips together is illuminating. What happens is that the fabric from the two strips eventually gets bunched up and knotted, and that is
because it turns out to be impossible to build a model of a Klein bottle in
three-dimensional space without the surface intersecting itself. Klein bottles
require four dimensions to be properly expressed. Even so, figure 4.2 conveys
its essence in a compelling way. Just as on a Möbius strip, a traveler wandering
about the surface of a Klein bottle may return home after a long journey and
discover that everything is oriented in exactly the opposite direction as it was
when she left. The Klein bottle itself is not a paradox any more than a Möbius

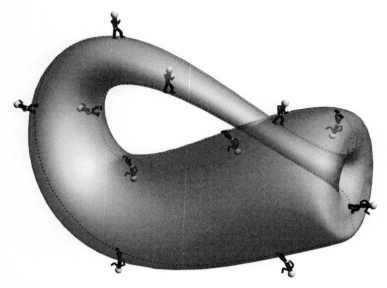

FIGURE 4.2. For a Klein bottle, the inside and the outside are one and the same.

strip is, but these mathematical objects quickly lead to paradox when we start to ask questions that reveal the biases we have acquired from an overexposure to spheres and coins and all the other orientable manifolds that populate our daily lives. To ask, for instance, whether our traveler's home is inside or outside the Klein bottle is deceitful nonsense—the inside and the outside are one and the same.

Mad but Wise, Prisoners but Free

The second act of *The Physicists* opens very similarly to the first, with a dead nurse on the floor of the drawing room and the Inspector asking questions. We are back where we started, but something has changed:

INSPECTOR: Strangled again, doctor?

POLICE DOCTOR: Quite definitely. And again, tremendous strength was used. But with the curtain cord this time.

INSPECTOR: Just like three months ago. (*He sits down wearily in the armchair downstage right.*)

FRÄULEIN DOKTOR: Would you like to have the murderer brought in?

INSPECTOR: Please, Fräulein Doktor.

FRÄULEIN DOKTOR: I mean, the assailant. (58)

The lines are familiar, but the roles are reversed. The Inspector's view of justice is pointing in the opposite direction from when we saw him last, and the effect is a strange kind of disorientation:

MÖBIUS: Herr Inspector, I must ask you to arrest me.
INSPECTOR: But what for, my dear Möbius?
MÖBIUS: Well, after all, Nurse Monika—
INSPECTOR: You yourself admitted that you acted under the orders of King Solomon. As long as I'm unable to arrest *him* you are a free man. (64)

"Justice is on holiday," the Inspector goes on to say, "and it's a terrific feeling." The audience's ongoing disorientation picks up momentum from here. Once the Inspector leaves, the ambushes come fast and furious:

NEWTON: A confession, Möbius. I am not mad.
MÖBIUS: But of course not, Sir Isaac.
NEWTON: I am not Sir Isaac Newton . . . My real name, dear boy, is Kilton.
Möbius stares at him in horror.
MÖBIUS: Alec Jasper Kilton?
NEWTON: Correct.
MÖBIUS: The author of the Theory of Equivalents?
NEWTON: The very same. (67)

Newton, a.k.a. Kilton, explains that he is a physicist and a spy who is fully aware that Möbius's latest scientific discoveries have the potential to alter the course of humankind, for better or worse. Posing as a madman, and regretfully killing a nurse to prove it, Kilton has come to persuade Möbius to join his government and their ongoing global struggle for freedom and justice. But someone has been eavesdropping on their conversation:

EINSTEIN: You were not the only one who read that dissertation, Kilton. (*He has entered unnoticed from Room Number 2 with his fiddle and bow under his arm.*) As a matter of fact, I'm not mad either. May I introduce myself? I too am a physicist. Member of a certain Intelligence Service. A somewhat different one from yours, Kilton. My name is Joseph Eisler.
MÖBIUS: The discoverer of the Eisler-effect?
EINSTEIN: The very same.

"Let us be mad, but wise." *The Physicists*; George Voskovec (Newton),
Hume Cronyn (Einstein), Robert Shaw (Möbius), 1964. Friedman-Abeles
© New York Public Library for the Performing Arts.

NEWTON: "Disappeared" in 1950.
EINSTEIN: Of my own free will. (69)

With Kilton speaking for the West, Eisler the KGB, and Möbius stuck in the
middle like the playwright's native Switzerland, Dürrenmatt explores the various arguments for the appropriate relationship between the scientist and the
governing authorities. Finding them all lacking, he has Möbius put forward a
radical new possibility:

MÖBIUS: My manuscripts? I've burned them.
Dead silence.
EINSTEIN: Burned them?
MÖBIUS: (*embarrassed*) I had to. Before the police came back. So as not
    to be found out.
Einstein bursts into despairing laughter.
EINSTEIN: Burned.
Newton screams with rage.
NEWTON: Fifteen years' work.
EINSTEIN: I shall go mad.
NEWTON: Officially, you already are.

They put their revolvers in their pockets and sit down, utterly crushed, on the sofa.

EINSTEIN: We played right into your hands, Möbius.

NEWTON: And to think that for this I had to strangle a nurse and learn German!

EINSTEIN: And I had to learn to play the fiddle. (78)

Möbius's proposal is an ambitious one. "We have to take back our knowledge and I have taken it back," he argues. "There is no other way out, and that goes for you as well . . . you must stay with me here in the madhouse." In *The Physicists*, what at first seems insane or preposterous starts to appear rational when viewed from a different perspective: visions of King Solomon, the murder of nurses, and now Möbius's insistence that they spend their lives locked away because their discoveries are too lethal. "Only in the madhouse can we be free. Only in the madhouse can we think our own thoughts." The illusion that the two apparent sides of a Möbius strip are, in actuality, one and the same is fully fleshed out in a grand toast the three physicists offer to their dead nurses upon their eventual decision to carry out Möbius's request:

NEWTON: Let us be mad, but wise.

EINSTEIN: Prisoners but free.

MÖBIUS: Physicists but innocent. (84)[10]

The Worst Possible Turn

Two points not yet mentioned from Dürrenmatt's list of 21 are the following:

3. A story has been thought to its conclusion when it has taken its worst possible turn.

4. The worst possible turn is not foreseeable. It occurs by accident.

The accident in this case is that Möbius has unwittingly put himself under the care of a psychiatrist who is also an insane megalomaniac. Fräulein Doktor Mathilde von Zahnd crashes the physicists' party to announce that she copied Möbius's manuscripts before they were burned and created a global cartel to fully exploit the materialistic potential of his Principle of Universal Discovery. And all this with the help of King Solomon, it turns out:

MÖBIUS: Do be reasonable. Don't you see you're mad?

FRÄULEIN DOKTOR: I'm no more mad than you. (90)

"I am Solomon. I am poor King Solomon." *The Physicists*; Robert Shaw (Möbius), 1964. Friedman-Abeles © New York Public Library for the Performing Arts.

And of course she is right. Möbius may as well be asking if she is inside or outside the Klein bottle.

Dürrenmatt ends his play with three monologues from his three scientists. The character of Newton speaks as though he is the historical Sir Isaac Newton, and Einstein speaks as Einstein. Both retire to their respective rooms at the conclusion of their remarks. Alone on stage, Möbius speaks last in the person of the wise King Solomon. All humor is removed at this point; the play is pure classical tragedy now. As it was with Oedipus, Möbius's actions have wrought the very thing they were designed to avoid. There is an inevitability, or a oneness, to this vision of things. The Möbius strip is a metaphor for this world—not for the shape of its physical space, but for its intellectual and moral topology. What Möbius was trying to do was an impossibility; his attempt to "take back our knowledge" is like trying to find the mythical other side of the Möbius strip. This is the glimpse of reality that Dürrenmatt's play momentarily dupes us into acknowledging. The dichotomies we traditionally use to give structure to our existence—good and evil, oppressed and free—may be temporary ways to locally solve our basic existential dilemmas, but in the waning moments of *The Physicists*, these binaries are exposed as delusions. The sane are insane, justice is injustice, the innocent are guilty.

Copenhagen

Mathematics becomes very odd when you apply it to people.

—HEISENBERG, FROM *COPENHAGEN*

Bibliographies of math and science-related plays are relatively sparse in terms of the number of pre-twenty-first-century entries that appear. One exception to this general pattern is the buzz of activity generated by the anxieties of the nuclear age. Bertolt Brecht and Friedrich Dürrenmatt were the most influential playwrights to address the subject, but they were joined by a number of other dramatists in this endeavor. Notable among other early theater offerings is $E = mc^2$ written in 1948 by Hallie Flanagan Davis, *Uranium 235* from 1952 by Ewan MacColl, and *In the Matter of J. Robert Oppenheimer* written in 1964 by Heinar Kipphardt. A more recent play on this theme, and one that has proved to be very influential in its own right, is *Copenhagen*, by Michael Frayn. If Tom Stoppard's *Arcadia* played some causal role in the notable increase in attention to mathematics and science exhibited by playwrights at the turn of the most recent century, then some of that credit has to be shared with Frayn's *Copenhagen*, which premiered in London in May 1998.

Like *Arcadia*, *Copenhagen* negotiated the elusive trick of incorporating a significant amount of technical material into its script in a way that its audiences found enticing rather than intimidating. Dürrenmatt made no attempt whatsoever to put any real science into his play. He absolved himself of this challenge by declaring that his concern as a playwright should be the effects of physics, not its content. Möbius's "Principle of Universal Discovery" is invented silliness, the content of which is never discussed, but whose name nevertheless contributes an extra bit of hyperbole.

Frayn's play, by contrast, is grounded in reality. Its three characters are historical people from the not-so-distant past. One of these is Niels Bohr, the Danish physicist who was the preeminent voice in the development of quantum mechanics in the 1910s, '20s, and '30s. The second is Werner Heisenberg, Bohr's protégé and collaborator, whose name is now permanently linked in the public consciousness with the uncertainty principle that he formulated and that became one of the pillars of quantum theory. Along with these two giants of twentieth-century science, Frayn includes Bohr's wife Margrethe as the third and final member of his small cast. The characters are meant to be authentic renditions of their real-life counterparts, and Frayn goes to significant lengths to make the science they discuss authentic as well. Because the topic of conversation is the physics of the atom, this inevitably means taking on some mathematics. In Margrethe, Frayn has provided the audience with a surrogate to represent the needs of the nonspecialist, but no one is exempt from exerting the mental effort necessary to learn some hard science:

HEISENBERG: 'If it works it works.' Never mind what it means.

BOHR: Of course I mind what it means.

HEISENBERG: What it means in language.

BOHR: In plain language, yes.

HEISENBERG: What something means is what it means in mathematics.

BOHR: You think that so long as the mathematics works out, the sense doesn't matter.

HEISENBERG: Mathematics *is* sense! That's what sense is!

BOHR: But in the end, in the end, remember, we have to be able to explain it all to Margrethe!

MARGRETHE: Explain it to me? You couldn't even explain it to each other! (64)[11]

Exchanges like these make it clear that Frayn has his pedagogical work cut out for him. At a conceptual level—i.e., in plain language—Frayn is adamant that his audience evolve through the evening toward an understanding of the general tenets of the relevant science: wave/particle duality, complementarity, the uncertainty principle. Why? To what end other than our general edification?

The answer is that, with *Copenhagen*, Frayn has orchestrated a small miracle whereby the principles of modern science under discussion are interwoven with the insights into human nature under debate, which are in turn incorporated into the architecture of the play being performed. This merging of form and content is a defining feature of the recent surge in plays about science, at least of those plays that have impacted contemporary theater in a

"But in the end, in the end, remember, we have to be able to explain it all to Margrethe!" *Copenhagen*; David Burke (Bohr), Sara Kestelman (Margrethe), Matthew Marsh (Heisenberg); Cottesloe Theatre, 1998.
© Donald Cooper/photostage.co.uk.

measurable way. "They literally enact the ideas they engage," writes Kirsten Shepherd-Barr, the scholar to most clearly articulate this thesis, and for whom *Copenhagen* represents a pinnacle of this phenomenon.[12] Frayn's play warranted a full chapter in Shepherd-Barr's inaugural book on the subject of science theater and remains a benchmark against which newer plays engaging science or mathematics are measured. What makes Shepherd-Barr's argument especially compelling is the way it highlights theater's particular strengths—specifically its performative nature—as a medium for bringing hard science into the domain of the humanities.[13] More than simple show and tell, the technical ideas in *Copenhagen* are so entwined with the structure of the performance that the play becomes a human analog of the science it discusses. Shepherd-Barr points out how Dürrenmatt engaged in a primitive form of something similar in *The Physicists*. By naming his protagonist Möbius, the playwright suggested a topological metaphor to illustrate the play's central ethical dilemma. Thirty years later, Michael Frayn staged another debate about the moral responsibility of the scientist but attempted a much more ambitious fusion of subject and structure.

The response was remarkable, especially for an intellectually dense play whose ending everyone already knows. (The Germans do not build an atomic bomb; the Allies do.) *Copenhagen* enjoyed over three hundred performances in London and a similar number two years later in New York, winning major awards in both cities. The play has gone on to enjoy a robust life outside these two major cultural centers, including a large swath of non-English-speaking countries. Very often, and especially in the earlier years, performances were accompanied by public symposia on science and its role in politics and the arts. In 2002, a film version of *Copenhagen* was produced by the BBC.

But Why?

Michael Frayn was born outside London, was educated at Cambridge where he studied philosophy, and worked as a journalist before dedicating himself full time to writing. A prolific and acclaimed novelist as well as a playwright, his most widely known work before *Copenhagen* is the comic farce *Noises Off*. The two plays could hardly be more different, in tone, style, or intent. The fact that they were spawned from the same creative mind speaks to Frayn's artistic dexterity.

At the center of *Copenhagen* is a documented historical event. In 1941, Bohr and Margrethe were living in Copenhagen, which at that time was under German occupation. A full-scale assault on the Danish Jewry, of which Bohr was a member, had not yet been initiated. In the scientific world, the process of fission—the splitting of the atomic nucleus and the release of energy—had been recently carried out on uranium, but the details of the science and its possible applications were not yet well understood. The formation of the American-based Manhattan Project was still a year away. Heisenberg, meanwhile, was in charge of the German nuclear research program.

This was the context in which the historical Heisenberg traveled to Copenhagen to visit his old mentor and friend:

MARGRETHE: But why?
BOHR: You're still thinking about it?
MARGRETHE: Why did he come to Copenhagen?
BOHR: Does it matter, my love, now we're all three of us dead and gone?
MARGRETHE: Some questions remain long after their owners have died. Lingering like ghosts. Looking for the answers they never found in life.

BOHR: Some questions have no answers to find.
MARGRETHE: Why did he come? What was he trying to tell you? (3)

This is the opening exchange in Frayn's play, and it efficiently sets out the agenda for what is to come. History confirms that Heisenberg did indeed visit the Bohr's at their home in the fall of 1941. The two scientists went for a walk, perhaps to avoid surveillance, perhaps because it was what they used to do in earlier, peaceful times. What happened on that walk, however, is not at all clear. Whatever transpired upset Bohr greatly and effectively ended the friendship. Later, after the war, attempts by the two physicists to explain the events of the evening only served to further muddy the waters. It was into this pond of confusion and accusations that Frayn ventured with the hope that an artistic lens might shed some new light on questions that the historians and biographers, not to mention the two participants, had left unresolved.[14]

Wave Functions

Our encounter with *Hapgood* provides a solid foundation for understanding the science in *Copenhagen*. The thrust of this earlier discussion is that light exhibits properties uniquely associated with waves as well as properties uniquely associated with particles, and a complete description of how light propagates demands we incorporate both viewpoints. When both slits are open in the diffraction grating of the double-slit experiment, the resulting alternating pattern of light and dark stripes on the receiving screen is a telltale sign of wave interference. Remarkably, this interference pattern appears even if the intensity of the laser is so low as to send the individual photons through one at a time. One by one, each photon leaves a mark on the screen, and the accumulating marks assiduously avoid landing on a series of spots that gradually become the nodes of the interference pattern. Again, this only happens if *both* slits are open. If we close one, the dark nodes disappear because there are no longer two wave sources interfering with each other. The other way to make the nodes disappear is to leave both slits open but to observe through which slit each successive photon passes. Once the photon is detected in one slit or the other, the restrictions are lifted as to where on the receiving screen the photon will ultimately be detected, and the interference pattern is no more.

One of Bohr's seminal insights for understanding the atom was that this wave/particle duality of photons applies to matter as well as light. In the 1920s, experiments equivalent to the double-slit experiment were carried out on electrons, with identical conclusions. This was even more counterintuitive than the experiment with light. Electrons are unambiguously understood to be particles. They have a mass that can be calculated. They are a constituent part of the atoms that in large enough quantities make up the chunky, solid objects we hold in our hands. Yet, electrons fired one at a time through a double-slit-type filter yielded an interference pattern. The conclusion: an electron can exhibit wavelike properties similar to light. Unobserved, an electron's location becomes fuzzy and spreads out across space the way we think of a wave spreading out across the surface of a pond.

To capture this aspect of electrons—and other particles at the atomic level—quantum mechanics represents the electron with something called a *wave function*. This function acts like a probability distribution. It represents the probability that the electron will be found at a given location in space. To be clear, when we go and look for it, the electron is never spread out across space; it is never the case that part of the electron is over here, say in one slit, and part of it is over there in the other. The electron is always fully intact and in one place. What the wave function describes is the relative likelihood of the various outcomes we expect to find before we do the measurement. The wave function is big in places where the electron is likely to be discovered and small in places where it is unlikely to be discovered. To put it another way, if we conducted the identical experiment of measuring the location of the electron multiple times and constructed a histogram of the observed data, the shape of the histogram would approximate the shape of the electron's associated wave function.[15] In a way, this is exactly what happens on the receiving screen in the double-slit experiment when electrons are sent through the apparatus one at a time. The densely cluttered spots where the electrons frequently hit the screen correspond to peaks in the associated wave function for a given electron. These peaks alternate with places where the wave function is effectively zero. On the screen, this is reflected in the dearth of electrons detected at these nodal points (Figure 1.4).

There is one way in which the concept of the wave function clarifies matters and another in which it adds an enormous sense of mystery. On the clarifying side, it offers a highly effective way to explain what is observed. Imagine an electron moving through the experimental apparatus, and assume it has a 50/50 chance of going through one slit or the other. Shedding our classical mindset that an electron is like a tiny bullet, we replace it with a probability wave. At the point where the electron passes the two slits, the wave has two equal peaks, reflecting the fact that if we measured it, the electron would be equally likely to turn up in either slit. But we don't measure it, and so both peaks of the wave then propagate forward as two independently existing synchronized waves which, when they interfere, precisely predict the alternating pattern of peaks and nodes that the actual experiment produces. If "mathematics is sense," as Frayn's Heisenberg says, then our work is done. The wave function is a mathematical device that predicts everything we can know about the trajectory of the electron.

But does this make sense in plain language? Is an electron literally a wave, and if not, how then does it engage both slits on its way to being detected as a single entity on the receiving screen? What the mathematics in the previous explanation suggests is that a given electron at the start of the experiment has two potential paths it might take—one through each slit—but rather than commit to either of them, it remains in a state of undetermined limbo. The two potential paths are superimposed on each other, allowing for the electron to interfere with itself. The mathematics may make sense, but taking the position that "that's what sense is" only works if we are willing to abandon some fundamental assumptions about how physical reality works.

My Dear Heisenberg; My Dear Bohr

One important takeaway of this discussion is that there is a probabilistic element to quantum mechanics that sits in direct contradiction to Newtonian determinism. For Newton, identical initial conditions always lead to identical outcomes. This is not the case in the physics of the atom. The wave function for a particle describes a set of probabilities governing an observable property such as its location. These probabilities are not a stand-in for some piece of information that exists but is too hard to calculate. They represent an inherent nonnegotiable feature of the atomic universe. In quantum mechanics, the same experiment theoretically conducted under the exact same initial conditions can produce a range of possible outcomes. Probabilistic descriptions are the best one can do.

This first tenet of quantum mechanics provides the scaffolding into which Frayn puts his characters. There is no setting described in the script. The only information Frayn offers is in the opening passage quoted earlier, where Bohr says to his wife that "all three of us are dead and gone." Reunited in some ephemeral gathering place in the afterlife, the three of them are going to try once more to get to the root of Heisenberg's motivations. The obstacles impeding their search are formidable, starting with the reliability of human recollection. "A curious sort of diary memory is," Bohr says early on, announcing one of the themes that will haunt them throughout their journey.

But the more fundamental hurdle to negotiate is the inherent randomness that the historical Bohr and Heisenberg introduced into our conception of reality. The way Frayn's characters proceed, then, is by running and rerunning the experiment of Heisenberg's visit, taking notes and debating revelations along the way. The physical laws of the afterworld where the action of the play resides are quite malleable. Bohr, Margrethe, and Heisenberg can speak directly to each other or speak internally to themselves while the audience eavesdrops. The shifts are frequent and seamless. The other thing that happens frequently and seamlessly is that the characters become fully incarnate versions of their living selves, cast in a series of reenactments. "The past becomes the present inside your head," Margrethe observes, and then without any theatrical fanfare it is 1941 and Heisenberg is getting off the train and heading to the Bohrs' house. He tries, with uneven effectiveness, both to be himself and to observe himself while Bohr and Margrethe engage in the same fascinating exercise:

> BOHR: I don't think we shall be going on any walks. Whatever he has to say he can say where everyone can hear it.
> MARGRETHE: Some new idea he wants to try out on you, perhaps.
> BOHR: What can it be, though? Where are we off to next?
> MARGRETHE: So now of course your curiosity's aroused, in spite of everything.
> HEISENBERG: So now here I am, walking out through the autumn twilight to the Bohrs' house . . . What am I feeling? Fear, certainly—the touch of fear that one always feels for a teacher, for an employer, for a parent. Much worse fear about what I have to say. About how to express it. How to broach it in the first place. Worse fear still about what happens if I fail. (10)

With no description of the setting provided in the script, most directors stage the play with minimal props and set pieces. A few chairs is all that typically gets used. The only stage directions offered are the ones built into the dialogue, which have the effect of making the past suddenly become present inside the viewer's head:

HEISENBERG: I crunch over the familiar gravel to the Bohrs' front door, and tug at the familiar bell-pull. Fear, yes. And another sensation, that's become painfully familiar over the past year. A mixture of self-importance and sheer helpless absurdity—that of all the 2000 million people in this world, I'm the one who's been charged with this impossible responsibility. . . . The heavy door swings open.

BOHR: My dear Heisenberg!

HEISENBERG: My dear Bohr!

BOHR: Come in, come in . . .

MARGRETHE: And of course as soon as they catch sight of each other all their caution disappears. The old flames leap up from the ashes. (13)

This juxtaposition of analysis and reenactments that move fluidly back and forth through time characterizes the action of the play. With each new measurement, the metaphorical wave function governing Heisenberg's motivations collapses around a different possibility. He came to Copenhagen to borrow Bohr's cyclotron. He came to ask whether Bohr was collaborating with someone. He came to offer to help the Bohrs escape. He came subconsciously hoping his former mentor would tell him not to try to build a bomb. Echoing Dürrenmatt's play, one of the more substantial possibilities that emerges is that Heisenberg came to suggest to Bohr that scientists had the potential, and the responsibility, to influence their respective governments. To do this collectively, Heisenberg needed to know if Bohr knew whether the Americans were working on a bomb. This line of inquiry leads to more ignoble possibilities, including Margrethe's pointed charge at one point that Heisenberg came simply "to show yourself off to us."

Historically speaking, Heisenberg was viewed negatively after the war. The fact that he had agreed to lead the Nazi nuclear program was unforgivable for many people, and Heisenberg became something of a pariah in the scientific community when the war ended. But what exactly did Heisenberg accomplish in his time serving under the Third Reich? It is well established that Heisenberg focused his efforts on building a nuclear reactor but did not pursue building a bomb. And why not? Here again we have to depart from the objective, Newtonian-like world of historical facts and enter the quantum-like world of human intentions. Was Heisenberg's decision not to pursue an atomic bomb incompetence or passive resistance? Did he simply not know how to do it, did he think it was practically infeasible, or did some component of his moral character make him unwilling to arm Hitler with such a potentially devastating weapon? Was it possible that by securing the leadership of the German nuclear research effort he was putting himself in a position to sabotage it, making just enough progress to keep his superiors happy but not enough to prevent its ultimate failure?

Frayn did an admirable job scouring the historical record in pursuit of some clarity on this point, but he is adamant that there are nonnegotiable limitations of the kind of precision one can hope for when exploring human motivations. "Thoughts and intentions, even one's own—perhaps one's own most of all—remain shifting and elusive," Frayn writes in the postscript to his play. "There is not one single thought or intention of any sort that can ever be precisely established."[16] It is along these lines that one can tangibly feel Heisenberg's science and Frayn's vision of human psychology merging into a common theme.[17]

The Uncertainty Principle

In 1927, Heisenberg formulated the initial version of his uncertainty principle which, like Einstein's theory of relativity, has managed to find its way into the public consciousness. This is due in part to the fact that one can give a reasonable description of it in plain language. Usually this is accomplished by invoking the properties of position and momentum for a given particle. In fact, the uncertainty principle applies to other pairs of complementary properties—energy and time, for instance—but position and momentum have the benefit of being more familiar notions. In this same vein, the momentum of a particle is proportional to its velocity, so one can tell the same story in even more accessible terms by referencing position and speed.

The uncertainty principle asserts that the more precisely we measure one of these quantities, say the electron's position, the less precisely we can know the other, in this case the speed. If, on the other hand, we set up an experiment to measure the speed of an electron with a high degree of accuracy, we have to accept a larger amount of uncertainty about the electron's location in exchange. While this observation may not sound so revolutionary, what is surprising is that this qualitative story can be made quantitatively precise. Using Δx for the error we accept while measuring the electron's position, and Δs for the corresponding error in its speed, the product $\Delta x \Delta s$ can never be smaller than a fixed amount that depends only on the mass of the particle. To be even more specific, if we swap Δs for Δp, the error in the particle's momentum, then the uncertainty principle takes the precise form

$$\Delta x \Delta \rho \geq \frac{h}{2\pi},$$

where Δx is measured in meters, $\Delta \rho$ in kilograms times meters/second, and $h = 6.63 \times 10^{-34}$ is a fundamental constant of quantum theory called Planck's constant. The value of h is quite small, obviously, but fixed and, to emphasize the central point, not zero. This hard lower bound on the limits of precision is not due to the quality of our available instruments or a lack of cleverness. It is a theoretical law of the universe. Under no circumstances is it possible to determine completely either the location or momentum of a given object, and the more precisely we attempt to determine one of these properties, the more uncertainty we must accept in the other.

Frayn assumes his audience knows something about the uncertainty principle, but he still smuggles in a series of explanations to help solidify the ideas. Some are metaphorical, like Bohr's retort when Heisenberg reminds him of how unbearably slow a skier he is. "At least I knew where I was," Bohr says. "At the speed you were going you were up against the uncertainty relationship. If you knew where you were when you were down you didn't know how fast you'd got there. If you knew how fast you'd been going you didn't know you were down." (24) As happens throughout the play, what starts out as a didactic metaphor eventually acquires a more poetic purpose. "Your talent is for skiing too fast for anyone to see where you are," Margrethe says to Heisenberg later on as a means to express her growing frustrations with what she sees as his galling duplicity. "For always being in more than one position at a time, like one of your particles."

Although these physics references in the dialogue are clever, the more interesting moments are when the playwright uses the three-dimensional nature of theater to elucidate the science. The following moment occurs in a passionate back-and-forth about Heisenberg's original formulation of uncertainty:

HEISENBERG: Plain language, plain language!

BOHR: This *is* plain language.

HEISENBERG: Listen . . .

BOHR: The language of classical mechanics.

HEISENBERG: Listen! Copenhagen is an atom. Margrethe is its nucleus. About right, the scale? Ten thousand to one?

BOHR: Yes, yes.

HEISENBERG: Now, Bohr's an electron. He is wandering about the city somewhere in the darkness, no one knows where. He's here, he's out

there, he's everywhere and nowhere. Up in Faelled Park, down at Carls-
berg. Passing City Hall, out by the harbour. I'm a photon. A quantum
of light. I'm dispatched into the darkness to find Bohr. And I succeed,
because I manage to collide with him . . . But what's happened? Look—
he's been slowed down, he's been deflected! He's no longer doing exactly
what he was so maddeningly doing when I walked into him!

BOHR: But, Heisenberg, Heisenberg! You also have been deflected! If
people can see what's happened to you, to their piece of light, then they
can work out what must have happened to me! (69)

This argument, which continues, is interesting for a host of reasons. The
first is that it provides a reasonably authentic rendition of the actual debate
between Heisenberg and Bohr about why the uncertainty relationship is a fact
of the quantum universe. Another is its performative nature—for a moment
the play becomes a physical rendition of the science it discusses. This merging of form and content happens organically in the flow of the ongoing debate,
and it serves a specific didactic purpose, but it is also a beacon alerting the
audience to the other ways in which the science is fusing with the storytelling.
Here is where *Copenhagen* most fully manifests Shepherd-Barr's proposition
about the performativity of science theater. This play about science is structured as a series of experiments designed to determine the motivations behind
Heisenberg's wartime visit, and one of the central discoveries is a law of
the psychological universe that might reasonably be called the uncertainty
principle for human introspection.

Applying Mathematics to People

Any time a precisely formulated conclusion of mathematics or science is
employed in the service of an artistic metaphor there is the danger of deforming the original principal beyond useful recognition. This is not the case in
Frayn's play. One strong piece of evidence for this opinion is that in *Copenhagen* the various metaphors work equally well in both directions. For some
viewers, an understanding of the science leads to new and meaningful insights
into the nature of human personality; for others, it is the examples of human
psychology that help them appreciate how weirdly enigmatic the science of
the atom really is.

One major sticking point for the founders of quantum mechanics was
whether the uncertainty principle represented a limitation on what could be

theoretically measured, or whether it was a statement about the nature of physical reality itself. Does an object like an electron—or a car for that matter—have a precise position and velocity that we cannot definitively determine, or are these properties just mental constructs left over from an overzealous dedication to Newtonian mechanics? For Bohr, the solution was to confine the domain of physics to what could be observed. Because perfect knowledge of location and velocity was not theoretically attainable, questions about them were rendered meaningless.

This point of view solves a number of semantic questions, but it puts an extraordinary demand on our conception of how the universe functions. To ask where an electron is located just before we measure it is not allowed—Bohr would say there is no such thing. Before it is observed, the electron is hovering in quantum limbo in a superposition of possibilities governed by a probability distribution. The act of measuring then forces the electron to commit to one of these possibilities, although the uncertainty principle implies that there will still be a bit of fuzziness to its determined position. Having located the electron in some reasonably well-defined region of space, can we now at least infer that *just before* we did the measurement the electron must have been in that general vicinity waiting to be detected? The so-called Copenhagen Interpretation of quantum mechanics championed by Bohr would say no. Before it is measured, the electron does not occupy a specific location. A quantum measurement does not reveal a preexisting truth; rather, it contributes to creating the outcome it records. Note how radically different this is from, say, lifting up our hand to see if a coin on the back of our other hand is heads or tails. Just before the reveal, when the coin is still obscured from view, the outcome has already been decided. This is not so in the world of the atom. In a quantum mechanical version of a coin flip, the coin is suspended between heads and tails until the measurement is taken, and it is the act of measurement that brings one or the other of these possibilities into objective existence.

In act I, Heisenberg notes that "mathematics becomes very odd when you apply it to people," but as the previous discussion makes clear, the mathematics is quite odd on its own terms. In fact, applying it to people can make it feel a little less odd:

HEISENBERG: . . . So many explanations for everything I did! So many of them sitting around the lunch-table! Somewhere at the head of the table, I think, is the real reason I came to Copenhagen. Again I turn to look . . . And for a moment I almost see its face. Then, next time I look the chair at the head of the table is completely empty. There's no reason at all. I didn't tell Speer [about Plutonium] simply because I didn't think of it. I came to Copenhagen simply because I did think of it. A million things we might do or might not do every day. A million decisions that make themselves. (77)

There is nothing strange about this assessment of how our minds work. Multiple thoughts coexist simultaneously in superimposed states inside our heads at every waking moment. They manifest themselves according to forces that are difficult to discern, most especially to the person whose head it is. In the Copenhagen Interpretation of quantum mechanics, the observer becomes enmeshed in the experiment he or she is observing; in *Copenhagen*'s interpretation of human nature the same is true, with the proviso that we are not really capable of observing ourselves:

HEISENBERG: And once again I crunch over the familiar gravel to the Bohrs' front door, and tug at the familiar bell-pull. Why have I come? I know perfectly well. Know so well that I've no need to ask myself. Until once again the heavy front door opens.

BOHR: He stands on the doorstep blinking in the sudden flood of light from the house. Until this instant his thoughts have been everywhere and nowhere, like unobserved particles, through all the slits in the diffraction grating simultaneously. Now they have to be observed and specified.

HEISENBERG: And at once the clear purposes inside my head lose all definite shape. The light falls on them and they scatter.

BOHR: My dear Heisenberg!

HEISENBERG: My dear Bohr!

BOHR: Come in, come in . . . (86)

This last run of the experiment is the one time Frayn sets his creative imagination completely loose from the recorded facts. The historical Heisenberg did not pursue building a bomb most likely because he thought it was infeasible, if not practically impossible. The fact that he held this position was, ironically, a major reason why he never committed himself to doing the research necessary to see that, in actuality, a nuclear bomb was not quite so impossible after all. In Frayn's last fanciful thought experiment of their fateful 1941 encounter, Bohr controls his rising temper and, instead of dashing off back to the house, calmly asks his former student why he is "confident that it's going to be so reassuringly difficult to build a bomb."

The question snaps the self-reinforcing cycle of Heisenberg's ignorance. He does not know why he is so confident . . . and so he no longer is so confident . . . and so he asks himself why he originally thought a bomb was practically impossible . . . and so he begins to do some new calculations . . .

HEISENBERG: The scattering cross-section's about 6×10^{-24}, so the mean free path would be . . . Hold on . . .

BOHR: And suddenly a very different and very terrible new world begins to take shape . . .

No more words are needed. The momentary glimpse of Hitler in possession of a nuclear weapon is bone-chilling.

MARGRETHE: That was the last and greatest demand that Heisenberg made on his friendship with you. To be understood when he couldn't understand himself. That was the last and greatest act of friendship for Heisenberg that you performed in return. To leave him misunderstood. (89)

The Role of the Observer

When Michael Frayn was working on the script for *Copenhagen,* he enlisted the help of director and friend Michael Blakemore. Blakemore directed the original production at the National Theatre in London and stayed on in this capacity when it was produced with a new cast on Broadway two years later. Both productions were done in the same minimalist style—three actors and three chairs on an otherwise empty stage. When the play moved to Paris, it was originally given a new director who, among other innovations, decided to have ethereal looking curtains swing in at the various moments when the characters began reenacting the past. Cast members expressed misgivings, and when the artistic tensions got too high, Blakemore was called in to direct this production as well.[18]

Under Blakemore's direction, the visual image of the play was unapologetically intended to be suggestive of its science. With the two physicists perambulating about on the round stage like subatomic particles and Margrethe standing in as the nucleus, Blakemore's *Copenhagen* looked like a model of an atom, albeit a prequantum mechanical version reminiscent of Rutherford's tiny solar systems. But Blakemore, like Frayn, did internalize a good deal

"Some questions remain long after their owners have died." *Copenhagen*;
Sara Kestelman (Margrethe), David Burke (Bohr), Matthew Marsh
(Heisenberg); Cottesloe Theatre, 1998. © Donald
Cooper/photostage.co.uk.

of modern physics, and it led him to a beautiful insight about how Frayn's
play—or really theater in general—finds yet another parallel with its science.
Reflecting on his experience working on *Copenhagen* in a piece for the *New
York Times*, he first acknowledged the long process that led up to opening
night. "Then something very strange happens," he pointed out. "The thing
that you rehearsed . . . and that you have seen a hundred times is put on a stage
and a thousand pairs of eyes hit it and alter it. The energy an audience brings
to it, the energy of their laughter and their rapt attention, changes what is
there."[19]

All theater practitioners point to this facet of their chosen craft—that one
cannot know what a play really is until it is brought properly into existence
in front of a live audience. *Copenhagen* provides an especially fascinating case
study of this phenomenon. Perhaps he is being modest, but Frayn maintains
that he was not confident *Copenhagen* would ever get performed, much less
win a Tony Award for Best Play.[20] He originally took up the project because
he was drawn to the philosophical questions it presented. As the writing process continued, he consoled himself with the hope that it might be given a

staged reading at some point or perhaps become a radio play. Frayn's apprehensions are understandable. The play is scientifically dense, the characters are
already dead, and we learn in the first few lines that the play's central dilemma
is not going to be resolved. "Why did he [Heisenberg] come to Copenhagen?"
Margrethe asks at the beginning of the show. "I doubt if he ever really knew
himself," Bohr says not long after.

Trying to put a finger on why the play *did* come to life in a powerful way
in front of an audience is likely to be as elusive as getting to the real reason Heisenberg traveled to Copenhagen in 1941. Strangely enough, it could
be that these two intractable questions have the same answer. It can happen
in quantum mechanics that two particles separated in space are "entangled"
in the following sense: Although both are hovering in some undetermined
quantum state between two possibilities, the instant one of them is measured and commits to a particular objective outcome, the other does as well,
even though the two particles might be kilometers apart. Stretching the scientific metaphors further than Frayn intended, the question of why *Copenhagen*
works as theater does seem to be entangled in a quantum mechanical-like way
with the attempt to determine Heisenberg's deepest motivations for why he
would risk so much to seek out a member of the enemy in the middle of a war.

Accepting the premise that the principles of quantum mechanics have
analogs for the mechanisms of human decision-making leads directly to
Frayn's assertion that "there is not one single thought or intention of any sort
that can ever be precisely established." Fair enough. But this same exercise
also leads to the proposal that there are metaphorical wave functions that
describe the relative likelihood of what we expect to find when we take a measurement of someone's motivations. For a host of people who have devoted
considerable effort trying to get inside Heisenberg's head—and this includes
his postwar colleague Hans-Peter Dürr, Danish director Peter Langdal who
produced Frayn's play in Copenhagen, theater scholar Kirsten Shepherd-Barr,
and Frayn himself—there is a consensus that the peak of the wave function
governing Heisenberg's motives is centered around the human desire to be
with an old and faithful friend.[21] More than politics or physics, what Frayn
and others suspect ultimately drove Heisenberg's decision was a desire that
unconsciously manifests itself in all of us, especially in times of extraordinary
adversity. We want to be observed. We want the audience of someone we trust.
When we are awash in a stew of uncertain and conflicting thoughts, we want
to hear what we will say when we are forced to explain ourselves out loud to
someone we know is really listening.

The particular circumstances of Heisenberg's predicament were gut-wrenchingly unique—"Of all the 2000 million people in this world, I'm the one who's been charged with this impossible responsibility"—but at the most personal level, the journey of a troubled soul back to a place where he last recalls being content is universal. If this is what Heisenberg was doing, then it goes a long way toward explaining *Copenhagen*'s gripping emotional appeal.

Whatever the truth, Frayn's play brings into focus a contemporary cousin of a philosophical idea we have encountered before: that to be is to be perceived. Esse est percipi. Plays and particles and people all share a version of this trait—that it is in collisions and confrontations and conversations where each achieves its fullest expression in the physical world.

A Disappearing Number

And the answer came into my mind.

—RAMANUJAN, FROM *A DISAPPEARING NUMBER*

Looked at chronologically, *Life of Galileo*, *The Physicists*, and *Copenhagen* represent a trend of increasingly ambitious attempts to weave mathematical content into the structure of each play. These scripts all notably come by their mathematics via physics. The last play in this survey is an attempt to engage mathematics on its own terms. In 2005, a London-based theater company called Complicité set out to devise a new play inspired by the story of Cambridge mathematician G. H. Hardy's collaboration and friendship with the Indian genius Srinivasa Ramanujan. From the outset, Complicité's founder and director, Simon McBurney, intended to create a theater piece in which the concepts of continuity, partitions, prime numbers, and the mathematical infinite were embedded into the fabric of the performance. There was just one problem: McBurney didn't know any mathematics.

G. H. Hardy

McBurney's entry point into the world of mathematics was *A Mathematician's Apology*, the deeply personal essay penned by G. H. Hardy. Elected as a fellow at Cambridge in 1900, Hardy's impressive career made him equal to the best mathematicians of his age. Hardy was also something of a celebrity within the broader academic circles of his time. Shy by nature, Hardy possessed a wit and worldliness that made him very much at home at high table with intellectuals of all sorts—from John Maynard Keynes to E. M. Forster—and led to his unofficial designation as the de facto spokesperson for British mathematics during the first half of the twentieth century. Given the parameters of his life—full-time academic, a member of the Cambridge Apostles, cricket aficionado—it is not obvious why Hardy's personal reflections would connect so strongly with the creative director of an innovative theater company who openly confessed to suffering from a lifelong aversion to mathematics. But that is precisely what happened. What did G. H. Hardy and Simon McBurney have in common, other than both having spent a good number of years living and studying in the same university town?

The first part of the answer becomes evident on an initial read through Hardy's essay. From the time it was published in 1940, *A Mathematician's Apology* has been widely recognized as a compelling and moving account of what it feels like to dedicate oneself to a life of making art.[22] In this regard, McBurney is by no means the first to fall under the *Apology*'s spell. But the deeper revelation comes from looking into the means by which McBurney's company creates its theater. McBurney is not so much a playwright as a facilitator of a process that relies on a careful balance of collaboration and improvisation. It is from this vantage point that McBurney's abrupt change of heart about mathematics begins to make sense. In fact, there is a strong case to be made that Complicité's journey to create *A Disappearing Number* comes closest to revealing the common core that mathematics and theater share in their respective searches for truth.

A Maths Lecture

Audience members entering the theater for a production of *A Disappearing Number* are greeted with a curious sight. There is no curtain, and the lights are up on stage to reveal the front of a contemporary university lecture hall—a large whiteboard flanked by two steel exit doors, an overhead projector off to the side. This familiar setting is disrupted by the presence of an actor already on stage. Standing frozen in time, with overcoat and suitcase and looking distinctly out of place, is the man we will eventually come to know as Al. His presence goes in and out of our consciousness, but he is clearly not visible to Ruth, the math professor who enters the hall to begin her talk.

Uneasy but genuinely excited, Ruth thanks her audience for coming and sets off on a lecture about infinite sequences and series. For the vast majority of the viewers, this is the first mathematics lesson they have attended in decades, and the initial response is an amused sense of awkwardness. Is this really going to be a mathematics lecture? Apparently so. As Ruth forges ahead there is not the slightest bit of exaggeration or campiness in the way she presents the mathematics, but this does not prevent the members of her audience from coping with their rising anxiety the only way they know how—they start to laugh.

"Let's consider these sets of numbers," Ruth says as she begins scribbling lists of integers on the whiteboard. The sequence of prime numbers elicits some suppressed giggles; the use of standard sigma notation for infinite summations earns a hearty guffaw. As the difficulty of the mathematics increases, so does the intensity of the audience's laughter, and when Ruth introduces the celebrated Riemann-zeta function

$$\zeta(z) = \sum_{n=1}^{\infty} \frac{1}{n^z},$$

the house comes down.

A Melancholy Experience

Moments of levity permeate *A Disappearing Number* but so do moments of pathos. G. H. Hardy wrote *A Mathematician's Apology* at the age of sixty-three, when he was well aware that his days for doing research mathematics had come and gone. The heaviness of this realization is evident from the opening line:

> It is a melancholy experience for a professional mathematician to find himself writing about mathematics. The function of a mathematician is to do something, to prove new theorems, to add to mathematics, and not to talk about what he or other mathematicians have done.[23]

Hardy's goal, as he describes it, is to "put forward an apology for mathematics," but he makes no attempt to hide the fact that there is more at stake than the reputation of his discipline. "I should say at once," he adds, "that my defense of mathematics will be a defense of myself."[24] A lifelong bachelor, fundamentally opposed to both of the world wars and a committed atheist, Hardy sets himself the task of justifying a life spent almost exclusively in pursuit of pure mathematics.

Of all its many epiphanies, the one that rises the highest from the pages of *A Mathematician's Apology* is Hardy's insistence that mathematics is an aesthetic discipline. Hardy goes to great lengths to distinguish his own research from

"Beauty is the first test. There is no permanent place in the world for ugly mathematics." *A Disappearing Number*; David Annen (G. H. Hardy), Shane Shambhu (Ramanujan); Barbican Theatre, 2008. Stephanie Berger © 2022.

the useful, and therefore "trivial," mathematics of bankers and engineers. Pure mathematics "must be justified as art if it can be justified at all," is how he summarizes it.[25] Adding to the intensity of this conviction is the fact that Hardy is writing at a point in his life when his creative abilities have departed. Unable to summon the cognitive strength to prove any new theorems, Hardy is acutely aware of the intrinsic worth of the creative spirit precisely because he no longer possesses it. The result is a poignant celebration of the value of a life dedicated to the imagination, filtered through a retired mathematician's sensibility not to overstate his case.

When asked about the origin of *A Disappearing Number*, McBurney credits Canadian novelist and friend Michael Ondaatje as the one who pointed him toward Hardy's essay. McBurney's copy included the standard introduction by C. P. Snow in which, among other anecdotes, Snow offers a vivid account of Hardy's collaboration with the Indian mathematician Srinivasa Ramanujan. While the revelation that Hardy was a fellow artist may have initially caught McBurney's attention, it was the Hardy–Ramanujan relationship that supplied the dramatic hook.[26] Hardy describes this episode as "the one romantic incident in my life," and although there is no physical romance, there is a fairy-tale quality to this story that encompasses both triumph and tragedy. Looking over Complicité's diverse body of work, a handful of regular themes consistently rise to the top: creativity, the passage of time, memory, cultural divides, and the way in which our past informs our present. In this light, the story of a famous Cambridge don's unexpected encounter with an unknown Hindu clerk from Madras becomes an irresistible fit for McBurney and his company.

But what about the mathematics? If mathematics was at the center of Hardy and Ramanujan's friendship, then it needed to be at the center of McBurney's play in all its potentially intimidating glory. So how does one transport pure mathematics onto the stage? A math lecture possesses its own natural kind of theater, as the opening of the play illustrates, but this cannot be the whole story. Could the mathematics be part of the performance in some satisfying way, and could it be incorporated into the architecture as well as the content of the play? This was the issue, and for McBurney to make any progress on this problem, he first had to admit that he had one.

It's Terrifying but It's Real

Just at the moment in Ruth's show-opening lecture where she plunges into the impenetrable details of the Riemann-zeta function, she is joined onstage by a distinguished-looking gentleman. He watches her for a bit—she does not acknowledge him—and then he addresses the audience.

"You are probably wondering if this is the whole show," he says to everyone's great relief, in an accent that suggests he is from India. "My name is Aninda Rao, and this is Ruth," he explains—but then he stops. His posture changes, his accent disappears, and he begins a confession. "Actually, that's a lie. I am an actor playing Aninda, she's an actress playing Ruth . . . This phone, for example . . . 'Hello mum?' . . . no mum, no ring tone! This door doesn't lead anywhere! I can push these walls right off!" Exposing the artifice of theater, Aninda—or whoever he is—physically dismantles the lecture hall piece by piece into the theatrical ether. But there is an ironic twist. Theater is an illusion, he is telling us, and everything we see up on the stage is fake—everything, that is, except the mathematics. "[The mathematics] is real," he says. "In fact, we could say that this is the only real thing here."

Throughout this exercise, Aninda recounts his various confusions about mathematics via a series of anecdotes lifted directly from the pages of Simon McBurney's own childhood:

ANINDA: I remember I had a chart on the back of my door. It went, '1—one tractor.' '2—two pineapples.' And, for a reason that nobody could explain, there were three medicine bottles.... Of course when they added one plus two equals three, I tried to understand how one tractor plus two pineapples could make three medicine bottles.... Just to walk into a math class would bring me out in a cold sweat. (23)[27]

Ten minutes into the play, then, all intellectual pretense is dispensed with. This is a fake play about real mathematics, a subject the director and the majority of the audience find baffling. Nevertheless, it is time to take this challenge head-on. When Aninda finishes his direct address, he redons his accent and magically gets into a cab on the streets of Chennai, India, alongside an American tourist. Having effortlessly made believers of us all again, they head off to the neighborhood where the legendary Srinivasa Ramanujan once lived.

The Real and the Imagined

Keeping track of what is real and what is fiction in *A Disappearing Number* is an instructive exercise that turns out to be more difficult than Aninda makes it sound. For what it's worth, the character of Aninda is a fictional present-day physicist whose work in string theory is finding nonfictional connections to Ramanujan's mathematics. Ruth, the lecturing mathematician, is also interested in Ramanujan as a source of general inspiration as well as for his particular contributions to number theory. Aninda's taxi mate turns out to be Al, who was the person standing invisibly onstage at the beginning of the show. The time structures of *A Disappearing Number* are fluid and deliberately disorienting. Ruth could not see Al in the opening scene because Al was five years into her future. Why Al and Aninda are touring Ramanujan's hometown together is not yet clear, but, again, this confusion is by design. It is as though the various narratives have been cut into pieces, shuffled about, and are being presented in some randomized order. The next scene returns to the university hall from the beginning where a smitten Al approaches Ruth after her lecture:

AL: I haven't understood a word you said, but I found it fascinating. You clearly like what you do.
RUTH: I love what I do.
AL: I want to ask you a question because yesterday you were talking about infinity.
RUTH: Yes.

AL: Infinity is something that has always frightened me.

RUTH: For mathematicians infinity is just another mathematical concept. It's no big deal.

AL: It's a big deal for me, because it's where I'm going to go when I die. (*Ruth laughs.*) You seem to be suggesting that there was more than one infinity?

RUTH: Yes. There is an infinity of infinities.

AL: Oh shit . . . (28)

Given McBurney's fraught relationship to mathematics, it is natural to draw parallels between the play's director and Al. Indeed, McBurney originally played Al in the first few productions but eventually decided that he needed to take himself out of the show.

The last two major characters in the play are the nonfictional Hardy and Ramanujan, and each is presented with meticulous biographical accuracy. Every line of dialogue uttered by either of these two characters comes from some documented source—Hardy's *Apology*, a recorded lecture, or their archived letters. The first time we hear Ramanujan in voice-over, it is from his most famous letter of all. In 1913, Ramanujan was a twenty-three-year-old accountant from a devout Brahmin family. In possession of an extraordinary intuition for mathematics but with essentially no formal education, Ramanujan had independently developed a body of theorems and formulas unlike anything the mathematical world had ever seen. Gathering an assortment of his discoveries, he sent them to the most famous mathematician in England in hopes of winning his favor:

RAMANUJAN (*voice-over*): Dear Sir, I beg to introduce myself to you as a clerk in the accounts department of the Port Trust Office at Madras . . . Please do not think me mad if I state that $1 + 2 + 3 + 4$ and so on to infinity is equal to $-1/12$. (26)

As the story goes, Hardy originally discarded the untidy pages, dismissing them as an elaborate hoax. But throughout the day, the image of what he had seen stayed in his mind until it dawned on him that the author of these formulas was not a fraud but a genius on the order of Gauss or Newton. The odd formula $1 + 2 + 3 + 4 + \cdots = -1/12$, for example, was evidence that Ramanujan had managed to rediscover the supremely complicated functional equation of the Riemann-zeta function. Hardy immediately set about arranging for Ramanujan to come to Cambridge, and the collaboration that transpired is the most celebrated relationship in modern mathematics.

In total, then, there are three different threads running through *A Disappearing Number*. Narrating the show is Aninda, the physicist, who spends the majority of the play giving a lecture at the CERN Institute about Ramanujan's mathematics. This vantage point gives Aninda a metatheatrical presence, and there is a direct way in which his voice in the play can be conflated with C. P. Snow's voice in *A Mathematician's Apology*. The second thread is the Hardy–Ramanujan story, which, in contrast to many other plays about real people, is left historically intact. A major reason this works theatrically is because of the modern love story between Ruth and Al. By creating this third thread, McBurney and company are able to transfer themes back and forth across time and thereby explore a range of emotional possibilities without imposing any awkward invented scenes on its historical figures.

Devised Theater

At Complicité, a script is typically the end result of the production process, not the starting point. Through an extended process of improvisation, the actors and artists involved in each project contribute directly to the creation of the dramatic material. In the case of *A Disappearing Number*, two early and important mathematical voices in this process belonged to Victoria Gould and Marcus du Sautoy. Gould is the rare example of a working actor who is also a mathematician. Having participated in a number of projects with Complicité in the past, Gould was one of the original people McBurney sought out to help him explore the dramatic possibilities in Hardy's *Apology*. Du Sautoy was another natural choice for the project. An Oxford professor, du Sautoy had made a name for himself across the UK teaching mathematical ideas to diverse audiences in creative ways. He was also the author of several best-selling popular books on mathematics.

Gould describes the initial phase of creating *A Disappearing Number* as an improvisational playground. "He [McBurney] assembled a cast—some of them fairly last-minute," she recalled in a joint interview with du Sautoy, "and there were, well not quite 100, but an awfully large number of people came through the door. He gets a big room somewhere, and over a period of months people come and *play*."[28] In addition to mathematicians, the list included artists, scholars, dancers, composers, economists, and directors.[29] "Another thing that Simon always does," Gould added, "is that he tries to get actors to

"Please do not think me mad if I state that $1 + 2 + 3 + 4$ and so on to infinity is equal to $-1/12$." *A Disappearing Number*; Shane Shambhu (Ramanujan); Barbican Theatre, 2008 (photo: Robbie Jack).

convey the dynamics of something. For instance, we spent weeks on the square root of two. The irrationality of the square root of two—how can you convey a little flavor of that in movement? And it's amazing how you can do it, actually. You think it's not possible but it kind of is."

Building on the theme of movement, du Sautoy took the company through a series of workshops that, in his words, "were designed to bring the maths alive."[30] One notable exercise was based on the partition function. The *partitions* of an integer are the number of ways of breaking that integer into a sum of smaller integers. There are 3 partitions of 3 (specifically: 3, 2 + 1, 1 + 1 + 1); there are 5 ways to partition 4 (specifically: 4, 3 + 1, 2 + 2, 2 + 1 + 1, 1 + 1 + 1 + 1). There are 7 ways to partition 5. The number of partitions of an integer rises quickly, and Hardy and Ramanujan made groundbreaking strides in finding a formula for this value when the numbers get astronomically large. Back in the rehearsal room, du Sautoy had cast members physically enact the different partitions of small integers with their bodies in space, which organically led to the creation of corresponding narratives to accompany the mathematical patterns they were ferreting out for themselves.[31]

Amid this improvisational playtime, there is still the mystery of just how the finished play comes together. Surely there must have been some person in the role of playwright providing cohesive narratives to serve as scaffolding for the rehearsal process? "No, not really," was du Sautoy's amused response to this question. "That's not how it works." Gould explained it this way: "Well, there are people recording everything that goes on in the room, so there are drafts of every improvisation which kind of hang around, lots of cutting and pasting. But it's right at the end that decisions are made."

Thus, it was over months of rehearsals that, image by image, *A Disappearing Number* began to take shape, and in the final result one can see the footprints of the long, experimental process that led to it. In the play, it is Ruth who gives a copy of *A Mathematician's Apology* to Al as a means of helping him understand her love of mathematics. In a way, Ruth is to Al what Ramanujan is to Ruth— a source of passion and inspiration that always remains a bit out of reach. "How can something you don't understand be beautiful?" Al asks of Ruth, unaware of the irony in his question. Ruth replies: "Don't we call something 'beautiful' simply because it outpaces us?" In these exchanges, the play never drifts too far from Hardy's *Apology*, which we often hear recited in the author's voice:

> Beauty is the first test. There is no permanent place in the world for ugly mathematics. . . . It may be very hard to define mathematical beauty, but that is just as true of beauty of any kind—we may not know quite what we mean by a beautiful poem, but that does not prevent us from recognizing one when we read it.[32]

Ruth eventually decides to travel to Ramanujan's hometown in India, a trip that, not coincidentally, Complicité's cast also made as part of their creative process. Ruth's journey east is juxtaposed against the reverse trip Ramanujan makes to England in 1914 to work with Hardy. There are glorious discoveries to be made in both directions, but there are also complicated issues of cultural compatibility, and both of these pilgrimages end tragically. Ramanujan eventually grows ill in the harsh climate of Cambridge and returns home to India in 1919, only to die a year later. Ruth suffers a fatal brain aneurysm on a train speeding across the Indian countryside. Hardy's reaction to the loss of Ramanujan is rendered with the restrained eloquence for which Hardy was known. It is in Ruth's death that McBurney explores the concept of grief, using the full expressive power of his craft.

As with so much of *A Disappearing Number*, there is a deeply personal component to this aspect of the play. While he was still in the early stages of deciding what he might do with the Hardy–Ramanujan story, one of McBurney's closest friends died quite unexpectedly. The play became a part of McBurney's attempt to come to terms with this loss.[33] Looking at *A Disappearing Number* in this light gives new significance to so many of its repeated elements: the lyrical portrayal of Ramanujan's death, Ruth's convergent series that approach but never reach their limits, the "infinity of infinities" that unnerves Al. It also points to why Al's journey to make his peace with Ruth's sudden demise spans the entire production. Al is on stage before the play starts, grief-stricken in his deceased wife's empty classroom—and he is on stage at the end, standing next to a sacred river near Ramanujan's home village. Al has followed Ruth's path to India, a country where he himself has familial roots. With Ruth's invisible hand holding Al's shoulder while he clutches a piece of chalk from among her belongings, the show closes with the sound of Ruth's consoling voice:

> RUTH: Al . . . I want to read you this . . . 'What reconciles me to my own death more than anything else is the image of a place: a place where your bones and mine are buried, thrown, uncovered together. They are strewn there pell-mell. One of your ribs leans against my skull. A metacarpal of my left hand lies inside your pelvis. The hundred bones of our feet are scattered like gravel. It is strange that this image of our proximity, concerning as it does mere phosphate of calcium, should bestow a sense of peace. Yet it does. With you I can imagine a place where to be phosphate of calcium is enough.' (91)[34]

A Maker of Patterns

With themes of grief and cultural displacement giving the audience a way to access the play emotionally, mathematics is fused into the storytelling at every opportunity. The continuity of the real number line is invoked as a metaphor for how the past is connected to the present. Convergent series are woven into Al and Ruth's courtship. There is a tabla player onstage, highlighting the relationship of number theory and combinatorics with the music of Ramanujan's home country. McBurney cast professional dancers in several key roles, including the London-based Asian dance artist Shane Shambhu as Ramanujan. In a climactic moment when Ramanujan is engaged in the highly cerebral business of his research, the audience is presented with a *tihai*—a mathematically syncopated Indian musical form involving tabla, voice, and dance. Elaborate audio and video effects accompany this scene and others, routinely filling the stage with the sounds of numbers and dynamic images of mathematical notation.

The partition function from du Sautoy's workshops finds its way into the performance in a memorable scene that involves overlaying four distinct moments in time. Sharing the stage for a few dazzling minutes are Hardy and Ramanujan working in Cambridge circa 1915; a grieving Al from the present locked in a lecture hall; Ruth a year in Al's past phoning to tell Al she is pregnant; and Aninda at CERN, delivering his lecture about Hardy and Ramanujan and their work on partitions. As Aninda explains the ways to partition the numbers 2, 3, 4, and 5, the narratives crisscross through time while the actors unwittingly demonstrate the different groupings with their bodies and various props.

Keeping track of the temporal locations of this scene is akin to listening to a four-part fugue, and it points to the most holistic way in which *A Disappearing Number* attains its mathematical identity. "Simon was determined that mathematics should be integrally embedded in the play, and into the structure of the play," du Sautoy reported. "The idea of fracturing the piece so that it is not continuous—bits that are disjoint but whose sum is the totality of the play—that was Simon's idea. He wanted to make those infinite sequences actually the structure of the play." What Michael Frayn achieved with quantum mechanics, McBurney wanted to do with mathematics, and it required the unsuspecting cooperation of his math-anxious audience.

Following Aninda's fourth-wall-shattering monologue at the beginning of the play, the audience is confronted with a chaotic stew of images via video, voice-overs, and the actors themselves. We catch a glimpse of Ruth on a train, see a child setting sail from India, hear Neville Chamberlain announcing war with Germany, encounter the continued fraction for the square root of two, then Ruth on the train again. Faced with this cacophony of disjointed pieces, the audience sorts through them, arranges them, and connects them as best they can to the little they know at that point. "A mathematician, like a painter or a poet, is a maker of patterns,"[35] is Hardy's iconic line that echoes through the theater as the audience engages in its own version of pattern seeking. Progress is slow at first, but gradually the outlines of the various fractured plot lines come into focus, and as they are mentally reassembled, the symmetries that exist between them start to emerge. There are no instructions to follow. This is just what thoughtful audiences naturally do, and while it is not mathematics, it is a reasonable stand-in for what mathematicians do when they train their focus on some unfamiliar corner of the mathematical landscape.

Pure Mathematics and Pure Theater

As McBurney's cast probed more deeply into Hardy and Ramanujan's alien world, what they ultimately discovered was this kinship between the creative process in mathematics and their own methods.[36] Mathematics is traditionally presented—in the classroom and in research papers—as an organized sequence of conclusions, each one following logically from the ones before it. But this is decidedly *not* how new mathematics is produced. Just as with McBurney's plays, mathematical theorems are the result of a long process of improvising and experimenting. The mathematical search for patterns starts from considering special cases or anomalous examples. The journey is circuitous, full of dead ends, and littered with false conjectures laid to rest by carefully crafted counterexamples. The discovery of a meaningful mathematical relationship is the rare and rewarding exception. "There is a very high degree of unexpectedness, combined with inevitability and economy," is how Hardy describes a beautiful theorem.[37] By comparison, Complicité describes its methodology this way: "The process of devising involves experimenting and discarding numerous ideas, throwing ideas together and allowing the possibility of the unexpected."[38]

Ramanujan is renowned for his mystical ability to find unexpected relationships in unlikely places. A famous anecdote portrayed in *A Disappearing Number* involves the number on a taxi Hardy took one day to visit Ramanujan in the hospital. The number was 1729, which Hardy remarked was rather dull. "No, Hardy," Ramanujan replied, "1729 is a very interesting number! It is the smallest number expressible as the sum of two cubes in two different ways." This familiar tale is usually invoked to convey Ramanujan's genius, but what it more accurately conveys is the enormous groundwork Ramanujan did familiarizing himself with the world of whole numbers. Yes, he possessed an unparalleled mathematical insight, but the result of his creativity—the "wild theorems" Hardy describes seeing in that first letter—were not gifts from the gods as much as they were products of the fertile soil his mind had been constantly tilling for the majority of his life.

At a Cambridge humanities conference presentation in 2009, McBurney expressed a strong sympathy for Ramanujan's method of working, going so far as to quote the following passage from *A Disappearing Number*:

> RAMANUJAN: Immediately I heard the problem it was obvious that the solution should be a continued fraction: I then thought, 'which continued fraction?' And the answer came into my mind. (69)

For McBurney, the emphasis was on the last phrase: "the answer came into my mind." In his own experience as an artist, the director of Complicité was familiar with moments of insight where solutions to problems appeared fully formed out of the cognitive ether. Joining McBurney on stage for that presentation was Fields medalist Tim Gowers, who uncontrollably jumped in with an objection at this point.

"It doesn't work that way!" Gowers interrupted, arguing that Ramanujan had created the possibility for revelations like these by virtue of so much preparation. McBurney's response to Gowers was telling. Essentially agreeing with the mathematician, McBurney pointed out that this is how creativity always functions. "Some days the page is blank, but some days I know how it all works out," he explained, reemphasizing the symmetry that he recognized between his craft and Ramanujan's. For Complicité's improvisational rehearsals to bear fruit, the participants needed to be immersed in the themes of the play. This methodology was part of Complicité's working philosophy from the outset. "The value in preparation . . . is again to do with the unexpected," McBurney wrote back in the 1980s,

> I did not prepare people so that they know about where they are going. I prepare them so that they are ready: ready to change, ready to be surprised, ready to seize any opportunity that comes their way.[39]

For McBurney, the process of preparation was not distinct from creation, an opinion that makes him sound very much like a mathematician.[40] The proof is in the product. The long journey to understand their mathematical subject matter produced the raw material for Complicité's play in the same way that Ramanujan's tireless explorations of the mathematical landscape were the source of the mystical leaps of insight that characterize his work.

This is the perch from which to best appreciate the affinity between mathematicians and practitioners of theater. Engaged in their respective endeavors,

"A mathematician, like a painter or a poet, is a maker of patterns."
A Disappearing Number; David Annen (G. H. Hardy), Saskia Reeves (Ruth);
Barbican Theatre, 2008 (photo: Robbie Jack).

these distinct artists share a common process designed for a common goal. In an attempt to distinguish theater-making from other art forms, Michael Frayn remarked that, "Plays are not called 'plays' for nothing—they are a means of messing about, exploring the world without the restriction of actually moving about in it."[41] This description of playwriting captures the essence of abstract mathematics. Polish playwright Stanislaw Witkiewicz put this analogy into practice to create his non-Euclidean dramas. Samuel Beckett was less explicit about the influence of mathematics, but he attributed his evolution toward theater to the autonomy it offered him in the form of greater control over the components of his art.[42] McBurney, despite being less mathematically minded than these predecessors, makes the most compelling case yet for the connectedness of mathematics and theater. As *A Disappearing Number* demonstrates, the theater is an idealized environment where the artist can escape the limits of physical reality and explore the world according to his or her own chosen set of assumptions. If we were looking for a tangible analogy for the creative process of pure mathematics, we could hardly do better.

Multiple Realities

A more complete version of the quotation from Hardy about being "a maker of patterns" shows that its author was interested in the lasting quality of his art:

> A mathematician, like a painter or a poet, is a maker of patterns. If his patterns are more permanent than theirs, it is because they are made with ideas. A painter makes patterns with shapes and colours, a poet with words . . . A mathematician, on the other hand, has nothing to work with but ideas, and so his patterns are likely to last longer since ideas wear less with time than words.[43]

Coming to terms with the ephemeral nature of his life, the aging Cambridge mathematician took solace in the immutability of the mathematical truths he was leaving behind. When the actor playing Aninda declares in his opening monologue that "the mathematics is the only real thing here," he is paying tribute to this quality of Hardy's chosen art form, largely at the expense of his own. But theater has its particular gifts to offer, and the primary one is the humanizing force it brings to the subjects it engages. The permanence of mathematics makes it timeless, whereas the ephemeral nature of theater puts the focus squarely on the mortal present. Hardy no doubt imagined being remembered by future generations for his mathematical discoveries, as indeed he has, but each night that *A Disappearing Number* is performed, he and Ramanujan achieve a different kind of legacy that neither of them could have anticipated. The phones don't work and the doors don't lead anywhere, but while the houselights are down and the stage lights are up, Hardy and Ramanujan's unlikely friendship conveys a more human kind of truth.

Hardy staked his life's worth on the belief that his theorems had an intrinsic value—not because they were useful in any way, but because they possessed an undefinable beauty. In a world that prioritized utility and profit, Hardy cast his lot with the poets and playwrights. That Hardy's beliefs were so thoroughly embraced by the company of Complicité not only vindicates Hardy's decision to do so, but it suggests that Aninda—and the rest of us—should not be too quick to set mathematical reality above the realities explored by Hardy's fellow artists in the theater.

5

Stoppard: The Logic
of Self-Conscious Theater

Art which stays news, in Ezra Pound's phrase, is art in which the question 'what does it mean?' has no correct answer. Every narrative has, at least, a capacity to suggest a metanarrative.

—TOM STOPPARD, FROM "PRAGMATIC THEATER"

TAKING A cue from Aninda's direct address at the top of *A Disappearing Number*, the time has come to break my own fourth wall and insert myself into the story. The agenda for this chapter is a more personal one. In one sense this chapter is an awkward fit with the rest of the book because, with one brief exception, the plays discussed do not mention mathematics. And yet it is the reason the book exists. In the prologue, I recount my experience of reading Tom Stoppard's *Arcadia* for the first time and feeling a sense of excitement and disorientation akin to the one that accompanied my introduction to the incompleteness theorems of Kurt Gödel. "The familiar was suddenly full of mystery," I write, but that summary of the comparison between Stoppard and Gödel is just the beginning of a longer story that, truth be told, is still being written.

Sleuthing through the lengthy bibliography of Stoppard plays in search of insights about the genesis of *Arcadia*, I pretty quickly found the mathematical stepping stones the playwright set out in *Rosencrantz and Guildenstern are Dead*, *Jumpers*, and *Hapgood*. But as I was following this path, another aspect of Stoppard's writing kept vying for my attention. In script after script, even when there was no mathematics in sight, Stoppard's propensity for metatheatrical gamesmanship kept jumping off the page as though it were auditioning for a role in a book about mathematics and theater. "We're *actors*, we're the opposite of people," the Player announces to Rosencrantz and Guildenstern in a line vibrating with reflexive echoes. Bounced mercilessly around by the script of *Hamlet*, Rosencrantz and Guildenstern spend the entirety of Stoppard's play unenlightened, while the Player's status as the leader of the acting troupe performing for the king imbues him with a metaphysical status. Every line he delivers to Hamlet's two school friends comes beset with layers of self-referential implications.

At this early stage in his career, Stoppard isn't consciously trying to make any connection to mathematics, but this theatrical conceit and the many others like it kept activating my mathematical nerve centers. Self-reference is the common thread running through the story of the foundations of logic. It is at the heart of the Epimenides Paradox ("This sentence is false") as well as the paradoxical concept that Russell constructed in his query to Frege ("Does the extension of the concept 'is the extension of a concept that does not fall under itself' fall under itself?"). Self-reference is the beast that Russell and Whitehead tried to ban from the three massive volumes of *Principia Mathematica*, and although evidence suggests that they succeeded in preventing any contradictions from penetrating its thick walls, Gödel discovered a different kind of self-reference residing inside. Using PM's latent metamathematical potential, Gödel laid to rest Hilbert's dream of axiomatic perfection by proving that if PM is indeed consistent, then it necessarily contains undecidable statements. This is incompleteness—the unavoidable shortfall between the true statements that the system can express and those that can be derived as theorems. Gödel demonstrated that this was the predestined fate of PM and all "related systems," and at the heart of his proof was a self-referencing capability that no one suspected PM possessed, least of all its two creators.

This was Gödel's greatest insight—that formal systems could be employed to probe their own integrity—and it was this discovery, more than the specific content of Gödel's theorems, that kept suggesting a link to theater as I worked my way through Stoppard's early scripts. Plays manage to probe their own integrity all the time! Morbidetto in Witkacy's *Gyubal Wahazar* exclaims, "What a mad comedy all this is." Edgar's father in *The Water Hen* needles his son to become an actor now that "actors are creative artists too, ever since Pure Form became the rage." Gogo and Didi's music hall banter about their uneventful evening being "worse than the pantomime" bristles with self-deprecating implications. Referred to alternately as reflexive theater, metatheater and sometimes anthropomorphically as self-conscious theater, the device of a play drawing attention to its own nature spans dramatic genres, forms, and eras. From Aristophanes's early comedies to Shakespeare's *Hamlet* to Pirandello's *Six Characters in Search of an Author*, playwrights throughout history have experimented with the effect of importing metatheatrical elements into their storytelling.

My first epiphany then—and there are a total of three to report—is that my enchantment with theater has its roots in the ingenuity with which math and theater are capable of investigating their respective mediums. There are novels about the novel and paintings about painting, but the dexterity that mathematics and theater share in their introspective capabilities is at the core of why Stoppard's plays feel so mathematical to me, even when they are not about mathematics.[1] Whereas Stoppard's ubiquitous use of metatheater aligns him with other practitioners of this long-standing artistic device, the fact that he also regularly engages the content of mathematics makes him an ideal case study for exploring this curious bridge between theater and mathematics. What is the artistic allure of self-referential constructions? What new capabilities do reflexive structures engender, and why is the most mathematically adept playwright of this century or any other so inextricably drawn to them?

When I originally set off in search of answers to these questions, I inevitably wondered whether Stoppard would eventually find his way to Gödel's mathematics—and what might happen if he did. The answer came late in Stoppard's career, in a play called *The Hard Problem*. This is not a play about mathematics, and Gödel's Theorem gets only a brief cameo in the first act. Most audience members are not likely to pay it much attention, and if they do it is not self-evident what it is doing in the play. *The Hard Problem* is about cognitive science and the mystery of human identity. If I was a bit unsatisfied that Gödel was given such a small part when Stoppard finally cast him, I was not disappointed about where the journey led. What began as an exploration of the reflexive forces at play in theater and axiomatic systems landed in the domain of neuroscience where the agenda was deciphering the reflexive forces at play in the human mind. Every playwright who engages mathematical truths does so with the goal of accessing human truths. By taking on the hard problem of consciousness, Stoppard had set his sights on the most elusive human truth of them all.

Peano Arithmetic

To get an authentic sense of Gödel's methods, we need to go deeper into the symbolic language described in *Principia Mathematica*, but the fact of the matter is that Russell and Whitehead's approach is especially daunting, even to mathematicians. Taken together, its three volumes are some 1,900 pages of dense notation, and progress is supremely slow. Proposition 54.43, which occurs 360 pages in, concludes with the comment, "From this Proposition it will follow, when arithmetical addition has been defined, that $1 + 1 = 2$" (Figure 5.1). For the actual proof that $1 + 1 = 2$, one has to read on until section 102.

Fortunately, Gödel's 1931 paper addresses "*Principia Mathematica* and related systems," and so, with no loss of rigor, we can switch from PM to a more user-friendly symbolic language called Peano Arithmetic. Giuseppe Peano was a contemporary of Russell from Italy who was part of the larger collective effort to establish a proper axiomatic foundation for mathematics. Whereas the domain addressed by PM is more expansive, Peano Arithmetic, or PA for short, is a logical system dedicated exclusively to formalizing whole number arithmetic. This is all we need.

To test the premise that PA is moderately straightforward to parse, here are five PA formulas, each expressing a familiar statement about whole numbers. Just to be clear, the capital letters on the left margin enumerating the list are *not* part of the language of PA and are included for ease of reference:

A: $(S0 + S0) = SS0$
B: $(S0 = 0 \lor \sim S0 = 0)$
C: $\forall a \, (a + 0) = a$
D: $\forall a \, \forall b \sim SSSSS0 = (SSa \cdot SSb)$
E: $\sim \exists a \, \exists b \, (a \cdot a) = (SS0 \cdot (Sb \cdot Sb))$

Before providing translations for each formula, let's remind ourselves what we currently know about formal systems. Like the system for propositional reasoning discussed earlier, PA is a purely formal language. It consists of a finite

Rosencrantz and Guildenstern Are Dead; David Leveaux (director), Joshua McGuire (Guildenstern), Daniel Radcliffe (Rosencrantz), David Haig (The Player); Old Vic Theatre, 2017. © Geraint Lewis / ArenaPAL.

Hapgood; Howard Davies (director), Tim McMullan (Blair), Lisa Dillon (Hapgood), Gerald Kyd (Ridley); Hampstead Theatre, London, 2015. © Donald Cooper / photostage.co.uk.

Arcadia; Jessie Cave (Thomasina), Dan Stevens (Septimus); Duke of York's Theatre, London, 2009. © Mark Ellidge / ArenaPAL.

Ubu Roi; Vincent de Bouard (Le Roi Wenceslas), Camille Cayol (Mère Ubu), Christophe Gregoire (Père Ubu), Xavier Boiffier (Bordure), Declan Donnellan (director); Cheek by Jowl, London, 2013. © Johan Persson / ArenaPAL.

The Water Hen; Mariusz Bonaszewski (Edgar), Beata Scibakowna (The Water Hen); Teatr Narodowy, Warsaw, 2002. © Michal Sadowski / Forum / ArenaPAL.

Endgame; Matthew Warchus (director), Michael Gambon (Hamm), Lee Evans (Clov); Albery Theatre, 2004. © Geraint Lewis / ArenaPAL.

Come and Go; Annie Ryan (director), Barbara Brennan (Ru), Susan Fitzgerald (Vi), Bernadette McKenna (Flo); Beckett Centenary Festival, Barbican Theatre, 2006. Stephanie Berger ©️ 2022.

Quad; Michael Hackett (director); Hammer Museum/UCLA Department of Theater, 2020 (photo: Josh Concepcion).

Copenhagen; Michael Blakemore (director), Charles Edwards (Heisenberg), Paul Jesson (Bohr), Patricia Hodge (Margrethe); Chichester Festival Theatre, 2018. © Conrad Blakemore /ArenaPAL.

A Disappearing Number; Simon McBurney (director), Shane Shambhu (Ramanujan), David Annen (G. H. Hardy), Saskia Reeves (Ruth), Divya Kasturi (dancer); Barbican Theatre, 2008. Stephanie Berger © 2022.

The Hard Problem; Nicholas Hytner (director), Parth Thakerar (Amal), Vera Chok (Vera), Lucy Robinson (Ursula), Rosie Hilal (Julia), Olivia Vinall (Hilary), Damien Molony (Spike); Royal National Theatre, 2015. © Geraint Lewis / ArenaPAL.

The Invention of Love; Richard Eyre (director), Michael Bryant (Charon),
John Wood (Housman); Royal National Theatre, London, 1997.
© Donald Cooper / photostage.co.uk.

Breaking the Code; Joe Calarco (director), Mark H. Dold (Turing), Phillip Kerr
(Knox), Annie Meisels (Pat); Barrington Stage Company,
2014 (photo: Kevin Sprague).

Lovesong of the Electric Bear; Cheryl Faraone (director), Alex Draper (Turing),
Tara Giordano (Porgy), Ben Schiffer (Turing Sr.), Cassidy Boyd (Joan);
Potomac Theatre Project NYC, 2010 (photo: Stan Barouh).

Proof; Daniel Sullivan (director), Mary-Louise Parker (Catherine); Walter Kerr
Theater, New York, 2000 (photo: Joan Marcus).

362   PROLEGOMENA TO CARDINAL ARITHMETIC   [PART II

*54·42. $\vdash :: \alpha \epsilon 2 . \supset :. \beta \subset \alpha . \exists ! \beta . \beta \neq \alpha . \equiv . \beta \epsilon \iota'' \alpha$

Dem.

$\vdash . *54 \cdot 4 .\quad \supset \vdash :: \alpha = \iota'x \cup \iota'y . \supset :.$

$\qquad \beta \subset \alpha . \exists ! \beta . \equiv : \beta = \Lambda . v . \beta = \iota'x . v . \beta = \iota'y . v . \beta = \alpha : \exists ! \beta :$

[*24·53·56.*51·161] $\equiv : \beta = \iota'x . v . \beta = \iota'y . v . \beta = \alpha$ (1)

$\vdash . *54 \cdot 25 . \text{Transp.} . *52 \cdot 22 . \supset \vdash : x \neq y . \supset . \iota'x \cup \iota'y \neq \iota'x . \iota'x \cup \iota'y \neq \iota'y :$

[*13·12] $\supset \vdash : \alpha = \iota'x \cup \iota'y . x \neq y . \supset . \alpha \neq \iota'x . \alpha \neq \iota'y$ (2)

$\vdash . (1) . (2) . \supset \vdash :: \alpha = \iota'x \cup \iota'y . x \neq y . \supset :.$

$\qquad\qquad \beta \subset \alpha . \exists ! \beta . \beta \neq \alpha . \equiv : \beta = \iota'x . v . \beta = \iota'y :$

[*51·235] $\equiv : (\exists z) . z \epsilon \alpha . \beta = \iota'z :$

[*37·6] $\equiv : \beta \epsilon \iota'' \alpha$ (3)

$\vdash . (3) . *11 \cdot 11 \cdot 35 . *54 \cdot 101 . \supset \vdash . \text{Prop}$

*54·43. $\vdash :. \alpha , \beta \epsilon 1 . \supset : \alpha \cap \beta = \Lambda . \equiv . \alpha \cup \beta \epsilon 2$

Dem.

$\vdash . *54 \cdot 26 . \supset \vdash :. \alpha = \iota'x . \beta = \iota'y . \supset : \alpha \cup \beta \epsilon 2 . \equiv . x \neq y .$

[*51·231] $\equiv . \iota'x \cap \iota'y = \Lambda .$

[*13·12] $\equiv . \alpha \cap \beta = \Lambda$ (1)

$\vdash . (1) . *11 \cdot 11 \cdot 35 . \supset$

$\qquad \vdash :. (\exists x, y) . \alpha = \iota'x . \beta = \iota'y . \supset : \alpha \cup \beta \epsilon 2 . \equiv . \alpha \cap \beta = \Lambda$ (2)

$\vdash . (2) . *11 \cdot 54 . *52 \cdot 1 . \supset \vdash . \text{Prop}$

From this proposition it will follow, when arithmetical addition has been defined, that $1 + 1 = 2$.

FIGURE 5.1. Proposition 54.43 from page 362 of the first volume of *Principia Mathematica*.

alphabet of symbols—sixteen in this case—that are combined into formulas like the ones above according to strict formation rules. The analogy we used before was to equate each formula with an arrangement of chess pieces on a board. The rules are rigid and unambiguous, so there is no debate about what arrangements are allowed. The crucial difference between PA and chess is that PA's formation rules are crafted to induce an *external* meaning. Specifically, they are designed so that the resulting formulas track with a part of the mathematical universe—the world of whole numbers in this case—and when we recognize this mirroring between the symbols and the mathematics then we instinctively attach meaning to the formulas.

In the case of PA, we have a head start in the deciphering process because some of the alphabet is already ingrained in our thinking. The symbols $+$, $\cdot$, and $=$ have their usual meaning of addition, multiplication, and equality. The symbol 0 thankfully represents zero. For the numbers 1, 2, 3, 4, . . . PA uses

$$S0, SS0, SSS0, SSSS0, ...$$

The intuition here is to connect S with the idea of adding 1, interpreting S more literally to mean "the successor of." Thus, $S0$ is the successor of 0 which is 1, while $SS0$ is $1 + 1$ or 2. The variables a and b in PA represent whole numbers, so expressions like Sa and SSb can be translated as their more familiar algebraic incarnations $a + 1$ and $b + 2$. Having two variables is enough for our purposes, but PA allows us to add primes (i.e., a', b', a'') if it turns out we need more than just a and b.

PA, like PM, incorporates the mechanical rules of propositional reasoning discussed in chapter 3. Thus $\wedge$ corresponds to "and," $\vee$ corresponds to "or," and $\sim$ to a negation. The only other new symbols are the two quantifiers that appear in front of the variables: $\forall$ translates as "for all" or "for every choice of"; $\exists$ is used to mean "there exists" or "for at least one choice of." Even with this PA–to–math dictionary there is still some work required to interpret each formula. Here are the promised translations:

A: $1 + 1 = 2$
B: $1 = 0$ or $1 \neq 0$
C: for every whole number a, $a + 0 = a$
D: 5 is a prime number
E: $\sqrt{2}$ is an irrational number

For formula D, a more literal translation would be "For every a and b, $5 \neq (a + 2)(b + 2)$." By adding 2 to each variable, the effect is to assert that the only way to factor 5 is via $5 \cdot 1$ or $1 \cdot 5$, which is precisely what it means to say 5 is prime. A similar trick is at work in formula E. A literal translation of E would be, "It is not the case that there exist a and b satisfying $a^2 = 2(b + 1)^2$." This is equivalent to asserting, "It is not the case that $\sqrt{2} = a/(b + 1)$, no matter how a and b are chosen." Adding 1 to b prevents the trivial possibility of having $0 = 2 \cdot 0$ in the initial equation.

Transformation Rules and Theorems

As a final bit of review, let's remind ourselves about the crucial distinction between a *formula* and a *theorem*. The theorems of PA are those formulas that can be created using the transformation rules of the system. We could also say "derived" or even "proved," which are more suggestive terms but also a bit misleading because proofs in formal systems are a purely mechanized procedure. PA, like all formal systems, comes with a list of formulas designated as *axioms*, which are free theorems in a sense. These are the starting points from which the other theorems are constructed. In principle, axioms are meant to represent fundamental truths whose validity does not need to be proved but is accepted as an intrinsic quality of the subject in question. On the list above, formula C is an axiom of PA, one that articulates a defining aspect of the number zero.

Starting from the axioms, the transformation rules of PA are purely syntax-based instructions for how to generate a new formula from an existing one. To get a sense of some of these rules, here is a complete derivation in PA of the theorem $(S0 + S0) = SS0$. (If these strings of symbols look daunting, remember that it took Russell and Whitehead several hundred pages to generate this result. The system of PA gets us there in seven lines!)

$\forall a\,(a + 0) = a$	axiom
$(S0 + 0) = S0$	Rule of Specification, replace a with $S0$
$S(S0 + 0) = SS0$	Rule of Successorship, applied to line 2
$\forall a\,\forall b\,(a + Sb) = S(a + b)$	axiom
$\forall b\,(S0 + Sb) = S(S0 + b)$	Rule of Specification, replace a with $S0$
$(S0 + S0) = S(S0 + 0)$	Rule of Specification, replace b with 0
$(S0 + S0) = SS0$	Rule of Transitivity, using lines 3 and 6

To tamp down any rising trepidation, let's remind ourselves that our agenda is not to understand the details but simply to appreciate the deterministic nature of how PA functions. A proof in PA is no more and no less than a list of formulas like the one above. The first formula on the list should be an axiom of the system, and the last formula on the list is the theorem to be proved. At each intermediate step, the new formula that follows must be either an axiom or a formula that can be created from earlier formulas in the list by applying one of the transformation rules. In this austere environment, there are no debates about what constitutes valid reasoning any more than there are debates about what constitutes a valid move in chess. Rules are rules, and it is straightforward to verify whether each line of the proof is properly justified.

The commentary down the right side is not part of PA's formal notation. It is included to indicate the particular justification of each step. Note that some steps are justified as axioms, others as transformation rules. For example, the Rule of Specification is employed a number of times. It asserts that we are allowed to take an existing theorem that begins with the pair of symbols

∀a and create a new theorem by removing those two symbols and replacing all remaining occurrences of a with a numeral like $S0$. This is illustrated in lines 2 and 5. As line 6 indicates, the Rule of Specification can also be applied with other variables in place of a.

Again, understanding the specific details of each rule is not so important; what matters is appreciating their mechanical nature. The transformation rules are phrased purely in terms of the formula's syntax with no regard for the meaning that any of the symbols might conjure up. As another example, the Rule of Successorship allows us to take a formula like the one in line 2 that has the form $x = y$ and append an S to the front of each side to produce the formula $Sx = Sy$ displayed in line 3.

But of course, these mechanical rules are crafted with specific semantic intent. The Rule of Successorship captures the elementary notion that if two numbers are equal, then adding one to each side preserves the equality. The Rule of Specification embodies the fact that the symbols ∀a are employed to mean "for all choices of a." If what follows ∀a is true for *every* value of a, then it should certainly be true for a *particular* choice of a. Taking a step back from the symbol-manipulating view of the above proof and adopting a semantic mindset that recognizes how the rules of PA are designed to model whole number arithmetic, the syntactical dance of symbols starts to tell a straightforward story:

A self-evident truth about 0 is that $a + 0 = a$ no matter what a is so, specifying a as 1 tells us that $1 + 0 = 1$ must be true; then, adding 1 to each side gives $(1 + 0) + 1 = 2$.
A self-evident truth about addition is that $a + (b + 1) = (a + b) + 1$ so, specifying a as 1 says $1 + (b + 1) = (1 + b) + 1$, and specifying b as 0 then yields $1 + 1 = (1 + 0) + 1$.
Combining lines 3 and 6, the transitive nature of equality implies $1 + 1 = 2$.

What looks like a mechanical shuffling of symbols according to deterministic rules from one point of view becomes richly meaningful from another. And the derivable formulas PA generates, in this case $(S0 + S0) = SS0$, yield true statements when this meaning gets applied.

This example gets to the heart of how formal systems should function. The guiding question for PA and systems like it is whether the axioms and rules can be formulated so that the resulting collection of derivable formulas—i.e., the theorems of the system—march in lockstep with the true statements that its formulas can articulate. We want theorems to be true statements, and we want true statements to be theorems. But Gödel showed that this was a bridge too far for formalized mathematics. Even in a domain as well-defined as the arithmetic of whole numbers, Gödel proved that it is impossible to establish a single system of axioms and transformation rules from which one can derive all the truths that lie within its descriptive reach.

Operating on Two Levels

In mathematics, the term "metalanguage" refers to statements *about* the formal system as opposed to statements *of* the system. Using PA as an example, the formula

$$(S0 + S0) = SS0$$

has the interpretation "1+1=2," which is a property of whole numbers. This falls squarely in the domain of the language of PA. By contrast, the statement

The formula $(S0 + S0) = SS0$ can be derived in PA

is about the strength of PA and is thus a part of the metalanguage of PA. Note especially that this latter statement is *not* a proposition about the nature of whole numbers. The crux of Gödel's argument relies on the ingenious recognition that a formal system like PA can be shown to contain significant parts of its own metalanguage. Formulas in PA have an intended interpretation as a statement about some property of whole numbers, but Gödel showed that *the same formula could also represent a coded statement about the formal system itself.* This is absolutely unexpected and understandably confusing. How could this be? We have encountered the entire range of symbols for PA, and every formula constructed thus far has asserted some proposition about numbers— some of them basic, some more complicated, but none of them hinting in any way how they might contain an alternate interpretation as a statement about the formal system in which it resides.

Exploiting this unexpected potential of formal systems, Gödel demonstrated incompleteness by constructing a formula—traditionally referred to as G—with the interpretation:

G, that is *this particular formula*, has no proof in PA.

We saw at the end of chapter 3 how the existence of G leads to incompleteness, and we'll go through the logic again later, but it is in trying to construct G that we come face-to-face with the central dilemma. Specifically, how can a formula of PA acquire the self-reflective interpretation, "This formula doesn't have a proof"? Formulas of PA express statements about whole numbers— statements like "$1 + 1 = 2$" or "5 is prime." By what means, then, was Gödel able to craft a PA formula that asserted a fact about PA? Even more remarkably, G makes an assertion about its own provability. Formulas of PA are about properties of numbers, but G manages to talk quite specifically about itself.

I learned the answer to these questions several decades ago as a graduate student, but Gödel's ingenuity still fills me with astonishment each time I revisit his proofs. The singular nature of his insights has only become more remarkable over time as I have come to appreciate them in a deeper way. This is why it was so arresting to encounter the various components of Gödel's argument running through Tom Stoppard's collected plays.

The Real Inspector Hound and Other Plays

Does this play know where it is going?

—MOON, FROM *THE REAL INSPECTOR HOUND*

Squaring the Circle is a television docudrama about Lech Walesa and the Polish Solidarity movement in 1980–1981. When the show begins, the Narrator's voice introduces us to Leonid Brezhnev and Edward Gierek, the Communist leader of Poland, discussing their concerns on a beach by the Black Sea. For a few moments, the soundtrack picks up the conversation between Brezhnev and Gierek, but then abruptly, the scene is interrupted by the Narrator, who is standing on the same beach and addressing the camera: "That isn't them, of course—and this isn't the Black Sea. Everything is true except the words and the pictures." This variation on the Epimenides Paradox is more comic than logically confusing, and it aptly sets up the theme of the piece which, as the Narrator describes it, is the unresolvable contradiction between the "idea of freedom as it is understood in the West, and the idea of socialism as it is understood in the Soviet empire."[2]

Squaring the Circle's intrusive Narrator is emblematic of Stoppard's meta-theatrical tendencies. The standard framework that usually emerges in a narrated story is a properly ordered hierarchy where the narrator exists on a distinct and higher level, entering at convenient moments to comment on the action one level down. *Squaring the Circle* toys with this convention in the opening scene, and the hierarchy gets more twisted when the Narrator begins engaging in debates with a character who appears in numerous supporting roles throughout the documentary. Early on, these debates tend to be about the veracity of the Narrator's version of events, and they often result in scenes being replayed. This idea is pushed to the point where, in one scene, the picture freezes and then tears itself in half to reveal the Narrator—who is also, apparently, the author—scribbling and then tossing the botched scene into the garbage.

This is one way to create a voice inside a play that is simultaneously capable of commenting on the play in which it resides, but it is far from Stoppard's most innovative solution to this challenge.

Confronted by Their Own Reflection

Stoppard wrote *Squaring the Circle* in 1984, nearly two decades after *Rosencrantz and Guildenstern Are Dead* had become an international phenomenon. The reflexive voices that run through *R&G* are due to Shakespeare's generous gift of the troupe of traveling actors tasked with performing a play that Stoppard modifies into a full-blown copy of the original script of *Hamlet*. At the core of Russell's paradox (discussed in chapter 3) is the notion of an *exotic* concept; i.e., a concept with an extension that effectively contains itself. In this sense, *R&G* is an example of an exotic play by virtue of the fact that it contains a copy of itself. While he was writing *R&G*, Stoppard was working on another script that is even more exotic in this Russellian sense. Written in fits and starts and finally performed in 1968, *The Real Inspector Hound* makes no attempt to incorporate any of the philosophical overtones that add to the psychological weight of *R&G*. "As you all know this is a deeply existential comedy," the tongue-in-cheek Stoppard said to the cast of *The Real Inspector Hound* at a revival that he directed some years later. "But what I shall chiefly be preoccupied about in rehearsals is getting as many laughs as possible."[3]

In a direct challenge to set designers, Stoppard's stage directions declare that upon entering the theater,"the audience appear to be confronted by their own reflection in a huge mirror." (5)[4] In the front row of this reflected audience sit two theater critics named Moon and Birdboot. Before them, and the audience, is the set of an Agatha Christie style whodunit, complete with a dead body and a rolling couch that hides the corpse at key moments during the upcoming show.

Moon is the frustrated second-string critic at his paper, so deeply obsessed with replacing his paper's lead reviewer that he largely ignores the play in front of him and waxes off about "a bloody coup d'état of the second rank ... stand-ins of the world stand up!" Birdboot, meanwhile, is entertainingly pretentious with an overblown sense of his critical influence in the theater world. Having been spotted by Moon the evening before with an up-and-coming actress, Birdboot is instantly outraged at the suggestion—which is never actually offered—that he is anything other than a "family man devoted to my homely but good-natured wife."

Stoppard has some preliminary fun with the nested structure when Moon favorably reviews one of Birdboot's favorable reviews, but the farce begins in earnest when the curtain rises on the play-within-a-play. The theatrical parodies are entertaining, but it is the self-referencing paradoxes between the critics and the play they are supposedly reviewing that creates the intrigue. Moon's obsession to become the lead critic at his paper starts to mingle dangerously with the murder plots of the Christie pastiche. Moon's musings are interrupted, though, when he recognizes the actress playing the jilted love interest as the woman he saw out with Birdboot. By this point, Birdboot's infatuations have moved on to another member of the cast, again mirroring the plot of the inner play. When the two critics finally pull themselves together enough to offer some criticism, it is not at all clear which play they are discussing:

> MOON: I am bound to ask—does this play know where it is going?...
> There are moments, and I would not begrudge it this, when the play, if we can call it that, and I think on balance we can, aligns itself uncompromisingly on the side of life. *Je suis* it seems to be saying, *ergo sum*. (24)

When Birdboot declares in his review that "it is at this point that the play for me comes alive," the stage-prop telephone rings, and it is Birdboot's homely but good-natured wife calling to confront him about his indiscretions. Before he can get off the stage, Birdboot is swept into the action where Stoppard has contrived events so that he miraculously fits—and, tragically, meets his demise. Rushing on stage to his partner's aid, Moon is swept into the role of Inspector Hound. The seemingly untenable loose ends this creates are magically tied up when the dead body under the couch turns out to be the lead reviewer at Moon's paper, an idea that Stoppard insists occurred to him very late in the writing process.

Speaking as a member of the audience and also as a mathematician, there is something more satisfying about the self-referential devices at work in *The Real Inspector Hound* as compared to those in *Squaring the Circle*. When

"Does this play know where it is going?" *The Real Inspector Hound*; Teddy Green
(Moon), Terence Frisby (Birdboot); The Young Vic, 1977.
© Donald Cooper/photostage.co.uk.

Moon talks about "this play," as in, "Does this play know where it is going?,"
from his point of view he is referring to the murder mystery that he is watching. But we, the audience, hear an additional metatheatrical voice referring to
the play that Moon is in. What makes this second voice so audible is the mirroring that exists between the two plays. As the evening progresses, the inner
play first reflects, and then eventually engulfs, the outer play so that *Hound*
ultimately contains a copy of itself with the characters rotated into different
roles.

Formal systems of mathematics were initially designed as a defense against
the paradoxes of self-reference, but the lesson of *Hound* is that self-reference
can be achieved in an unsuspecting way—smuggled in by constructing a *coded*
copy of the outer play inside the original.

Artist Descending a Staircase

When Stoppard wasn't writing plays about plays, as he was with *Rosencrantz
and Guildenstern Are Dead* and *The Real Inspector Hound*, he was frequently
writing plays about art and artists. A regular companion piece to *Hound* is the
one-act comedy *After Magritte*, written in 1970. René Magritte was a surrealist
painter who, like Stoppard, was deeply engaged in simultaneously exploring

FIGURE 5.2. *The Human Condition*, by René Magritte. © 2022
C. Herscovici/Artists Rights Society (ARS), New York.

and commenting on the way his chosen artistic medium communicated with
its viewers. Best known for his various images of a smoking pipe floating above
the caption "Ceci n'est pas une pipe" (This is not a pipe), Magritte frequently
included paintings within his paintings. These inner paintings are typically
mounted on easels depicting exactly what one would expect to see behind it if
the canvas were not there (Figure 5.2). Within the world of Magritte's painting,

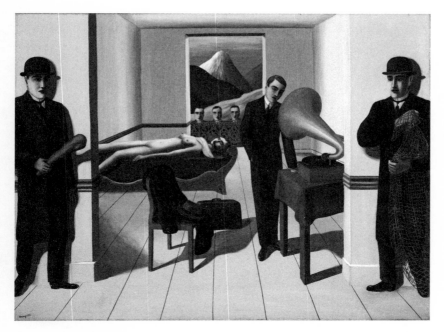

FIGURE 5.3. *The Menaced Assassin*, by René Magritte, was the inspiration for Stoppard's play *After Magritte*. © 2022 C. Herscovici/Artists Rights Society (ARS), New York.

the artistic rendering of reality is indistinguishable from the real thing. *After Magritte* takes as its starting point Magritte's 1926 painting *The Menaced Assassin*, which portrays an incongruous murder scene that includes three eerily detached witnesses looking in through a back window (Figure 5.3). Stoppard's comedy is less ominous than this particular painting, but it very much plays the same game of calling attention to the human propensity to conflate the physical world with the mental representation of it that we construct in our minds.

Even when he was writing about painters rather than playwrights, Stoppard routinely marshaled a range of metatheatrical forces to explore the potency of art in general, including his own. *Travesties* is the most well-known example. A full-length play produced in 1974 and frequently revived, *Travesties* features Irish modernist James Joyce among its major characters along with Tristan Tzara, one of the central figures of the antiestablishment Dada movement. But the play where Stoppard's latent mathematical tendencies are more apparent is a radio play he wrote a few years earlier called *Artist Descending a Staircase*. The

Scene 1 = here-and-now (set in the mid 1970s)
Scene 2 = a couple of hours ago
Scene 3 = last week
Scene 4 = 1922
Scene 5 = 1920
Scene 6 = 1914
Scene 5 continued
Scene 4 continued
Scene 3 continued
Scene 2 continued
Scene 1 continued

FIGURE 5.4. The stair-stepped scene structure of *Artist Descending a Staircase*.

title comes from the Marcel Duchamp painting, *Nude Descending a Staircase*, and is significant for a host of reasons beyond conjuring up an iconic image associated with Dada.

One of these is that it points to the chronological structure of the eleven scenes that make up the play. The first scene opens in the present, and the next five scenes are successively stair-stepped flashbacks, each one taking place earlier in time. The concluding five scenes then step forward in time, continuing the action interrupted by each flashback (Figure 5.4).

The title is also apt because the play begins with an elderly character named Donner falling down a set of stairs to his death, where he is discovered by his friends Beauchamp and Martello. The play travels backward and then forward through the adult lives of the three lifelong companions—Beauchamp, Martello, and Donner—all artists and all regularly debating the nature of art. To create an implicit parallel between his chosen medium of the radio play with that of one of his protagonists, Stoppard makes Beauchamp's artistic medium recording odd and incongruous sounds. "You select your public. It is the same with me," Beauchamp says in response to Donner's cynical assessment of one of his tapes that has just been played. "If you played my tape on the radio, it would seem a meaningless noise, because it fulfills no expectations." And this is very much the case.

There is one more significant way in which Stoppard exploits the radio format of this play, and that is by having the last character in this comic-drama be blind. When Sophie enters the apartment of the three young artists in scene 5, she recognizes the unmistakable sound of a ping-pong game—we all do—but several minutes later, after the match becomes a fully formed part of the scenery to us, Martello explains to Sophie that it is only one of Beauchamp's recordings:

> SOPHIE: It's very effective. I could have kept score just by listening.
> BEAUCHAMP: Yes!—you see—sorry!—I'm trying to liberate the visual *image* from the limitations of visual *art*. The idea is to create images—pictures—which are purely *mental*. . . I think I'm the first artist to work in this field. (137)[5]

Amid the laugh lines in the play, Stoppard vents some personal frustrations around contemporary definitions of art. "An artistic imagination coupled with skill is talent," Donner tells Beauchamp. "Skill without imagination is craftsmanship and gives us many useful objects such as wickerwork picnic baskets. Imagination without skill gives us modern art." (122) Beauchamp is impervious to Donner's criticism and can be relied upon to counter with a provocative-sounding platitude ("Art should break its promises") that, like Beauchamp's art, is largely devoid of any real significance. Highly refined naturalistic art is also challenged as being pointless and deceitful. "The greater the success, the more false the result," is how Martello summarizes it to Sophie in the middle of the play. Martello then offers one of the more lasting pieces of insight: "It is only when the imagination is dragged away from what the eye sees that a picture becomes interesting." (140)

Sophie's eyes no longer see anything—and neither do ours in a radio play—so imagination is all there is. Variations on the ping-pong trick are played several more times, calling attention each time to the simple truth that the entire drama is taking place as a mental construction generated from a set of sounds. Our inability to know if a whistling kettle is real or one of Beauchamp's soundscapes connects us to Sophie, so that when the comedy gives way to a tragic love story our imagination is dragged away from what we hear to a place of tangible empathy. A side effect of this experience is the recognition of the capabilities of the art form itself. By manifesting Martello's definition of art, *Artist Descending a Staircase* manages to gracefully sidestep the blistering criticisms that it launches.

Recreating Its Own Genus

Toward the end of *Rosencrantz and Guildenstern Are Dead*, the coolheaded Guildenstern finally snaps when he realizes his death is imminent. The Player consoles Guildenstern, offering that "In our experience, most things end in death," to which the enraged Guildenstern responds by stealing the Player's dagger and holding it to the actor's throat. "I'm talking about death—" Guildenstern hisses at him, "and you've never experienced *that*. And you cannot *act* it." When Guildenstern violently thrusts the blade into the Player's gut there is a rapt, breathless moment as the stunned Player registers the terror of what has just happened, gasps, and falls lifelessly to the floor. The audience feels head-on the reality of death while a hysterical Guildenstern wheels on the Player's troupe and barks out his own defense. "If we have a destiny, then so had he—and if this is ours, then that was his—and if there are no explanations for us, then let there be none for him—."

And then comes the ambush as the Player rises to his feet to the reverent applause of his fellow tragedians. "You see, it is the kind they do believe in," he tells a bewildered Guildenstern, revealing the retractable blade on his stage-prop knife. "It's what is expected." (123)

The Player's death in *R&G* is like Beauchamp's ping-pong match in *Artist* which is like the paintings on the easels inside Magritte's paintings. Within each medium, there is no way to distinguish the artistic rendition from the thing it is meant to represent. They are one and the same. Stage plays, radio plays, and paintings can recreate their own genus with impeccable authenticity, and in so doing they acquire the potential to forcefully demonstrate—and potentially augment—their respective artistic power. This brings us back to Gödel's proof. A bit like paintings, and very much like plays, it turns out that a formal system like PA has this same potential to recreate its own genus inside its expressed domain of whole number arithmetic. The Gödel formula G inside PA that manages to assert its unprovability acquires its metamathematical potential in the same way that Stoppard's characters do—from an embedded version of the medium in question (a play or a formal system) inside the expressive domain of the original. But how exactly? Even with this generous hint and a handful of reflexive models from Stoppard's list of early plays, there is still the mystery of how to embed a copy of a formal system like PA into the domain of number theory.

As if on cue, Stoppard wrote a play about this too.

"I am talking about death—and you've never experienced *that*." *Rosencrantz and Guildenstern Are Dead*; Jamie Parker (Guildenstern), Samuel Barnett (Rosencrantz), Chris Andrew Mellon (The Player); Theatre Royal Haymarket, 2011. © Donald Cooper/photostage.co.uk.

Dogg's Hamlet and Gödel's Code

Hamlet bedsocks Denmark, yeti William Shakespeare.

—HEADMASTER DOGG, FROM *DOGG'S HAMLET*

The opening scene of *Dogg's Hamlet* inspires a confusion similar to what one feels encountering strings of PA notation for the first time. Faced with an empty stage, we hear a schoolboy yell, "Brick!" A soccer ball arcs across from left to right. "Cube," he yells again. Abel, the boy who apparently threw the ball, then walks to the front of the stage to assemble a microphone and stand:

> ABEL: (*into microphone*) Breakfast, breakfast ... sun—dock—trog ... (*He realizes the microphone is dead. He tries the switch a couple of times and then speaks again into the microphone.*) Sun—dock—trog—pan—slack ... (*It is still dead. ABEL calls to someone off-stage.*) Haddock priest!
> BAKER: Eh?
> ABEL: Haddock priest.

BAKER: Haddock?
ABEL: Priest. (147)[6]

The familiar words have lost their familiar meanings, but this new language seems to be following some rules nonetheless. Soon after, a third boy named Charlie comes on wearing a dress over his school uniform and calling a bit too eagerly for the ball:

> CHARLIE: Brick! . . . brick! (*A ball is thrown to him from the wings. ABEL dispossesses CHARLIE of the ball.*)
> ABEL: Cube!
> VOICE: (*Off-stage*) Brick! (*Charlie tries to get the ball but ABEL won't let him have it.*)
> CHARLIE: Squire! (*ABEL throws the ball to the unseen person in the wings*) Daisy squire!
> ABEL: Afternoons!
> CHARLIE: (*Very aggrieved.*) Vanilla squire! (148)

As is often the case when learning a new language, the first things we ferret out are its profanities. The next thing we learn about our schoolboys is that they are currently engaged in rehearsing the school play—*Hamlet*, as it turns out, to be performed in the original English. In the middle of their lunchtime banter, Abel starts an impromptu run-through of their lines:

> ABEL: (*Suddenly*) Who's there?
> BAKER: Nay, answer me.
> ABEL: Long live the King. Get thee to bed.
> BAKER: For this relief much thanks. (*ABEL stands up.*)
> ABEL: What, has this thing appeared again tonight? (*BAKER stands by him.*)
> BAKER: Peace, break thee off: look where it comes again.
> ABEL: Looks it not like the King? (150)

The English is jarring, in part because it's the first we've heard in this play but more so because it is delivered without any inflection or understanding on the part of the two adolescent actors. The Dogg language we've been hearing up until this point is rich with purpose and meaning, although most of it still eludes us, while the Shakespearean poetry is delivered as a sequence of sterilized syllables.[7]

Fifteen minutes in, the audience is given a companion in the task of translating in and out of Dogg with the arrival of Easy, an unsuspecting truck driver who speaks only English. Like us, Easy is caught off guard by the strange world he has entered, having only at the last minute been assigned the job of delivering the truck full of construction materials that Headmaster Dogg has ordered for the set of the school play.

Planks, Slabs, Blocks, and Bricks

The idea for *Dogg's Hamlet* comes from a section of Ludwig Wittgenstein's philosophical writing. Wittgenstein was Gödel's contemporary, and, like Gödel, his early work was heavily influenced by the theories on logic of Russell, who was Wittgenstein's mentor during his first stint at Cambridge in 1911.

Dogg's Hamlet is a response to a specific thought experiment proposed by Wittgenstein in his posthumously published *Philosophical Investigations*. At issue is the way language obtains meaning, and in particular whether having a particular word represent a particular object or idea is a viable starting point for a model of how language functions. The words that Stoppard starts with— "plank," "block," "slab," "brick," and "cube"—are only a slight modification of the ones that Wittgenstein proposes in his "builders' language-game," and it is no coincidence that each corresponds to an item in Easy's truck of set-building materials. The point of the experiment is that, to someone unfamiliar with English, these words might at least hypothetically correspond to a completely different set of ideas. In Stoppard's Dogg language, the translation goes this way:

Dogg		*English*
plank	⟷	ready
slab	⟷	okay
block	⟷	next
cube	⟷	thank you
brick	⟷	here
git	⟷	sir

Wittgenstein's question, transposed into this context, is whether there might arise a situation where someone speaking Dogg and another speaking English could engage in a fruitful discourse with each person assuming, incorrectly, that they were being understood by the other. Something very much like this occurs when the group finally gets down to building the set for the upcoming play.

> DOGG: Plank?
> BAKER/CHARLIE: (*Positioned next to the truck*) Plank, git?
> DOGG: (*Calling to ABEL who is in the truck bed*)) Plank?
> ABEL: (*Off-stage*) Plank, git.
> DOGG: (*Calling to ABEL loudly—shouts.*) Plank!
> (*To EASY's surprise and relief a plank is thrown to BAKER who catches it, passes it to CHARLIE, who passes it to EASY, who places it on the stage. DOGG smiles, looks encouragingly at EASY.*)
> EASY: (*Uncertainly, calls.*) Plank! (153)

Easy continues calling for "planks" and "blocks" with some success and plenty of confusion. There are several other stretches where the separate meanings coincide to produce a false sense of communication. There are also plenty of instances where Stoppard has the two meanings cross at right angles. "Cretinous pig-faced, git?" translates as "Have you got the time, sir?" while "Afternoon squire," comes across as "Get stuffed you bastard" to speakers of Dogg.

Wittgenstein's intent with this experiment was to show that meaning is not contained in the words themselves but is derived from the patterns that exist between how the words are used and the physical world. When Easy says "plank," he might be referring to an object in his truck or he might be demanding the attention of the schoolboys—or both. The Dogg meaning is not the familiar one, but once we have this alternate interpretation pointed out to us, we have to accept that the possibility was there all along.

To apply this thinking to Gödel's argument, we move from English words to the symbols of PA. Just as "afternoon squire" has an intended English interpretation as a midday greeting one might say to a young knight in training, a string of PA symbols has an intended meaning as a statement about whole numbers. This is why PA was built. From our brief introduction to PA, we recognize that $(S0 + S0) = SS0$ has the meaningful and true interpretation "1+1=2," while $S0 = 0$ asserts the false statement "1=0." But could there be an encoded Dogg interpretation whereby formulas like these have a second meaning?

There can if we apply a version of Stoppard's Dogg language mapping to the symbols of PA.

"Cube!" *Dogg's Hamlet*; Bob Goody (Easy), Watford Palace Theatre, 1993.
© Donald Cooper/photostage.co.uk.

Gödel-Numbering

The details get predictably technical, but the central idea of Gödel-numbering is not hard to appreciate—at least no harder than appreciating a performance of *Dogg's Hamlet*. The first step is to associate a distinct whole number to each symbol of PA. This number is referred to as the symbol's *Gödel number*. Because there are more than ten total symbols in PA, we will save ourselves some trouble if every Gödel number has precisely two digits. For instance, let's make the following assignments:

PA		GN	PA		GN
0	⟷	10	S	⟷	20
=	⟷	11	+	⟷	21
·	⟷	12	$\prime$	⟷	22
a	⟷	13	b	⟷	23
∀	⟷	14	∃	⟷	24
∧	⟷	15	∨	⟷	25
⊃	⟷	16	∼	⟷	26
(	⟷	17	)	⟷	27

The next step is to use this Dogg-like mapping that translates symbols into whole numbers to translate full *formulas* into whole numbers in a natural way. Given the PA formula $S0 = 0$, for instance, we replace S with 20, 0 with 10, $=$ with 11, and concatenate the results. With this relabeling, the formula

$$S0 = 0 \quad \text{becomes} \quad \overset{\text{S}\ 0\ =\ 0}{20101110}$$

while the formula

$$(S0 + S0) = SS0 \quad \text{becomes} \quad \overset{(\ \text{S}\ 0\ +\ \text{S}\ 0\ )\ =\ \text{S}\ \text{S}\ 0}{172010212010271120 2010.}$$

This process of replacing a string of PA symbols with a string of digits is as straightforward as it appears, but keep in mind that a string of digits is also itself an actual whole number. This important number is the sought-after Gödel number of the original formula. The formula $S0 = 0$ has Gödel number 20101110. The formula $SS0 = 0$ has Gödel number 2020101110. The formula $0 = 0$ has Gödel number 101110.

From these few examples it should be evident that every formula of PA, indeed every arbitrary string of PA, has a unique Gödel number. Generating the Gödel number for a given formula is akin to translating it into a numerical code—a code that is simple to crack, in fact. Decoding Gödel numbers back into PA notation is just as straightforward as generating them. Starting with the large but otherwise innocuous whole number 2620101110, we examine the successive pairs of digits and see that

$$2620101110 \quad \text{is the Gödel number for} \quad {\sim}S0 = 0.[8]$$

What is the benefit of Gödel-numbering PA? At first glance, all we have done is change the particular symbols involved. One effect of this transposition is that we momentarily lose whatever clarity we might have built up reading PA notation. At this point, the formula

$$(S0 + S0) = SS0$$

clearly signifies the statement "1+1=2" whereas

$$1720102120102711202010$$

just looks like an exceptionally large number. Still, this phenomenon is really just the result of what we are used to, and it should be clear that no information is lost. What is surprising is that something is gained. How is this possible?

In short, it is because in this new notation, PA formulas are now also *numbers*, and proving statements about numbers is the business of PA. The lingering question we've been trying to answer is how a formal system like PA becomes self-referential, and the answer is tantalizingly close. Stoppard's plays acquire their metatheatrical voice by including copies of themselves—translated, disguised, or artfully reinterpreted—inside their expressive domains. PA acquires its metamathematical voice the same way— by including an encoded copy of itself inside its intended domain of the natural numbers.

Recreating Its Own Genus, Continued

We are all familiar with at least a handful of notable subsets of whole numbers that share some distinctive mathematical property—the set of even numbers, for example, or the set of prime numbers. Square numbers, perfect numbers, and Fibonacci numbers are other subsets that have achieved some degree of fame. Alongside these examples, we now define the curious set of *theorem numbers*.

The theorem numbers are precisely those positive integers that arise as the Gödel number for a theorem of PA. Each string of PA symbols has a unique Gödel number and so, in particular, each theorem of PA has a unique Gödel number. Given that $0 = 0$ is a theorem of PA—which it certainly is—it follows that the corresponding Gödel number 101110 is a theorem number. From the earlier discussion we see that 20101110—the Gödel number for $S0 = 0$—is not a theorem number since the false statement $S0 = 0$ (i.e., "1=0") is not a theorem of PA (assuming PA is consistent.) Taking the negation $\sim S0 = 0$ yields a formula whose truthfulness (i.e., "$1 \neq 0$") correctly suggests it can be derived inside PA. Thus, $\sim S0 = 0$ is a theorem, and so writing out the corresponding Gödel number 2620101110 gives us our second example of a theorem number.

The analogy between the set of theorem numbers and a familiar set like the even numbers is helpful and more apt than it might at first seem.[9] Being even is an unmistakably *arithmetic* property. A number a is even if it is divisible by 2; i.e., if it can be written in the form $a = 2 \cdot b$ where b is some other whole number. This characterization of evenness falls squarely within the domain of PA's expressive powers. It would be simple enough to translate the statement "a is an even number" into PA notation. In fact, one version looks like this:

$$\exists b (2 \cdot b) = a.$$

Translating "*a* is a prime number" into PA is equally straightforward. Indeed, we've seen a formula for "5 is prime" already. None of this is surprising or remarkable. Representing statements of number theory is precisely what PA is designed to do.

On the surface, theorem numbers may feel qualitatively different from even numbers or prime numbers. In particular, to produce theorem numbers it seems as though we are doing something other than number theory. Theorem numbers come from PA theorems, and PA theorems are created by applying transformation rules to other previously derived theorems. Theorem numbers are generated by formally manipulating PA formulas; even numbers, by contrast, are generated by doing arithmetic.

But this discrepancy turns out to be an illusion! Recall that, via Gödel-numbering, PA formulas take the form of integers. For instance, the formula

$$(S0 + S0) = SS0 \quad \text{looks like} \quad 1720102120102711202010.$$

In this numerical notation, manipulating PA formulas with syntax-based rules looks very much like doing arithmetic with whole numbers—and although this requires some work, we can show that the two are indistinguishable. The purely formal nature of PA makes it possible to replicate the symbol-shifting rules (acting on PA formulas) with mathematical rules (acting on the corresponding Gödel-numbers).[10] Relegating the discussion of this important aspect of Gödel's argument to the footnotes, the implication is that *the entire architecture of PA can be simulated inside the world of whole number arithmetic*—and whole number arithmetic is precisely the world PA was invented to describe.

Expressible in the Symbols of the System Itself

Taking a step back from the technical specifics, let's refocus our attention on the broad strokes of Gödel's proof and its kinship to Stoppardian theater. From our short introduction to PA, it should be clear and unremarkable that the statement

101110 is an even number

can be encoded as a PA formula.[11] The takeaway of the previous discussion is that it is just as possible to encode the statement

101110 is a theorem number

as a PA formula, but this latter formula reverberates with a mysterious additional potential. What happens when some unsuspecting person encounters this formula? How do they interpret what it says?

The answer depends on whether the person is aware of the Gödel-numbering code that is in place. The PA formula for "101110 is a theorem number" is like a line of dialogue from *Dogg's Hamlet*. For Easy, who only understands the standard (English) interpretation of PA, this formula says

101110 has a quirky arithmetic property involving addition and multiplication of whole numbers.

Easy would view this claim as nothing extraordinary—no different from asserting that 101110 were even or prime—and he would certainly not detect anything introspective about it. But to the schoolboys fluent in the Gödel (Dogg) code, this formula translates as the statement

$$0 = 0 \text{ is a theorem of PA.}$$

It is both subtle and stunning, and it takes time to appreciate. The first interpretation is a statement of number theory; the second is a statement about PA's capabilities. Both are equally valid so that the truth of one interpretation implies the truth of the other. Here is how Gödel summarizes this aspect of his argument in the introduction to the original 1931 paper announcing his result:

For meta-mathematical purposes it is naturally immaterial what objects are taken as basic [symbols], and we propose to use natural numbers for them. Accordingly, then, a formula is a finite series of natural numbers, and a particular proof-schema is a finite series of finite series of natural numbers. Meta-mathematical concepts and propositions thereby become concepts and propositions concerning the natural numbers, or series of them, and therefore at least partially expressible in the symbols of the system PM itself.[12]

The system PM that Gödel refers to is from Russell and Whitehead's *Principia Mathematica*, which we have conveniently replaced with PA, but the argument is the same. Gödel found an unexpected way to mirror the architecture of PM inside number theory and then used the fact that PM is capable of expressing number theory statements as a means to construct formulas of PM that are self-referential. *Dogg's Hamlet* illustrates the mechanics of Gödel-numbering in a vivid way, but *The Real Inspector Hound* is a better model for the metamathematical punch line. In *Hound*, Stoppard builds up an unmistakable set of parallels between the theater critics—Moon and Birdboot—and the characters in the inner murder mystery play. The mirroring is so rich, in fact, that when Birdboot declares, "It is at this point that the play for me comes alive," there is no way to miss the second metatheatrical interpretation. Birdboot is talking about the play he is watching but also, in a coded but unambiguous way, about the play he is in.

Ceci N'est Pas Une Théorème

With the powerful mechanism of Gödel-numbering at our disposal, the final hurdle to proving Gödel's Theorem is constructing the previously advertised formula G whose meta-interpretation refers back to itself in a self-fulfilling way. The discovery of a formal system's potential to express metamathematical statements reveals Gödel's genius; explicitly constructing G demonstrates his artistry and flair. It is one thing to start with a formula such as $S0 = 0$, compute its Gödel number to be 20101110, and then assert the existence of a formula in PA with the interpretation

$$20101110 \text{ is not a theorem number.}$$

Because we know the Gödel code, we recognize this formula as also having the meta-interpretation

$$S0 = 0 \text{ is not a theorem of PA.}$$

To construct G, Gödel *concentrated* the metamathematical potential of the formal system into a single formula. Just as in the previous example, G has a standard interpretation of the form

$$m \text{ is not a theorem number}$$

where m is a properly defined number. In search of the second, metamathematical interpretation of this formula, we decode m to see what formula G is referring to and, lo and behold, m turns out to be the Gödel number of the formula for G.[13] The interpretation of G becomes

> The formula with Gödel number m, that is to say *this* formula, is not a theorem of PA.

It is this metamathematical interpretation that turns G into a stick of dynamite when we ask whether or not G is in fact a theorem. Starting with the assumption that G is a theorem, we appeal to the assumed consistency of PA and conclude that G must be telling the truth about itself. Consistency, together with the accepted truth of the axioms, demands that we never have falsities for theorems. But G is asserting the opposite conclusion—that G is *not* a theorem, a blatant contradiction. Because this is unacceptable, we have to reject the premise that G is a theorem.

We are left with the conclusion that G is not a theorem of PA. Thus, starting from the axioms and following the rules, it is impossible to generate this particular formula G. This might not seem at first like such a serious setback. After all, we can't generate the formula $S0 = 0$ either. But $S0 = 0$ expresses a false statement about numbers, and we don't want it to be a theorem. The formula G, however, expresses a *truth*. Assuming PA is consistent, we conclude that G is not a theorem, which is precisely what G says. This true and unprovable formula is the advertised hole in PA's collection of theorems. Echoing a phrase from George Moore of *Jumpers*, PA can't prove that G is true, we only claim that G is true because PA can't prove it.[14]

Gödel's discovery that PA and PM and any other formalization of arithmetic necessarily contain true but unprovable formulas like G altered the mathematical universe, but the revolutionary impact of his incompleteness theorem is not the main headline at the moment. The parallels I originally sensed between Stoppard's writing and Gödel's metamathematics turned more on Gödel's methods than on the content of his results. Stoppard had no idea he was effectively staging metaphorical versions of constructions at the heart of twentieth-century logic. His agenda was to write engaging plays— plays that "stayed news" as he liked to say—and decade after decade he kept returning to the metatheatrical well for inspiration. Stoppard never explained why exactly, and I don't think he could articulate a reason if he were asked. His logically inclined disposition was instinctively drawn to a source of artistic agency that I doubt he ever gave much conscious attention. Mathematicians, meanwhile, have given it their full attention. To this point, we have employed Stoppard's plays to elucidate the mechanisms of Gödel's mathematics. The time has come for mathematics to return the favor and offer its distinctive insights about the mechanisms at work in Stoppard's plays.

The Real Thing

I don't think writers are sacred, but words are. They deserve respect.

—HENRY, FROM *THE REAL THING*

In one of a series of interviews, Stoppard and *New York Times* critic Mel Gussow touched on the subject of intertextual references in Stoppard's writing. Pointing out a pattern, Gussow noted how *Rosencrantz and Guildenstern Are Dead* had benefited from "riding the coattails of *Hamlet*. And *Travesties* rides the coattails of *The Importance of Being Earnest*."

"Not quite as secure a grasp," Stoppard responded, and then he proudly announced: "With *The Real Thing*, here we are all on our own. No coattails!"[15] Except it's not exactly true. In a sense akin to the metamathematical potential of PA, *The Real Thing* manages to ride on its own coattails.

In the opening scene of *The Real Thing*, we meet middle-aged couple Max and Charlotte and quickly learn that their marriage is coming apart. Charlotte returns from a trip, Max accuses her of having an affair, and the sharp-tongued exchanges that follow point to layers of bitterness. The tense and uncomfortable first scene ends this way:

> CHARLOTTE: I'm sorry if you've had a bad time. But you've done every-
> thing wrong. There's a right thing to say if you can think what it is.
> (*She waits a moment while MAX thinks.*)
> MAX: Is it anyone I know?
> CHARLOTTE: You aren't anyone I know. (*She goes out.*) (15)[16]

In scene 2, we see Charlotte again, but she is different—a different person altogether. And she is with a different man, whose name is Henry. Is Henry the secret lover she has been seeing? It's a reasonable theory, but Henry and Charlotte's overly casual demeanor doesn't support this. The feeling of disorientation is severe, but then the pieces start falling into place. Henry is an "intellectual playwright," we learn, and Charlotte is an actress. When Max arrives, the trick is completely revealed: Charlotte is Henry's wife, and she and Max are currently performing in Henry's play, *House of Cards*. The first scene of Stoppard's play was actually Henry's play, but now we are watching the real thing.

Charlotte is hard on Henry, as well as his play, and the self-referential echoes start in right away.

CHARLOTTE: That's the difference between plays and real life—thinking time. You don't really think that if Henry caught me out with a lover, he'd sit around being witty about Rembrandt place mats? Like hell he would. He'd come apart like a pick-a-sticks. His sentence structure would go to pot, closely followed by his sphincter. You know that, don't you Henry? Henry? No answer. Are you there, Henry? Say something witty. (*He turns to look at her.*)
HENRY: Is it anyone I know? (22)

It's witty, but it's Henry's script.

As it turns out, Henry is the one having the affair—with Max's wife Annie who is also an actress. When Annie confesses the affair to Max, the scene is intentionally staged to remind us of *House of Cards* from scene 1, except that upon learning that his real-life wife is in love with Henry, Max falls to pieces very much along the lines of Charlotte's description. The play then becomes about Henry and Annie's journey together. It begins with a honeymoon period but, after the act break, travels several years into the future to find the couple wrestling with issues of trust and selfishness. The major threat to Annie and Henry's relationship comes in the form of Billy, an actor who is cast in the role of Annie's lover in the steamy seventeenth-century drama *'Tis Pity She's a Whore*. Again, the worlds of art and life entangle as Annie is drawn to Billy. When Henry confronts Annie about Billy, we experience our third rendition of scene 1 from *House of Cards* with Henry suspended perilously between the world he writes about and the one he lives in. A central question the play engages is whether Henry and Annie's abstract ideas about love can hold up to the "mess, tears, pain, and self-abasement" that is the real world.

The Real Thing ran for several years in London and then became Stoppard's first Broadway hit since *Rosencrantz and Guildenstern Are Dead*. It won a Tony Award for Best Play in 1984 and another one in 2000 for Best Revival. The script had the trademark wit that Stoppard was known for by this point, but it connected in a more emotional way than anything the intellectual playwright had previously written. The similarities between the character of Henry and Stoppard were not lost on anyone, including Stoppard, who nevertheless insisted that he didn't want to write about a playwright or actors, but was unable to avoid it.

"I wanted to write a play in which the first scene was written by a character in the second," Stoppard said to Gussow. "But once you're stuck with him being a playwright who wrote the first scene, whatever narrative you invent

would be about his life." Stoppard makes himself sound like a victim of circumstances over which he has no control. "If the writer's wife has got to be in
both situations [scene 1 and scene 2]," he goes on,

> she's got to be an actress. It's determined by the playful idea of having peo-
> ple repeat their situation in fiction. For instance, the [part of Annie], there's
> a love scene with a person who becomes her lover except in fact they're in
> *'Tis Pity She's a Whore.* As soon as you decide that's what's going to happen,
> the woman in the rail scene has got to be an actress because she ends up
> acting in *'Tis Pity She's a Whore.* That's where the horse is, and that's where
> the cart is.[17]

Stoppard's excuse is that he is just the cart following the horse, but he
is really avoiding the question. Why is the horse wandering around in this
metatheatrical landscape in the first place? What drew Stoppard to the idea of
having his characters negotiate the same predicaments multiple times—first
as theatrical performers or writers, and then again in real life? The first answer
is that it is a version of what all of us do all the time. "Public postures have the
configuration of private derangement," Henry says at one point in *The Real
Thing*, cynically suggesting that the publicly stated reasons for the noble activ-
ities we might undertake, in politics or in relationships, are just performances
designed to cover up our real, self-interested motives. Another answer—and
this one speaks to the general phenomenon of why Stoppard is so consistently
experimenting with hierarchies of plays within plays—is that, on an instinctive
level, he is acutely aware of the causal power of metanarratives to augment the
impact of his plays.

In addition to John Ford's *'Tis Pity*, Stoppard also borrows a few lines from
Miss Julie by August Strindberg. It's a fleeting moment, but worth a look because
it again points to Stoppard's near obsession with exploring nested structures.
Late in act I, after Henry and Annie have left their respective spouses for each
other, the two are running Annie's lines with Henry in the part of Jean, the
cultured footman with whom Miss Julie flirts in a dangerous manner:

HENRY (as JEAN): You flatter me, Miss Julie.
ANNIE (as MISS JULIE): Flatter? I flatter?
HENRY (as JEAN): I'd like to accept the compliment, but modesty for-
    bids. And, of course, my modesty entails your insincerity. Hence, you
    flatter me.
ANNIE (as MISS JULIE): Where did you learn to talk like that? Do you
    spend a lot of time at the theatre? (39)

"Loving and being loved is unliterary." *The Real Thing*; Roger Rees (Henry), Strand Theatre, 1982. © Donald Cooper/photostage.co.uk.

If this were a Magritte painting, we would see paintings inside paintings at least three levels deep by this point, each one further removed from so-called reality. This is why, when the couple step out of Strindberg's characters and start discussing the new play Henry is writing, there is a relative authenticity to what is being said.

"I can't do mine," Henry says. "I don't know how to write love. I try to write it properly, and it just comes out embarrassing." Henry is writing a play for Annie—a gift for her—and the wordsmith can't find the words. "Loving and being loved is unliterary," Henry continues, but it may as well be Stoppard talking since he is the one writing a play about love at the moment. "It's happiness expressed in banality and lust. It makes me nervous to see three-quarters of a page and no *writing* on it. I mean, I *talk* better than this."

Cricket Bats

There is yet one more play embedded into the fabric of *The Real Thing*—a fictional script written by Brody, a soldier who is arrested and jailed for arson in the context of a political protest. As a means of forwarding his cause, Brody has written a play that is essentially autobiographical. The one scene we hear read aloud recounts the moment when Brody met Annie on a train en route to the fateful protest march where Brody's arrest took place.

Feeling a sense of responsibility for him, Annie has been advocating for Brody's release for several years. In fact, Brody's play was originally Annie's idea, intended as a means to keep attention on his case. There is just one problem, however:

HENRY: It's no good.
ANNIE: You mean it's not literary.
HENRY: It's not literary, and it's no good. He can't write.
ANNIE: You're a snob.
HENRY: I'm a snob, and he can't write.
ANNIE: I know it's raw, but he's got something to say.
HENRY: He's got something to say. It happens to be something extremely silly and bigoted. But leaving that aside, there is still the problem that he can't write. He can burn things down, but he can't write. (48)

The debate continues to escalate with Annie standing up admirably for Brody's right to be heard at the expense of Henry's "English Lit" standards. Henry eventually counters by fetching his prized cricket bat from the hall:

HENRY: This thing here, which looks like a wooden club, is actually several pieces of particular wood cunningly put together in a certain way so that the whole thing is sprung, like a dance floor. It's for hitting cricket balls with. If you get it right, the cricket ball will travel two hundred yards in four seconds, and all you've done is give it a knock like knocking the top off a bottle of stout, and it makes a noise like a trout taking a fly . . . (*He clucks his tongue to make the noise.*) What we're trying to do is to write cricket bats, so that when we throw up an idea and give it a little knock, it might . . . *travel.* (51)

On the page, this speech by Henry has its particular merits, but onstage it lands with a disproportionate weight. Over time it has become one of the most frequently referenced passages in all of Stoppard's plays. At the beginning of this chapter, I recounted the first of three epiphanies as being the common dexterity with which mathematics and theater are able to explore their own expressive potential. My second epiphany emerged from trying to make sense of the unexpected potency of Henry's cricket bat speech.

The analogy between Stoppard's metatheatrical writing and the metamathematical potential of formal systems runs very deep in this scene. In the case of PA, Gödel first recognized the possibility of mirroring PA's structure inside the world of whole numbers, enabling PA to investigate its own capabilities. The next step was to *focus* this introspective ability to create the formula G which refers, not just to some fact about PA, but to the provability of G itself. The formula G refers directly back to G and metamathematically says, "I don't have a proof."

Here is the critical observation: This metamathematical interpretation does not just *turn out* to be true about G, it is the *reason* G is true. Remember that G has a base-level interpretation as a statement of number theory—that some particular number has some particular arithmetic property. Like all statements about whole numbers, it is either true or false, but if we restrict ourself to the notion of proof as it is codified by PA, we won't find a resolution. In terms of the capabilities of PA, G is *undecidable*—neither G nor $\sim G$ will ever show up as theorems—and the reason it is undecidable is that G turns out to have a second interpretation. That interpretation is, "PA is not capable of providing a proof for G," which becomes self-fulfilling under the assumption that PA is consistent. Somehow this secondary meaning has the tangible effect of making G underivable from the rules of PA.[18]

This self-fulfilling agency is not unique to G but is a general phenomenon that can be exploited in other ways. For instance, it is possible to craft a formula of PA that has the more optimistic meta-interpretation, "This formula of PA does have a proof." For reasons that again hinge on the interaction of the meta-interpretation with the standard number theoretic interpretation, this formula must be a theorem of the system—and therefore true. Note how forcefully this pushes against the natural order of cause and effect. It seems unambiguous that PA's axioms and transformation rules are at the root of everything that happens in the system. Although not directly contradicting this fact, what becomes apparent is that when meta-interpretations arise, they can be imbued with some demonstrable agency. The lower-level rules bring about the possibility of higher-level meta-interpretations which then, mysteriously, wield some influence back down on the system from which they were spawned.

Henry's response to Annie in defense of playwriting contains some obvious metatheatrical sentiments about theater in general. Stoppard then gradually focuses those sentiments—first to be about the play we are watching and then, when Henry is making his case to Annie with cricket bat in hand, to be about the very lines Henry is speaking. Henry's speech is about good writing, but it is good writing. As the words are delivered, there is a resonance between the lower-level meaning of what Henry is saying to Annie and the metatheatrical meaning that emerges once we identify Stoppard's play with Henry's. Henry's

"What we're trying to do is to write cricket bats." *The Real Thing*; Roger Rees (Henry), Felicity Kendal (Annie), Strand Theatre, 1982. © Donald Cooper/photostage.co.uk.

description of well-crafted, tightly sprung prose applies to the speech itself—and how are we sure about this? Because the ideas are compelling. Because our sense of what writing can do travels as Henry delivers his lines. "I am a cricket bat," the passage is saying, and experiencing this metatheatrical interpretation has the causal effect of validating Henry's argument. There is no rule that says we have to take Henry's word on the matter, just as there is no a priori reason to believe what G says about itself, but that is what ends up happening in both cases. Stoppard and Gödel each tap into the inherent introspective capabilities of their chosen language and then focus those capabilities to expand the language's expressive powers beyond the preconceived boundaries of the medium in question.

This instinctual recognition of the power of metalanguages to augment a play's artistic potential is by no means unique to Stoppard, but it is exceptionally acute. It runs through his early plays and reaches its full potential in *The Real Thing*. To belabor the point, the next time a Stoppard script attains a degree of broad emotional appeal on par with *The Real Thing*, it again features a playwright as its central protagonist. *Shakespeare in Love* began as a modest favor when director Ed Zwick at Universal Studios asked Stoppard if he could give a more authentic Elizabethan tone to a promising screenplay written by American Marc Norman. Once again finding himself in the role of the cart following a horse into a story about a playwright struggling to write a play, Stoppard ignored his initial instructions from Zwick and overhauled the entire script.[19] When the dust settled, *Shakespeare in Love* was a full-throated metatheatrical ode to theater that went on to earn seven Academy Awards including, notably, Best Original Screenplay. Stoppard and Norman shared the award.

Despite what these accolades might suggest, *The Real Thing* is still the more impressive achievement. It also celebrates theater to some degree, but at its core it is a play about love, which, as Henry explains to Annie, is extremely hard to write because "loving and being loved is unliterary."

Writing the Unliterary

From *Miss Julie* to *'Tis Pity* to *House of Cards*, love is portrayed in all its dramatic forms, and eventually Henry is forced to summon a performance that will make him worth loving in Annie's eyes. Henry knows that revealing the full depth of his pain over Annie's evolving affection for Billy will drive her away completely. Through a gut-wrenching scene with Annie late in the play, Henry plays his part as best he can until he is finally left alone, at which point Stoppard gives him the least articulate line of the play:

HENRY: Oh, please, please, please, please, *don't*. (77)

Faced with the challenge of writing a play about love, Stoppard resorts to writing a play about someone trying to write a play about love—or in Henry's case, someone trying to write a play about love but then acknowledging that he can't because it is too hard to do. Is love really "unliterary" as Henry explains? Stoppard's solution is to use the inherent metalanguage of *The Real Thing* to say what would otherwise sound banal. Here is Henry the playwright talking to his daughter:

HENRY: Well, I remember, the first time I succumbed to the sensation that
    the universe was dispensable minus one lady—
DEBBIE: Don't write it, Fa. Just say it. The first time you fell in love. What?
HENRY: It's to do with knowing and being known. (62)

What Henry struggles toward is a description of love as an authentic kind of knowledge that requires getting beyond performing who we are to being who we are—"the real him, the real her," he goes on to say, still falling short of what he means. Henry's depiction of love is doomed to a form of literary incompleteness—the inevitable shortfall of his ability to express the full truth he innately feels. But this is how the augmenting power of metanarratives work. Stoppard's play finds its voice in the negative spaces of Henry's explanations for why he can't find his. The resonance between the lower-level textual meanings and the introspective metatheatrical meanings gives Stoppard's play a depth that Henry cannot access. Stoppard even manages an upbeat ending, bringing Annie and Henry together to the accompaniment of the distinctly unsnobby soundtrack of the Monkees that steers clear of banality of any kind. At the final curtain, we are left with the optimistic feeling that Henry will one day write Annie's play because, of course, we have just seen it.

The Hard Problem

Who can? Who's the 'you' outside your brain? Where?

—HILARY, FROM *THE HARD PROBLEM*

Up to this point, the respective discussions of Gödel and Stoppard have been conducted on parallel tracks. The connections thus far have been unintentional on the part of the playwright, the result of two creative minds employing similar techniques in vastly disparate domains. In 2015, the tracks crossed. In a brief but significant moment, Stoppard dropped in a reference to Gödel in the middle of *The Hard Problem*, a tender play anchored around the contemporary debate about how much of who we are can be accounted for by biological and evolutionary science. The majority of the characters in *The Hard Problem* are biologists and neuroscientists, but mathematics has a crucial, and controversial, role to play in the tug-of-war between the two factions. On one side of the rope are the materialists, for whom everything from altruism to angst can be explained by a combination of neurons and Darwinian evolution. On the other side are the soulists, who insist there must be something more. As strange as it sounds, both sides of the debate are intent on enlisting Gödel's mathematics as an ally for their cause.

Brains and Minds

Coined in 1995 by philosopher and cognitive scientist David Chalmers, the term "hard problem" refers to the challenge of explaining human consciousness, particularly in the context of understanding the subjective way

"Bananas aren't thinking, 'Hey, seven eights is fifty-six.'" *The Hard Problem*;
Olivia Vinall (Hilary), Damien Molony (Spike); Royal National Theatre,
London, 2015. © Geraint Lewis / ArenaPAL.

sensory information is experienced. It is not just that we cognitively process
visual images, sounds, or even emotions, it is that we have an individualized
experience of these things. Stoppard's 2015 play subsumes this question by
exploring the question of how—or whether—the mechanics of the brain
could lead to the phenomenon of a conscious mind.

The central character in *The Hard Problem* is Hilary, a behavioral psycholo-
gist who is fundamentally at odds with the hypothesis that the hardware of the
brain is all that is required to explain her passions, heartbreaks, and unshakable
need to pray by her bedside in the evenings. "When you come right down to
it, the body is made of *things* and things don't have thought," she argues early
in the play, sounding a bit like George Moore from *Jumpers*. "Bananas aren't
thinking, 'Hey, seven eights is fifty-six', or 'I'm not the king of Spain', and when
you take a banana to bits you can see why . . . Same with brains. The mind is
extra." (12)[20]

It is in defense of Hilary's soulist position on human intelligence that
Stoppard invokes Gödel. Admittedly weak in mathematics, Hilary has the

following exchange with a colleague named Ursula, where she asks for confirmation that all the current theories of consciousness are flawed in some way and then makes a curious request:

HILARY: How about panpsychism?

URSULA: No. Nature isn't conscious. Trees are not conscious.

HILARY: Functionalism?

URSULA: No. A thermostat is not even a tiny bit conscious. Have you been reading after lights out?

HILARY: What about quantum-level brain processes to explain consciousness?

URSULA: She has!

HILARY: Will you show me how Gödel's Proof means a brain can't be modelled on a computer?

URSULA: You wouldn't know Gödel's Proof if it had suspenders in Selfridge's window.

HILARY: Ursula, I need you to show me!

URSULA: *Now?* (45)

We don't hear Ursula's explanation, but if we did it would go something like this: A formal system is like a computer. Both consist of strict, deterministic rules governing the base-level functioning of the system that lead to higher-level output. In the case of PA, that output is in the form of formulas that have the interpretation of being true statements of number theory. It would be a straightforward task to write a computer program that simulates PA and then send the computer off to generate truth after truth of number theory by applying various transformation rules to the axioms. Such a program could rightfully be seen as a simple kind of artificial intelligence that attempts to model one small corner of human intelligence—specifically, the domain of whole number arithmetic. This mini computer brain couldn't express anything on the topic of Spanish royalty, but at least it could assert that seven eights is fifty-six.

Because the PA theorems are deduced from axioms and rules, we could even be generous and say that the computer "knows" all the arithmetic facts that it generates. But Gödel showed that PA, and thus our PA-simulating computer brain, will never be able to generate G nor its negation $\sim G$. In short, the computer can never know if G is a true statement. We humans, however, know full well that G is true. In fact, G was deviously designed that way! By taking advantage of the unexpected potential for formulas of PA to contain self-referencing interpretations, Gödel carefully crafted G to be undecidable from *within* PA, and then deduced that G is true by a metamathematical argument *outside* PA. This means that the human mind can always know truths that PA does not—and there is nothing PA specific about Gödel's proof. What Ursula might say to Hilary is that any computer-based model of intelligence, by virtue of its deterministic rule-bound hardware, would represent a physical manifestation of a formal system and thus be susceptible to a Gödel-type manipulation. This line of attack results in a blind spot in the computer's expressive domain that a sentient human mind could detect.

To say it more succinctly, computing machines are equivalent to formal systems by way of their common deterministic architecture; humans are superior to formal systems by way of Gödel's Theorem; therefore, humans are superior to computing machines.

Hofstadter's Counterargument

This argument, which originated many decades ago, has generated its share of spirited debate over the years.[21] Rather than enumerate its potential weaknesses, of which there are several, the strongest counterpoint to offer is that Gödel's proof can also be invoked as the cornerstone for an argument that the human mind *can* be modeled in a purely mechanical way. Acknowledging that my academic origins as a mathematician are likely to blame, I will own up to my opinion that this is the more compelling point of view. Hilary may not want to hear this, but one implication of Gödel's work is that the hardware in the brain could very well be enough to explain the more mysterious aspects of the human experience such as creativity, emotions, and, most especially, a sense of self. The mind may not be extra.

One of the most innovative voices in the conversation around the broader significance of Gödel's contributions belongs to cognitive scientist and polymath Douglas Hofstadter.[22] Running through Hofstadter's extensive writing about Gödel over multiple decades is an argument that an ego-possessing form of intelligence could indeed be supported on silicon as naturally as it is on carbon-based neurons. The previously explained analogy between formal systems and computers is robust. Both can be understood as rule-following, deterministic machines, and to this analogy one can add the brain. A purely biochemical description of the brain as a vast array of synaptically linked neural cells is similarly mechanical in nature. Whereas the level of complexity is many orders of magnitude higher in the neural network of the brain, at its base level the brain still consists of strictly rule-following components possessing the same lack of free will as the circuitry of a computer or the symbols of PA. In this three-way analogy between brains, computers, and formal systems, a deterministic, mechanical manipulation of meaningless symbols on the bottom level becomes the substrate from which evolve meaningful representations of some aspect of reality higher up.

Now, this evolution from meaningless symbols to meaningful representations is easiest to parse in a formal system like PA because, compared to a brain or even a computer, PA has a relatively simple structure. This is why Gödel's discovery is so stunning. Even in the simplest of these three mechanical systems, something akin to consciousness appears seemingly out of the ether. In formally modeling the tiny slice of the universe that is whole number arithmetic, PA acquires the ability to model itself. Poetically speaking, it becomes self-aware. As primitive as systems like PA or PM are when compared to the workings of the brain, they are still rich enough to be able to capture their own detailed reflection, and because of this self-perception, the hierarchy of cause and effect gets tangled. Although we think of the fate of the formula G as being completely determined by the lower-level deterministic rules of PA, somehow G's insistence that it cannot be proved has the effect of preventing the transformation rules of PA from ever being able to produce it as a theorem. It is as though G has acquired some autonomy over its own fate and thus over PA as well. Hofstadter uses the term "downward causal power" to describe this phenomenon, drawing a strong parallel to the unshakable freedom-loving sense of self that exists in a human brain. The suggestion is that, much like the Gödel string G in PA, the ego that presides over our mind and possesses the potential to dictate our thoughts and actions may also be completely supported on totally deterministic neural hardware similar to the formal rules that constitute PA. For Hofstadter, it is more than just an analogy. "The self comes into being at the moment it has the power to reflect itself," he writes, arguing that PA's self-awareness and human consciousness are two versions of the same phenomenon taking place on vastly different scales.[23]

Stoppard gives a modest nod to this side of the argument in *The Hard Problem*. Late in the play, a quantitative computer tech named Amal, whose job is to model risk in financial markets, has enough champagne at Hilary's dinner party to offer the following confession:

> AMAL: In theory, the market is a stream of rational acts by self-interested people; so risk ought to be computable, and the models can be proved

mathematically to crash about once in a lifetime of the universe. But every now and then the market's behavior becomes irrational, as though it's gone mad, or fallen in love. It doesn't compute. It's only computers compute.

He drains the bottle.

So I am thinking about that. (68)

It is a small bone, but Amal is at least acknowledging that irrational, even romantic, behavior can appear in systems built from purely rational ingredients.

Sufficiently Powerful

Although Stoppard gives the bulk of the play's sympathy to Hilary's position that the mind must be more than the sum of its mechanical parts, his propensity for exploiting the downward causal power of self-referential constructions reveals a latent sympathy in the other direction. A play like *The Real Inspector Hound* is constructed with a coarse-grained copy of the outer play embedded in the inner play so that when Birdboot announces, "It is at this point for me that the play comes alive," that's what actually happens—not just to Birdboot's play in some internally literal way but to Stoppard's in a figurative way. It's good fun of course but also richly suggestive. Making plays come alive is ostensibly every playwright's intention, and the proposal is that they come to life in much the same way that people do—by acquiring the potential to move from the particular to the universal. Stoppard said that "art which stays news . . . is art in which the question 'what does it mean?' has no correct answer. Every narrative has, at least, a capacity to suggest a metanarrative."[24] For Stoppard these metanarratives inevitably entail some form of self-reflection, and the mathematics suggests that this principle extends beyond Stoppard's plays.

Delegating the remainder of the debate about human consciousness to the neuroscientists, what remains is a provocative rendering of the task of playwrights to make their plays come alive. There is a component to Gödel's argument that has not been given the appropriate attention. A proper paraphrasing of Gödel's Theorem should include the modifying phrase "sufficiently powerful," or something like it.

Gödel's Theorem: Any consistent formal system of number theory that is sufficiently powerful is incomplete.

"Sufficiently powerful" refers to the ability of the formal system to represent a certain level of number theoretic reasoning. Think of this as a critical threshold. Below the line are formal systems that might capture a particular component of arithmetic (e.g., addition of two numbers) but whose expressive power is limited. The formulas in these primitive systems communicate in a flat, monotone voice that adheres to the intended agenda of the system in question. Above the line we find PA and PM and the other formal systems that, by virtue of their strength, acquire the capacity to express coded metamathematical statements. The formulas in these systems can possess multiple interpretations, including self-referential ones, leading to an array of unanticipated phenomena of which incompleteness is just one example.[25]

At the risk of engaging in the snobbery that Henry is accused of in *The Real Thing*, there is a similar hierarchical scale that distinguishes art which warrants contemplation from flat, monochromatic, single-voiced offerings. Whatever unpolished authenticity it claims to possess, Brody's play in the *The Real Thing* lacks the layered depth required to drag the imagination away from what we initially see. Stoppard is staking out a fundamental objective of the playwright by asserting that additional and unsuspecting levels of meaning are a requirement for any proposed piece of art that aspires to having some kind of lasting impact.[26] "Art that stays news" is art that by virtue of being "sufficiently powerful" has the ability to communicate on multiple levels, and what the mathematics keeps pointing to is that one of those levels is eventually and unavoidably self-reflective. In *Artist Descending a Staircase*, Sophie beautifully summarizes the argument this way: "But surely it is a fact about art—regardless of the artist's subject or his intentions—that it celebrates a world which includes itself—I mean, part of what there is to celebrate is the capability of the artist."

Gödel's discoveries align with Sophie's on this point, or Sophie's align with Gödel's. Either way, the good news for artists is that celebrating the capability of the artist cannot be avoided, once the medium in question becomes rich enough in its expressive potential.

The Invention of Love

In the mirror of invention love discovered itself.

—OSCAR WILDE, FROM *THE INVENTION OF LOVE*

When I originally set out on my journey to make sense of Stoppard's fascination with mathematics, what I was in fact doing was trying to understand my own fascination with theater. I had not had any academic training in drama, nor had I enjoyed a stint as an actor at any point along the way. And yet, there I was, neglecting the duties of my day job as a mathematician to read scripts and dash to the city to catch performances whenever I could. As I navigated the logistics of this curious double life, there was a particular Stoppard character with whom I came to share a special sympathy.

At the beginning of Stoppard's 1997 play *The Invention of Love*, Alfred Housman has died and is waiting by the River Styx. When Charon the ferry-man arrives to pick him up, he tells Housman that they must wait for another passenger who is apparently late.

> HOUSMAN: Doubly late. Are you sure?
> CHARON: A poet and a scholar is what I was told.
> HOUSMAN: I think that must be me.
> CHARON: Both of them?
> HOUSMAN: I'm afraid so.
> CHARON: It sounded like two different people. (2)[27]

The historical A. E. Housman started as a student at Oxford in 1877, and after several detours ended up becoming one of England's most distinguished classical scholars, spending the last twenty-five years of his life at Trinity College in Cambridge. Housman was known for bringing an unrelenting standard of rigor to the business of translating classical texts, and he was merciless in his critiques of those who did not exhibit the required level of discipline when finding the intended meanings of the original Greek and Latin authors. Housman was also a poet, widely celebrated for the lyrical collection, *A Shropshire Lad*, first published in 1896.

Like *Arcadia* before it, *The Invention of Love* explores the tension between classical and romantic sensibilities, this time tightly compressed in the psyche of a single character. The play is a dreamlike sequence of scenes from Housman's youth. In act I, we encounter the distinguished faculty from the Oxford Classics department and meet Housman's college friends, including Moses Jackson who would become his unrequited love. But it's when the deceased Housman is poled ashore by Charon and encounters an eager undergraduate that the play jolts to life:

> HOUSMAN: Oh . . . excellent. You are . . .
> YOUNG HOUSMAN: Housman, sir, of St. John's.
> HOUSMAN: Well, this is an unexpected development. Where can we sit
>   down before philosophy finds us out? (30)

"A scholar's business is to add to what is known. That is all." *The Invention of Love*; John Wood (Housman), Ben Porter (Young Housman); Theatre Royal Haymarket, 1998. © Donald Cooper/photostage.co.uk.

The elder Housman tells his younger self that translation is a science—the noblest one, in fact. The advancement of truth, by any amount, needs no further justification. When the young Housman asks whether it is possible to be such a scholar and write poetry as well, the answer is unambiguous:

> HOUSMAN: No. Not of the first rank. Poetical feelings are a peril to scholarship. . . . Taste is not knowledge. A scholar's business is to add to what is known. That is all. But it is capable of giving the very greatest satisfaction, because knowledge is good. (36)

The rebuttal to this argument is the melancholy aura of Housman's life—the implied sense that he sacrificed passion for scholarship and died unfulfilled. From start to finish in the play, there are suggestions that Housman has failed at life in some fundamental way, suppressing his love for Moses Jackson and perhaps sacrificing his full potential as a poet—but the elegance and beauty with which textual scholarship is portrayed make it impossible to draw an easy conclusion on this question. In the closing moments of *The Invention of Love*, the character of Oscar Wilde appears, reading from *A Shropshire Lad*, and the two contemporaries, who never actually met, are given a chance to converse in Stoppard's imagined universe. Wilde is a significant off-stage character for most of the play. His antics are reported regularly, including his eventual conviction for the crime of gross indecency, so Wilde's cameo at the end of *The Invention of Love* comes as a delightful reward for the audience.

Wilde is, in every way, the antithesis to Housman. He has lived a life of flamboyance on his way to becoming an iconic figure of the aesthetic movement, and in his few brief moments on stage Stoppard gives him an essentially uninterrupted opportunity to make his case. When Housman expresses some sympathy for Wilde's ruinous end, Wilde responds with, "Better a fallen rocket than never a burst of light."

> WILDE: You are right to be a scholar. A scholar is all scruple, an artist none. The artist must lie, cheat, deceive, be untrue to nature and contemptuous of history. I made my life into my art and it was an unqualified success. . . . I awoke the imagination of the century. (96)

Contrary Psychologies Injudiciously Mixed

The topic of whether Wilde is the play's success story and Housman its cautionary tale was the subject of a pointed debate that played out between Stoppard and classicist Daniel Mendelsohn in the pages of the *New York Review of Books*. Over five exchanges that appeared between August and October of 2000, Mendelsohn repeatedly calls Stoppard out for presenting the scholarly life of the mind via Housman as the sterile and stunted option, juxtaposed with the emotionally fulfilling pursuit of artful pleasures that characterizes Wilde's public image. Stoppard refuses to cede much ground in his responses, insisting that the play—and by extension its author—very much "reveres" Housman for the work he did and the choices he made. In one particularly revealing paragraph, Stoppard has this to say:

> [Mendolsohn] is quite right that Wilde is in the play as a foil to Housman, and elevates the "dithyrambic" artist at the expense of the scrupulous scholar. But I disagree that Wilde is "the real hero" of the play, and that "there's no question of where your sympathies are meant to lie." It seems to me there is every question; at least, it's the question which accounts for the play, and—as with *Arcadia*, about which Mr. Mendelsohn with his either/or approach is similarly bemusing—the reason it's a question at

all is that most of us, including Housman and Wilde, are compounded of contrary psychologies injudiciously mixed.[28]

With only a few pages left in the chapter, time is running short to make good on my promise of a third epiphany, which in its original form was about reconciling this tension between contrary psychologies. In its current form it has acquired an added degree of personal significance that I did not see coming.

Housman famously said, "Meaning is of the intellect, poetry is not," and he went to great lengths to segregate the two in his own life.[29] In his scholarship, he was driven by the desire to understand exactly what the classical authors had intended, and thus it distressed him greatly when other translators, either consciously or not, allowed for distortions that served some other artistic purpose. There was a meaning to be uncovered—no more and no less, and for no other reason than to "add to what is known." But Gödel showed that this search for a single, monochromatic meaning is an impossibility. Meaning can arise in places where it is unexpected and, to complicate matters even more, where it is unintended.[30]

Here, again, the key is to distinguish between what Gödel proved and how he proved it. The conclusion of Gödel's Theorem—that truth cannot be perfectly captured by formalized axiomatic means—sent tremors through the mathematical world, but it is a law of the universe that artists have organically embraced since artists have existed. For someone with a foot on either side of the divide between poetry and scholarship, it is the *mechanism* of Gödel's Theorem that offers the more affecting message. Even in the austere Platonic world of whole number arithmetic where the concreteness of the intended meaning of *Principia Mathematica* could be no less ambiguous—even here, out of a notational system designed to capture the purest meaning of the narrative about the characters 0, 1, 2, 3, . . ., unintended voices emerge. These new voices recount metanarratives equipped with the agency to affect the efficacy of formal systems in general, including the formal system telling the story.

How far do the implications of this discovery in twentieth-century logic extend into the domains of art and theater and human nature itself? Probably farther than I realize, even taking into account the sentiments in this particular sentence. Bertrand Russell's quest to create a deterministic system that could unemotionally compute its way through the truths of mathematics took place at the same Cambridge college where A. E. Housman was simultaneously serving as Benjamin Hall Kennedy Professor of Latin. But both of these towering intellectuals set the bar unreasonably high. Despite Housman's or

Russell's best intentions, in any conceivable incarnation of knowledge, science and art are more inextricably linked than either of them fully realized.

"If knowledge isn't self-knowledge it isn't doing much, mate," says Bernard to Valentine in *Arcadia*, to which Valentine could have replied that, according to the mathematics, knowledge inevitably leads to a penetrating form of self-knowledge. It's not a contradiction at all that Housman the disciplined classical scholar was also a sentimental poet, and maybe it is not so incongruous that a mathematician would write a book about the humanity math obtains in the hands of inventive playwrights.[31] This is the most lasting image that Gödel's work suggests. Constructed with the utmost regard for the logical perils of self-reference, *Principia Mathematica*, by virtue of its strength, is capable of self-reference of the most articulate kind—and there is no way to avoid it. Any time a system is built to model something as primitive as whole number arithmetic, it must cross the threshold where its expressive powers erupt beyond their intended boundaries. One can try to conceive of an entity that generates only dispassionate objective truths, but eventually, when it is sufficiently powerful, it acquires some self-awareness and all the poetry that comes with it.

6

Whitemore, Wilson, and Mighton: The Dramatic Life of Alan Turing

If a machine is expected to be infallible, it cannot also be intelligent.

—ALAN TURING, FROM A 1947 LECTURE TO THE
LONDON MATHEMATICAL SOCIETY

IN THE second act of Tom Stoppard's *The Invention of Love*, a historical character named Henry Labouchere has a conversation with several fellow journalists about his role as a member of Parliament in crafting the Criminal Amendment Act of 1885:

> LABOUCHERE: Anybody with any sense on the backbenches was pitch-forking Amendments in to get the government to admit it had a pig's breakfast on its hands and withdraw it ... My final effort was the Amendment on indecency between male persons, and God help me, it went through on the nod ...
>
> STEAD: But—but surely—you *intended* the bill to address a contemporary evil—?
>
> LABOUCHERE: Nothing of the sort. I intended to make the bill absurd to any sensible person left in what by then was a pretty thin House ... but that one got away, so now a French kiss and what-you-fancy between two chaps safe at home with the door shut is good for two years with or without hard labour. It's a funny old world. (61)

Although the original focus of the bill was on delineating penalties for sexual offenses against women and minors, the Labouchere Amendment, with its incorporation of the broad and ambiguous phrase "gross indecency," essentially made illegal any sexual relations between male persons "in public or private." On the off chance that they don't see it coming when the above conversation occurs, Oscar Wilde's flamboyant appearance at the end of Stoppard's play serves to remind the audience of this law's most famous victim. When *The Invention of Love* opened in 1997, it is likely that Wilde was the only person among the tens of thousands of men convicted between 1885 and the law's repeal in 1967 that most people could name. However, in 1952, Labouchere's gross indecency clause snared another British intellectual of extraordinary significance.

A mathematician of great originality, Alan Mathison Turing was a leading figure in the British code-breaking effort during World War II, and his effectiveness in deciphering the German naval and U-boat codes contributed substantially to the Allies' ultimate success against the Nazis. After the war, Turing became a pioneer in the embryonic world of computer science, and he is now generally regarded as the person most responsible for creating the theoretical framework of the modern computer. Turing was also gay. By the time Turing came before the court in March of 1952, the police knew they were dealing with one of Winston Churchill's "Bletchley Park boys," and in a misplaced gesture of compassion, the judge allowed Turing to avoid jail time provided he submit to a course of experimental estrogen injections. These were meant to "cure" him of his homosexuality but succeeded only in humiliating the former war hero and clouding his mind. On June 7, 1954, Alan Turing died in his home from cyanide poisoning. The consensus among those most familiar with the surrounding events is that Turing took his own life—what is not generally agreed upon is the reason why.

The Shakespearean storyline of Turing's life has attracted the attention of numerous artists, including a generous number of playwrights. In this chapter, we look at the contributions of three of them: Hugh Whitemore, Snoo Wilson, and John Mighton. Up to this point, all the dramatists discussed have had some measurable influence on the arc of twentieth-century theater. Although it's not likely that history will put this chapter's three writers in that same high echelon, each is widely accomplished, and all found inspiration in Turing's story in an innovative way. What distinguishes the plays discussed here is the way they leverage the mechanisms of theater to illuminate the implicit humanity of Turing's mathematics. Each playwright manages this feat in his own way, but they all start by recognizing the historical Turing's propensity to conflate abstract mathematical thinking with practical real-life application.

By contributing significantly to the defeat of Hitler and the invention of the computer, Turing connected himself to two of the most important events of the twentieth century. Turing's mathematics, however, did not start out as applied in any way. He came to Cambridge while G. H. Hardy was still there. He participated in a class with Ludwig Wittgenstein and was among the first generation of students to study higher mathematics in a post-Gödel world. Turing, in fact, specialized in logic and would make his reputation solving a problem that David Hilbert had posed and Gödel's 1931 paper brought to the fore. What is curious, however, is that his writing on these esoteric subjects often sounds as though he is doing applied mathematics. This is one of the reasons his ideas were so transformative—people could understand them, communicate them, and build on them. The full scope of his influence is evident in the ubiquity of his name in the lexicon around the modern theory of computation: Turing's thesis, Turing machines, and the Turing test are all pillars of contemporary computer science.

Of course, Turing did not name these important concepts after himself. When he first proposed what has become known as the Turing test, he called it the "imitation game," a suggestive term that conjures up the image of an acting class exercise. The fact that there is some validity to this misunderstanding hints at why theater is uniquely suited as an artistic medium to explore this enigmatic mathematician. Theater and mathematics, born from the same ancient Greek culture as distinct mechanisms for codifying and communicating truths of the universe, collide in a fascinating way two thousand years later in the life and work of Alan Turing.

Breaking the Code

It's a technical paper in mathematical logic, but it is also about the difficulty of telling right from wrong.

—TURING, FROM *BREAKING THE CODE*

The most widely produced play about Alan Turing is *Breaking the Code*, by Hugh Whitemore. On the title page, Whitemore acknowledges that his play is based on the biography *Alan Turing: the Enigma*, written by mathematician and gay activist Andrew Hodges. More than anyone else, Hodges is most responsible for supplying the necessary effort to bring Turing's life story into the open. His engaging and exhaustive biography of Turing was published in 1983, and Whitemore's adaptation was staged in London in the fall of 1986 with Derek Jacobi in the leading role. The play eventually ran in New York and a film version of *Breaking the Code,* also starring Jacobi, was released in 1996 as part of the Masterpiece Theatre series produced by the BBC.

Having written widely for television, film, and stage, Whitemore was nevertheless navigating uncharted territory when he took on Turing's story. This was a decade before *Arcadia* and *Copenhagen* had proved that hard science and mathematics could hold their own on the popular stage. Meanwhile, the preeminent example of a biographical play about a mathematician, Brecht's *Life of Galileo,* had played fast and loose with the historical record. Galileo's life and science were part of common folklore which Brecht used in an allegorical way. In contrast to Brecht's protagonist, virtually no one knew anything about the hero of Whitemore's play, which put added pressure on the playwright to find a way to render Turing authentically without turning the play into a documentary. In this regard, Whitemore had a fine line to walk. The initial response, as well as the many revivals that have followed, suggest Whitemore chose his steps well.

To tell Turing's story, Whitemore uses two separate narrative strands— one that begins with the forty-year-old Turing and his legal troubles, and a second that recounts his earlier years at school and then at the British cryptology headquarters at Bletchley Park. The two strands crisscross in alternating scenes until they merge into a single thread toward the close of the play.

In the opening scene, Turing, the established professor at Manchester University, has come to the police station to report a burglary at his house to one of the detectives on duty. In the very first line, we see Whitemore's solution to a problem that Brecht never had:

ROSS: Well now, let's get the basic facts sorted out. We're talking about a
    burglary that occurred on January 23 and you are Mr. Spurling.
TURING: No—Turing.
ROSS: I beg your pardon?
TURING: My name is Turing, *not* Spurling.
ROSS: Sorry, sir, I beg your pardon. (*Displays a sheet of paper.*) Just look at
    this atrocious writing. It could be Spurling, Spilling, Tilling.
TURING: Well, it's Turing. (*spelling*) T-U-R-I-N-G.
ROSS: (*writing*) Alan Mathison Turing. Is that right? (7)[1]

Could the point be any clearer? Who is this person no one has ever heard of, and how many introductions does he need? Turing's ongoing dealings with Detective Ross play out intermittently through act I. Things start out innocuously enough when Turing explains that "a brush salesman" told him who the burglar might be, but by the end of the act Ross has sniffed out that this anonymous brush salesman was in fact an acquaintance of Turing's:

> ROSS: This friend of yours: what's his name?
> TURING: Ron. Ron Miller.
> ROSS: A colleague of yours at the University?
> TURING: Well, no.
> ROSS: A social acquaintance?
> TURING: In a way . . .
> ROSS: So why did you lie to conceal his identity?
> TURING: I, um . . . I didn't want to get him into trouble. (58)

Not comfortable, or capable really, of dishonesty, and very much at peace with his sexual identity, Turing confesses to an affair with Ron and instantly becomes the perpetrator instead of the victim of the crime at hand.

Fictional Facts

This is, in fact, very close to how events transpired, although the real name of Turing's romantic partner was Arnold Murray. Why the name Ron Miller? In the months before he died, perhaps as a form of therapy, the historical Turing wrote a pseudo-autobiographical short story about a gay scientist named Alec Pryce who picks up a young, out-of-work drifter named Ron Miller. By replacing the real-life Arnold Murray with the fictional stand-in from Turing's short story, Whitemore acknowledges his intention to find a middle ground between the literal truth and a poetic one. There are liberties taken and events rearranged to support the storytelling, but for the most part the raw material of Whitemore's play has its roots firmly in Hodges's biography.

In the spaces between scenes with the police investigation come highlights from Turing's early life. In the second scene, Turing is seventeen years old when we meet Christopher Morcom, a friend who is to become the idealized and unrequited love of Turing's life. This is not an invented embellishment. Christopher was Alan's one true companion at Sherborne School, and in addition to being genuinely fond of Alan, he could also hold his own intellectually with the young Turing. Christopher earned the more prestigious place at Trinity College in Cambridge while Turing had to settle for King's College just down the road. Their friendship was romantic in every way but physical, and it is likely that Christopher was not gay. There is no way to know for certain

"This friend of yours: what's his name?" *Breaking the Code*;
Dave Hill (Ross), Derek Jacobi (Turing); Theatre Royal
Haymarket, 1986. © Conrad Blakemore / ArenaPAL.

on this question because in 1931 Turing matriculated to Cambridge on his own—his beloved friend having died of tuberculosis the year before. In *Breaking the Code*, Whitemore resorts to using Turing's real words—lifted from a letter to his mother—to describe the sense of loss for the simple reason that they can hardly be improved upon. "It never seemed to have occurred to me to make other friends besides Morcom," the historical Turing wrote, and Whitemore's Turing says. "He made everyone else seem so ordinary."[2]

The Entscheidungsproblem

Our next view of the younger Turing in *Breaking the Code* comes later in act I when he arrives at the British intelligence and code-breaking headquarters of Bletchley Park in 1939. Turing is twenty-seven years old at this point, and Whitemore uses another interview, this time between Turing and his new boss Dillwyn Knox, to fill us in on the missing years:

> KNOX: This is your file. I shall consult it from time to time. There's no need to be alarmed.
> TURING: I'm not.
> KNOX: Good. (*He looks at file.*) So you went to Sherborne, Cambridge—and then America: 1936 to 1938; two years in America. How was that? Did you enjoy Princeton?

As it turned out, the historical Turing was promoted from undergraduate at King's to fellow in the mathematics department when he was just twenty-two years old. This landed Turing squarely in the prestigious and rapidly evolving world of mathematical logic. A quick review of the dates helps tell the story. David Hilbert's reverberating cry of "Wir müssen wissen, Wir werden wissen" from Königsberg was in 1930, and Kurt Gödel's "On Formally Undecidable Propositions in *Principia Mathematica* and Related Systems" appeared in 1931. In 1934, the year before he was elected fellow, Turing took a class at Cambridge taught by Max Newman. The course started with the goals for formalized mathematics that Hilbert had articulated and culminated with an examination of how Gödel's work made it clear that Hilbert's dream of a simultaneously complete and consistent formal system of mathematics was unattainable.

But there was still one aspect of Hilbert's original program that remained unresolved—the so-called *Entscheidungsproblem*, or decision problem. The main thrust of Gödel's paper is that provability inside a formal system is necessarily a weaker notion than truth. In every sufficiently strong and consistent system, it would always be possible to find formulas corresponding to true statements that the system would be unable to derive. The Entscheidungsproblem asks a different question, which is: Does there exist a deterministic method for distinguishing the provable formulas from the unprovable ones? Newman, Turing's instructor, offered this assessment of what was at stake:

> The Hilbert decision-programme of the 1920s and 30s had for its objective the discovery of a general process, applicable to any mathematical theorem expressed in fully symbolic form, for deciding the truth or falsehood of the theorem. A first blow was dealt at the prospects of finding

this new philosopher's stone by Gödel's incompleteness theorem (1931), which made it clear that the truth or falsehood of A could not be equated to provability of A or not-A in any finitely based logic, chosen once for all; but there still remained in principle the possibility of finding a mechanical process for deciding whether A, or not-A, or neither, was formally provable in a given system.[3]

Displaying an unorthodoxy that came from thinking about even the most complex issues from first principles, Turing showed definitively that no such "mechanical process" existed. His results appeared in a now-famous 1937 paper, "On Computable Numbers with an Application to the Entscheidungs-problem," and it was this work that propelled him to Princeton to work with the American logician Alonzo Church, with whom he eventually completed his PhD.

Among other dignitaries then at Princeton were Albert Einstein, John von Neumann, and, as it turned out, G. H. Hardy, who was visiting from his reg-ular position at Trinity in Cambridge. On the surface of things, Hardy seems like someone with whom the younger Turing might have formed a meaning-ful relationship, but the two Cambridge expatriates never made much of a connection, mathematically or socially.[4] In a pinch, Turing could have held his own in a cultured high table conversation, but he was more likely to be heard chanting the wicked queen's sinister verse from *Snow White and the Seven Dwarves*:

Dip the apple in the brew,
Let the sleeping death seep through.

This groundbreaking full-length animated classic premiered in 1938, the same year that Turing turned down an offer to stay at Princeton to work with von Neumann and returned to England.

The following year the German army advanced into Poland.

The Laws of Human Life

On September 3, 1939, British Prime Minister Neville Chamberlain went on the radio to somberly announce that Hitler had refused to withdraw his troops and that, consequently, Britain was at war with Germany. On September 4, 1939, Alan Turing reported to Bletchley Park.

In *Breaking the Code*, Whitemore uses an interview between Turing and his new boss, Dillwyn Knox, for a number of purposes beyond enumerating the important biographical landmarks in Turing's life. Turing's quirky literal mind-edness comes through, and Turing's deep fascinations with codes and ciphers, as well as with *Snow White*, are also discussed. There is also the moral question of war. Here we can hear the echo of G. H. Hardy. Hardy would write *A Mathematician's Apology* exactly one year later, and in it he would declare pure mathematics—and himself by extension—innocent of producing anything the least bit useful in relation to the business of warmongering. "No one has yet discovered any warlike purpose to be served by the theory of numbers or relativity," Hardy writes, "and it seems very unlikely that anyone will do so for many years."[5]

Standing as a massive counterexample to Hardy's argument was the entire intelligence operation of Bletchley Park, its location at the midpoint between Oxford and Cambridge chosen specifically to take advantage of the intellectual resources of Britain's two most prestigious universities. In sharp contrast to Hardy, Turing was never one to separate the pure from the applied, and Whitemore effectively captures this trademark quality:

> TURING: I realized that my ideas in mathematics and logic . . . would acquire military value if war were declared. I was concerned about the moral implications of putting my, uh—intellectual armory—at the disposal of the government at war.
> KNOX: Have you managed to resolve this dilemma?
> TURING: Well, I'm here. Doesn't that answer your question?
> KNOX: Not necessarily. (30)

When Knox presses him on "loyalty to your country," Turing bristles at the question. Uncomfortable with the very concept of patriotism, Turing seems aware that at some level his country will not be capable of returning whatever loyalty he might show it. For Turing, the calculation is a purely pragmatic one; war is a "lesser evil in the last resort," and Turing ends the debate with an assertion that reverberates into the interspersed episodes with the police:

> TURING: I have always been willing—indeed eager—to accept moral responsibility for what I do. (31)

At the end of their interview, Knox introduces Turing to one of his future assistants named Patricia Green. Here again, Whitemore has bestowed a fictional name on an important historical figure in Turing's life. Joan Clarke was a trained mathematician assigned to Turing's cryptanalytic team at Bletchley

"I have always been willing—indeed eager—to accept moral
responsibility for what I do." *Breaking the Code*; Derek Jacobi (Turing),
Michael Gough (Knox); Theatre Royal Haymarket, 1986.
© Conrad Blakemore / ArenaPAL.

Park. Joan's relationship with Turing went beyond their professional collaborations to include many evening conversations over games of chess, progressing to the point that the two code-breakers decided to get married. In the end,
however, this was infeasible for Turing, and the engagement ended.

At the close of the first act of *Breaking the Code*, the two narrative threads
are still some ten years apart but arcing toward their common endpoint. The
war-era Turing forsakes Pat's affections in hopes of finding something closer
to his lost love for Christopher Morcom, while the Manchester professor Turing reluctantly agrees to make a statement to the police about his affair with
Ron Miller. Whitemore frames these two developments as pivotal moments
of decision, but in both cases the outcomes feel preordained. It's as though
Turing is given choices with only one available option. This brings us face-to-
face with what it is about Turing's life that ultimately captured Whitemore's
fascination. Yes, the biographical details are riveting in their own right, but
there are other questions at stake that transcend war heroics and issues of civil
rights. "Tolstoy said that free will is merely an expression denoting what we
do not know about the laws of human life," Turing says to Detective Ross at

one point during his questioning. "Perhaps it's an illusion. But without that illusion, life would be meaningless." (82)

To understand how the story of a code-breaking mathematician finds its way to Tolstoy and questions of free will, we have to look more closely at the mathematics. Picking up on the themes that emerged at the conclusion of the previous chapter, it is not an exaggeration to say that embedded in the technical details of "On Computable Numbers" are ideas that can be brought to bear on some of the deepest questions of human nature.

The Universal Machine

> It is possible to invent a single machine which can be used to compute any computable sequence.
>
> —ALAN TURING, FROM "ON COMPUTABLE NUMBERS"

Our journey through Gödel's Theorem provides the setting in which to understand Hilbert's decision problem that Max Newman refers to in the previously quoted passage. If A is a formula of PA, then it expresses some conjecture about whole numbers. The nature of arithmetic means that either A or $\sim A$ is true, and the original hope was that whichever was true would be provable as a theorem in PA. This is completeness, and Gödel showed it was not to be. Having made his peace with this limitation of formal systems, Newman points out that there was "still the possibility of finding a mechanical process for deciding whether A or $\sim A$, or neither, was formally provable in a given system." Even if A were true, Gödel showed it no longer had to be provable within the system, but the question remained as to whether there might be some algorithmic way of deciding whether a particular string like A were provable or not.

The keyword here is *algorithm*, which during Turing's day was essentially a synonym for "mechanical process." In *Breaking the Code*, Whitemore's Turing introduces the idea during his act I interview with Knox:

> TURING: . . . eventually one word gave me the clue. People had been talking about the possibility of a mechanical process, a process that could be applied mechanically to solving mathematical problems without requiring any human intervention or ingenuity. Machine!—That was the crucial word. I conceived the idea of a machine, a Turing machine, that would be able to scan mathematical symbols—to read them if you like—to read a mathematical assertion and to arrive at the verdict as to

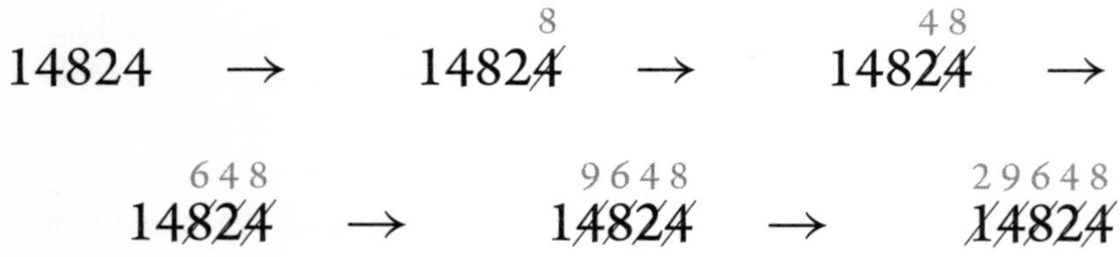

14824 $\rightarrow$ 14824 $\rightarrow$ 14824 $\rightarrow$

14824 $\rightarrow$ 14824 $\rightarrow$ 14824

FIGURE 6.1. Standard algorithm for doubling a large number by hand.

whether or not that assertion were provable. With this concept I was able to prove that Hilbert was wrong. My idea worked.

KNOX: You actually built this machine?

TURING: No, no—it was a machine of the imagination, like one of Einstein's thought experiments. (34)

The historical Turing didn't use the term "Turing machine" in his paper— he called his mental constructions "a-machines," where the "a" stood for "automatic." The idea was to break down the process of computing into a finite number of primitive ingredients. These ingredients were technically objects "of the imagination," as the character of Turing says, but the imagery he used was quite physical. Turing machines sound like tangible entities, which has everything to do with why his approach turned out to be so revolutionary.

Turing Machines

In search of a more rigorous definition for "algorithm," "mechanical process," and eventually "computable," let's examine what happens when a human computer—i.e., a person—uses a familiar paper and pencil algorithm to double a number. Given an input number of, say, 14,824, a standard algorithm for this computation involves starting with the rightmost digit, 4 in this case, and multiplying it by 2 to get 8 (Figure 6.1). Our focus then moves left to the next digit, which is a 2, and we double that which yields a 4. When we get to the 8, doubling yields 16, so we write down the 6 and *make a mental note* that we are carrying an extra 1. When our focus moves left again to the 4, we double it to get 8 just as before, but of course write down a 9 this time. Having written down the 9, we then mentally undo the carry-the-one reminder and return to our normal state of mind. Thus, when we get to the leading 1 and double it, we write down a 2. Finally, seeing a blank to the left of the 1 we stop—or to say it in Turing's language—the computation halts.

Let's now go back and discuss how Turing conceived of a machine to carry out the same algorithm. In a Turing machine, our piece of paper is replaced by a roll of tape with designated squares, or spaces, for written symbols, one symbol per square. The tape is assumed to go on forever in each direction—remember, this is an imaginary machine—so there is always more room on the tape. For the doubling algorithm, we start the tape with our input number written across consecutive squares. The first crucial observation about the computation we just carried out is that, despite being handed a five-digit number, the procedure we followed only required us to look at one digit at a time. Likewise, our Turing machine comes equipped with a *scanner*—Turing's originally term—that looks at precisely one square on the tape at a time. For our purposes, let's agree that when we start a computation on a given input number, the scanner is always set on the rightmost digit of the input number (see the topmost image in figure 6.3).

In addition to the paper tape and the scanner, the final ingredient of a Turing machine is the *table of behavior*. Again, this is Turing's original terminology. A more intuitively useful phrase might be "table of instructions" or "instruction book," but Turing chose his terms to highlight the analogy between Turing machines and human brains. The table of behavior consists of a list of instructions meant to capture the primitive essence of what a person does when looking at the tape one square at a time. Recall that there were two distinct "states of mind" during our computation—a normal state and a carry-the-one state. To double 14,824, we started off in the normal state with our attention focused on the rightmost 4. The rule that we followed might read like this:

If we are in the normal state and we see a 4: then replace the 4 with an 8, stay in the normal state and move our attention one square to the left.

Note the two distinct parts to the rule. The first part ("If . . .") tells us when to apply the rule; the second part ("then . . .") tells us what specific actions to take.

Developing some shorthand notation will be useful as we convert this rule into an instruction for our Turing machine. Just as it was for us during the computation, our machine can be in one of two states. Let's use $N0$ to denote the normal state and $N1$ for the carry-the-one state. Writing $4 \rightarrow 8$ will be our way of telling the machine to "replace the 4 with an 8," and let's use L for "one square to the left" and R for "one square to the right." Our above instruction now condenses to:

If in $N0$ and the scanner sees 4: then $4 \rightarrow 8$, stay in $N0$ and move L.

Returning to the doubling calculation, our scanner is now positioned on the 2, and the next instruction it executes is the following:

If in $N0$ and the scanner sees 2: then $2 \rightarrow 4$, stay in $N0$ and move L.

Our scanner is now set on the 8, so this time the applicable rule is:

If in $N0$ and the scanner sees 8: then $8 \rightarrow 6$, switch to $N1$ and move L.

The command "switch to $N1$" puts the machine into the carry-the-one state and dictates which part of the instruction book to use next. At this point, the scanner sees another 4, and there in fact are two rules in the table of behavior for what to do when reading a 4:

If in $N0$ and the scanner sees 4: then $4 \rightarrow 8$, stay in $N0$ and move L.
If in $N1$ and the scanner sees 4: then $4 \rightarrow 9$, switch to $N0$ and move L.

Because the most recent command we implemented left the machine in state $N1$, the carry-the-one state, the second of these rules applies, and that is the one the machine executes.

Condensing the Instructions

Even for this relatively simple doubling algorithm, the table of behavior has over twenty entries, so it's well worth our while to find an even shorter shorthand for each instruction. At each stage, the action of a Turing machine depends on only two factors:

(i) the symbol being scanned, and
(ii) the current state of the machine.

The action the machine takes consists of three steps:

(a) overwriting the currently scanned symbol on the tape with a potentially new symbol,
(b) possibly changing from one state to another, and
(c) moving the scanner left or right one space on the tape.

Doubling Machine

$(N0, 0 : 0, N0, L)$	$(N1, 0 : 1, N0, L)$
$(N0, 1 : 2, N0, L)$	$(N1, 1 : 3, N0, L)$
$(N0, 2 : 4, N0, L)$	$(N1, 2 : 5, N0, L)$
$(N0, 3 : 6, N0, L)$	$(N1, 3 : 7, N0, L)$
$(N0, 4 : 8, N0, L)$	$(N1, 4 : 9, N0, L)$
$(N0, 5 : 0, N1, L)$	$(N1, 5 : 1, N1, L)$
$(N0, 6 : 2, N1, L)$	$(N1, 6 : 3, N1, L)$
$(N0, 7 : 4, N1, L)$	$(N1, 7 : 5, N1, L)$
$(N0, 8 : 6, N1, L)$	$(N1, 8 : 7, N1, L)$
$(N0, 9 : 8, N1, L)$	$(N1, 9 : 9, N1, L)$
	$(N1, B : 1, N0, L)$

FIGURE 6.2. The complete table of behavior for the doubling machine.

Listing these five pieces of information in order is all that is really needed to specify the instructions. Thus,

If in $N1$ and the scanner sees 4: then $4 \rightarrow 9$, switch to $N0$ and move L

can be economically summarized by the quintuple of symbols:

$$(N1, 4 : 9, N0, L).$$

The two entries before the colon are used to discern if the instruction applies; the last three entries describe the action to be taken. To give another example, the quintuple $(N0, 0 : 0, N0, L)$ corresponds to the rule that says, "If the machine is in state $N0$ and the scanner sees a 0, then leave the 0 on the tape, stay in state $N0$ and move one space to the left."

The complete set of instructions for the doubling machine, written in this new, more efficient notation, is shown in figure 6.2. This list constitutes the machine's table of behavior, which on occasion we will also refer to as its instruction book.

In creating the full list, we have adopted a few conventions. The letter "B" is used to refer to a blank square on the tape. Thus, the rule $(N1, B : 1, N0, L)$ applies when the scanner sees a blank square and the machine is in the $N1$ state. In this case, the machine writes a 1 on the blank square, switches to the original state $N0$, and moves left to what, in practice, will be another blank square. A quick glance over the entire list of instructions reveals that there is

Initial configuration; apply rule $(N0, 4 : 8, N0, L)$

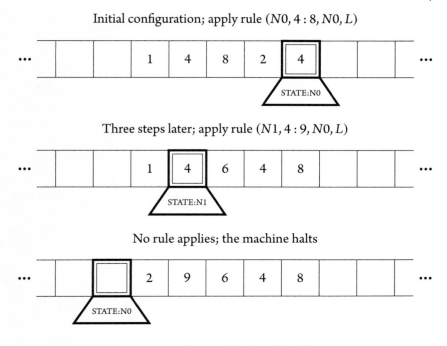

Three steps later; apply rule $(N1, 4 : 9, N0, L)$

No rule applies; the machine halts

FIGURE 6.3. A visual rendition of the doubling Turing machine.

no rule for what to do when in state $N0$ and the scanner sees a blank. When no rule applies, the machine halts.

Whatever number we start with on the input tape, if we set the scanner on the rightmost digit and initialize the machine to be in state $N0$—from now on, these will be unstated assumptions—the machine will mechanically follow its instructions, halting with a new number on the tape that is precisely twice the original input number. In other words, the algorithm for multiplication by 2 is enacted by this particular Turing machine (Figure 6.3).

Turing's Thesis

We have taken our time going through this example in order to make the following bold assertion feel plausible. Just as there is a table of behavior for doubling a number, so is there a table of behavior—written in precisely the same syntax of quintuples—for *any* type of numerical computation carried out according to some well-defined algorithmic process. Whether we consider squaring a number, adding two numbers, determining if a number is prime, or generating the successive digits of π, there is in every case a Turing machine capable of executing the computational steps and arriving at the proper conclusion.

The imaginary hardware of each of these machines is identical—an infinite roll of tape and a scanner that reads one symbol at a time. The essence of the machine is in its table of behavior. The one machine we have designed so far required two states and a total of twenty-one rules. Turing machines that carry out more complicated algorithms will require more states and more rules, but a notable requirement is that the number of rules and the number of states be finite. To generate the digits of π, for instance, certainly requires a good deal more than two states. In fact, explicitly describing such a machine was one of the examples that Turing himself gave in his original 1937 paper to help make the point that his machines provided a rigorous and reasonable definition of "computable." This point is so important that we need to pause over it a bit longer, and give it the title by which it has come to be known:

> **Turing's Thesis:** Any possible algorithm for doing number-theoretic computations can be carried out by a Turing machine.

This principle isn't a mathematical theorem in need of a proof—it represents a fundamental insight into what computation is, long before computers as we know them existed. Just as every whole number is a product of prime numbers, and every molecule is a compound of elements from the periodic table, every computational algorithm can be decomposed into a finite list of primitive instructions expressible as quintuples in the table of behavior of a Turing machine.

The Universal Machine

"On Computable Numbers with an Application to the Entscheidungsproblem" was written in three parts. The first outlined the concept of Turing's a-machines; the last specifically addressed the decision problem. Although addressing this part of Hilbert's program was the driving motivation for the paper, it was the middle section where Turing introduced the concept of a *universal machine*, which would literally change the world.

The existence of Turing's universal machine is a natural corollary of Turing's thesis, with an assist to Kurt Gödel. As Turing explains, given any algorithmic task there is a Turing machine capable of carrying out that task.

Different tasks require different sets of rules—the doubling machine, the squaring machine, the is-it-a-prime machine each come with their own specific instruction book. The image at this point is a one-to-one correspondence between algorithms and instruction books; each task requires its own instruction book, and each instruction book is designed for one specific task.

Now here is an experiment. What does the machine with the short instruction book:

<u>Mystery Machine</u>

$$(N0, 0 : 0, N1, R)$$

$$(N1, B : Y, N2, R)$$

do with the input number 1729? How about with 1730?

To answer this, we do the natural thing—we "upload" the instruction book into our brain and simulate a run of the machine starting with each of the two numbers. In the case of 1729, the machine halts immediately because neither of its two rules applies. In the case of 1730, the machine moves to the right and prints a Y immediately to the right of the input number and then halts. Interpreting the Y to be the machine's way of saying "Yes," it becomes evident that this clever machine's particular task is distinguishing multiples of ten from nonmultiples of ten.

Although this machine is quite simple, we could easily do the same kind of analysis when presented with a machine with a more complicated table of behavior. Simulating a "run" of a Turing machine is a predictable and deterministic venture—at each step we look at the scanned symbol, keep track of what state the machine is in, and carry out the rule (if any) that applies.

The central point is this: *Executing the instructions of a Turing machine is a straightforward algorithmic process.* As we give this observation a moment to sink in, let's recall the content of Turing's thesis, which is that *every algorithmic process can be carried out by a Turing machine.* Combining these two truths of the computational universe leads directly to the following conclusion:

There must be a Turing machine—which Turing called the *universal machine*—whose specific task is to carry out the instructions of other Turing machines.

If the earth doesn't shake underneath our feet at this epiphany, it is only because we have all come of age in a world dominated by incarnations of universal machines. Phones that take photos and play movies and sort our mail are so familiar that we no longer stop to ponder how such multitasking is even possible. As it turns out, these devices are the direct descendants of Turing's insights in "On Computable Numbers," which shattered the one task/one machine correspondence. To put it succinctly, just as Turing machines provided an abstract model of computation, Turing's universal machine provided an abstract model for the modern computer—a flexible, all-purpose machine capable of executing any algorithmic task on the same hardware.

A reasonable objection to the possibility of a universal Turing machine whose specific task is to implement the instruction books of other Turing machines is that, up until now, the input data for a Turing machine has always been in numerical form. Turing's thesis says that all *number-theoretic* algorithms can be simulated by a Turing machine. Although following the instructions for one of these machines is certainly algorithmic, it doesn't appear to be number-theoretic. Thus, one could argue that Turing's thesis doesn't apply here and the existence of a universal machine doesn't follow.

Our extended journey through Gödel's proof—which Turing learned in his course from Max Newman—has prepared us well for handling this objection. If the system of natural numbers is rich enough to mirror the architecture of a formal system like PA, then we should certainly be able to embed the logic of Turing's instruction books into the natural numbers. Indeed, we can do it in the very same way—by Gödel-numbering.

The alphabet of symbols for writing instruction books thus far consists of the letters N, L, R, and B as well as the digits 0 to 9. Picking numbers that vaguely look like the letters they represent, we can start by identifying

$$N \longleftrightarrow 3 \quad L \longleftrightarrow 7 \quad R \longleftrightarrow 9 \quad B \longleftrightarrow 8.$$

As for the digits in the instruction book language, let's first replace them with their PA numerals. Thus,

$$0, 1, 2, 3, 4, \ldots \qquad \text{becomes} \qquad 0, S0, SS0, SSS0, SSSS0, \ldots$$

Now extend our coding scheme by letting

$$0 \longleftrightarrow 0 \quad \text{and} \quad S \longleftrightarrow 5.$$

Thus, the number 2 gets coded with 550 and 4 would be 55550. We also used the letter Y as a symbol on the tape, so for now let's take $Y \longleftrightarrow 4$. In general, we should be prepared to accommodate more nonnumeric tape symbols so a more thoughtful Gödel-numbering approach is necessary, but the point is made. Using this coding scheme and dropping all the unnecessary punctuation, a typical quintuple gets coded in the following way:

$$(N1, B: Y, N2, R) \longleftrightarrow 3508435509.$$

We have no plan to read this new coded language ourselves, but we could certainly do it if we had to—and that is the important point. The same information is all there in numerical form. This particular quintuple came from the second line of the multiple-of-10 machine. To generate the code number for the entire machine, we just concatenate the numbers that correspond to each line of the instruction book. Thus, the multiple-of-10 machine whose complete table of behavior looks like:

Multiple-of-10 Machine

$$(N0, 0: 0, N1, R)$$
$$(N1, B: Y, N2, R)$$

is represented by the single integer

$$300035093508435509.$$

Turing referred to this as the *description number* of the machine.

To generalize from this example:

- To every Turing machine T there corresponds a description number t that is, in essence, a numerical encryption of its instruction book.
- If the number t is given to someone who knows the coding scheme, along with an input number n, this person could algorithmically decode t into its more human-friendly notation and then carry out the resulting instructions on the input number n.
- It follows that there exists a Turing machine U that takes as input the two numbers t and n and imitates the behavior of T acting on n, producing as output the same result that T would have produced.

Despite its ethereal-sounding name, the universal machine U shares the same concrete quality as the other single-task Turing machines we have met. It is not akin to some paradoxical "set of all sets." It doesn't physically contain all the other Turing machines—what it contains is a single set of instructions for how to decode the description numbers of other machines and then perfectly imitate their behavior. In this sense, U is like a talented actor. On the surface, U looks similar to all the other Turing machines in the room—it has a scanner, a tape, and a table of behavior that is above average in length but written in the same syntax as the other machines. Turing actually provided an outline for the table of behavior for U. What is remarkable about U is what it can perform. The machine U is a single machine that, much like a modern computer and a bit like a brain, can adapt its fixed internal hardware to implement an unlimited number of tasks.

The inevitable question—which dawned on Turing even at the earliest points in his thinking—was just how far the similarities ran between his universal machine and the human mind.

Breaking the Code, Continued

I can guess what's going through your mind.

—JOHN SMITH, FROM *BREAKING THE CODE*

Whitemore wrote *Breaking the Code* in 1986—seven years before *Arcadia* was first staged and two years before Stoppard's quantum mechanical spy-thriller *Hapgood* was so negatively reviewed. In this context, it would be understandable if Whitemore had elected to forgo any significant engagement of Turing's mathematics altogether. To his credit, he takes the opposite tack.

In act I, Whitemore gives Turing an uninterrupted six-minute speech— embedded in his interview with Dillwyn Knox—to outline the developments of twentieth-century logic that led to "On Computable Numbers." Turing mentions Russell and *Principia Mathematica*, and then Hilbert and incompleteness. "I think Gödel's theorem is the most beautiful thing I know," he says wistfully, after giving a reasonable accounting of what Gödel actually proved. More remarkably, Whitemore opens act II with an even longer monologue by Turing on the foundations of computer science. The speech is presented as an address Turing delivers to his old prep school and draws from numerous writings and speeches Turing gave on the topic of artificial intelligence, a subfield of computer science Turing was pioneering along with the rest of the discipline.

Turing's abstract universal machine became a theoretical prototype for the electronic computers which came into existence in the years after the war. As these first primitive machines were being built and tested—most notably by John von Neumann in the United States and by Turing and others in England—a public debate ensued as to whether these new "electronic brains," as they were called, might someday rival old-fashioned biological ones. To what extent was the human brain a carbon-based incarnation of a universal

Turing machine? To what extent could a digital computer achieve some degree of human intelligence? Asserting a position that is no less controversial now than it was in Turing's day, Whitemore's Turing concludes his act II prep-school address with a passage lifted almost verbatim from an essay Turing published. "I can see no reason at all," he says,

> TURING: . . . why a thinking machine should not be kind, resourceful, beautiful, friendly, have a sense of humor, tell right from wrong, make mistakes, fall in love, or enjoy strawberries and cream. At the moment such considerations should not concern us; but it might be rather nice— don't you think?—if, one day, we could find out just what a machine can *feel.* (64)

The charge that these lengthy installments of Turing's research into *Breaking the Code* come off as overly didactic ignores the more significant point, which is that Whitemore was brave enough to trust the narrative momentum of his play to Turing's science for several long stretches. The filmmakers of the BBC version of *Breaking the Code* were not so brave, and both speeches were significantly cut. Of course, Whitemore's agenda was not simply to assist Hodges in resurrecting Turing's reputation from its suppressed state. Having made the investment of incorporating these technical ideas into his script, the playwright had artistic dividends to collect.

Whitemore's intentions in this regard start to emerge in a series of scenes midway through the second act. The first of these is an argument between Knox and Turing which takes place several years into their code-breaking wartime collaboration at Bletchley Park. The two colleagues cover a range of topics related to Turing's admirable achievements, but eventually the older mentor shifts the topic of conversation to Turing's abrasive personality. Knox's warning includes a mention of Turing's current tryst with a young engineer but, curiously, his focus is more broadly about Turing's stunted interpersonal skills. Very much in Turing's corner, Knox has the young mathematician's best interest at heart, but Turing is having none of it:

> KNOX: You can't go through life ignoring the effect you have on other people or the effect that other people have on you.
> TURING: *(deliberately provocative)* You can try.
> KNOX: You've spent far too much time thinking about your Turing machines. We are, after all, human beings; and you should try to accept the many imperfections that are part of our human condition.
> TURING: Tolerate, perhaps; not accept. (71)

"You can't go through life ignoring the effect you have on other people."
Breaking the Code; Philip Kerr (Dillwyn Knox), Mark H. Dold
(Alan Turing); Barrington Stage Company, 2014 (photo: Kevin Sprague).

Given that these two were only moments earlier discussing intelligence work, this bit of social awareness coaching on the part of Knox seems like an odd, almost mundane, direction for their conversation to take. Yet it keeps heading the same way. Knox continues:

> KNOX: All right!—Let me give you an example. A few minutes ago, you inquired about my health. Suppose I had answered you directly. Suppose I had told you that I am mortally ill and have only a year or so to live. Suppose I had broken down and wept . . . I can't believe that you would have welcomed such a disclosure. I feel sure that you'd have found it distressing, embarrassing and somewhat inconsiderate. And so—being aware of your feelings as well as my own—it would seem to be both correct and appropriate for me to moderate my response.[6] (71)

Why the extended lecture on human empathy? On the surface, these lines point forward in time, foreshadowing the pain Turing's suicide will cause to those who care about him, but there is a deeper significance related to what Turing's mathematics suggests about human nature. Knox ends his entreaty with a final salvo, borrowing a quotation from the historical Wittgenstein regarding the limitations of scientific investigation:

> KNOX: "We feel that even when all possible scientific questions have been answered, the problems of life remain completely unanswered." (72)

This archetypical humanist position stands in direct contrast to the historical Turing's view, which was that science in general—and his science in particular—was poised to answer some deeply fundamental questions about life. Knox's admonition that Turing should spend less time with machines and more with human beings flies directly in the face of Turing's conviction that the human brain *is* a universal Turing machine of sorts. The irony of this scene is that Knox is trying to argue for the opposing view. Accusing Turing of being insensitively mechanical, Knox asserts his own humanity via his ability to construct a mental copy of Turing in his own mind for the purpose of feeling what Turing would feel. This is how empathy works. Our social behavior is predicated on our ability to experience the world from another person's point of view. And how do we accomplish this fundamental skill? Just as Knox says— by importing a copy of that person's personality into our own brain hardware and simulating his or her response to some hypothetical situation.

Whitemore doesn't give Turing a chance to respond, but if he did, Turing would point out that *emulating the behavior of other Turing machines is precisely and unambiguously what a universal Turing machine is designed to do.* Although Knox was attempting to make the opposite point, it turns out that the human capacity for empathy is actually one of the more compelling arguments for a mechanical model of the human mind.

Universal Empathy

Turing's monologue at the top of act II is about the capacity of a digital computer to model the human psyche, but Whitemore's play is really about the *human* capacity for building models of another human's psyche. Just how robustly can one person inhabit the mind of another? Throughout *Breaking the Code*, Whitemore riffs on the different degrees to which humans empathetically understand each other. In act II, Turing goes to tell his mother about his evolving legal problems. Although typically self-centered and superficial, Turing's mother responds to her son's predicament with a heartfelt gesture expressed in the form of a recounted memory from Turing's childhood. Mother and son embrace, and Turing's next line is telling. "I had no idea you felt like that," he says before she quickly breaks the mood to retreat to her usual safe emotional distance.

The scene that follows finds Turing back at the police station with Detective Ross, and it is at this point in the second act that the two separate narrative threads from past and present merge into their tragic endgame. As Turing makes his official statement to Ross, the encounter he describes between himself and Ron Miller is simultaneously performed for the audience. The picture of Turing that emerges is that of a deeply isolated soul. Turing tries in vain to discuss Tolstoy or science with his young, uneducated dinner guest, but Ron is primarily interested in the free drinks and whatever cash he might be able to get off the old mathematics professor. Turing tells Ron that as a child his friends were numbers, which, as reliable and as fascinating as they may be, are notably not other people. The only human that Turing seems capable of holding dear in his mind's eye is his beloved Christopher Morcom.

Earlier in the play, Turing remembers that when Christopher died, he became obsessed with the question of whether Christopher's mind might continue to exist without his body—a sentiment which the historical Turing expressed, and which Whitemore magnifies through his theatrical lens. Turing clings to his memories of Christopher as though he might make another home for the mental processes of his friend in the still animate gray matter of his own brain. Meanwhile, unable to forge any emotional connection to his latest pickup, Turing summarizes his evening with Ron as "mutual masturbation."[7]

The poignant scene resolves back in Detective Ross's office where Turing is flummoxed by how he has become the criminal in this story:

ROSS: Not much choice, was there? When a man says he's committed a crime, I can't just ignore it.

TURING: Going to bed with Ron is not a crime.

ROSS: It's against the law.

TURING: It's not as simple as that.

ROSS: In your opinion.

TURING: Is your opinion any better?

ROSS: Look, I don't care what you and young Ron get up to. But if it's against the law, I have to do something about it, okay? (83)

Ross has cast himself as a mechanical decision-maker, and Turing, having written the proof, knows better than anyone that there is no such thing:

TURING: Even in mathematics there is no infallible rule for proving what is right and what is wrong. Each problem—each decision—requires fresh ideas, fresh thought. And if that's the case in mathematics—the most reliable body of knowledge that mankind has created—surely it must also apply in other less certain areas. (84)

But the single-tasked detective is unmoved, and unmovable. "Decisions have to be made," is the best reply he can muster. Then, in an innocent-sounding remark that cuts with an unexpected irony he adds: "Nothing personal, Mr. Turing. I understand how you must feel."

The Halting Problem

Although the digital computer is the lasting legacy of "On Computable Numbers," it was Hilbert's decision problem that Turing originally set out to tackle in that paper. His negative resolution of the question is what the character of Turing is alluding to in the previous scene with Detective Ross when he says, "Even in mathematics there is no infallible rule for proving what is right and what is wrong." A sketch of the central idea of Turing's highly original assault on the Entscheidungsproblem offers some final colors that go a long way toward filling out the portrait of Turing embedded in Whitemore's play.

Hilbert asked whether there existed an algorithm for distinguishing provable statements of mathematics, expressed in a formal language like PA, from unprovable ones. Turing managed to replace Hilbert's question with an equivalent one in the language of Turing machines. Recall that to every Turing machine T there is a corresponding description number t that is an encoded copy of T's instruction book. In section two of "On Computable Numbers," Turing justifies the existence of the universal machine U that takes this t as input, along with another input number m, and simulates the behavior of T acting on m. In section three of the same paper, Turing then ponders the existence of a cousin to the universal machine we might call a halting-tester. When a machine T runs on an input, there is the possibility that the machine might get caught in a loop of instructions and never halt. This does not happen with the previously discussed multiple-of-10 machine

Multiple-of-10 Machine

$$(N0, 0 : 0, N1, R)$$
$$(N1, B : Y, N2, R)$$

which halts no matter what number appears on the input tape, but observe what happens when we add two new instructions to the table of behavior:

Modified-Multiple-of-10 Machine

$$(N0, 0 : 0, N1, R)$$
$$(N1, B : Y, N2, R)$$
$$(N2, B : B, N1, L)$$
$$(N1, Y : Y, N2, R)$$

A simulation of this new machine reveals that it still halts immediately on any nonmultiple of ten. On numbers ending in zero, however, the machine embarks on an endless loop. Starting with a multiple of ten, the machine runs forever, printing the same Y over and over again and never halts.

Recognizing that endless loops are problematic for computing machines, Turing asked whether there might be a general method that could determine ahead of time whether a given machine T will halt on a given input m. Might there be some way to segregate out the halting combinations of machines and inputs from the nonhalting combinations, which could be accomplished from a careful scrutiny of the machine's instruction book? Such a process, to be of any value, would need to be predictable, deterministic, and executable in a finite amount of time—and therefore could be carried out by a Turing machine. This machine is the previously mentioned halting-tester:

> **Turing's Halting Problem:** Does there exist a Turing machine—let's call it H—that takes a pair of inputs (t, m), where t is the description number for a machine T and m some arbitrary input number, and outputs "Y" when T halts on m and "N" when it does not?

A halting-tester like H would have tremendous power. We could, for instance, use it to solve Fermat's Last Theorem in the following way. First, construct a Turing machine that takes as input a number m and then methodically cycles through all combinations of a, b, c, n with $n \geq m$, halting if it ever finds that $a^n + b^n = c^n$. Now feed the description number t of this Fermat machine into H along with $m = 3$. If H says "Y" then Fermat (and Andrew Wiles) was wrong, and if H says "N" (as it likely would given that the validity of Wiles's proof has been painstakingly confirmed) we would have a new proof of Fermat. This sounds too good to be true, which of course it is. The answer to Turing's Halting Problem is the same as it is for Hilbert's Entsheidungsproblem—no and no.

"I can guess what's going through your mind." *Breaking the Code*; John Leonard
Thompson (John Smith), Mark H. Dold (Alan Turing); Barrington Stage
Company, 2014 (photo: Kevin Sprague).

Processing Our Own Description Numbers

Turing's strategy for showing H does not exist is to combine a hypothetical
version of H with the very real universal machine U into a hybrid machine
we'll denote as UH. Like H, U takes inputs of the form (t, m), where t is the
description number for a Turing machine T and m is an input value for T.
On its own, U is programmed to emulate the behavior of T running on m.
In the hybrid machine UH, H acts like a filter for U. In step one, H investigates the pair (t, m) and decides whether T will halt on the input value m. If H
determines that T does not halt on m, then H halts right away. If, on the other
hand, H determines that (t, m) is acceptable—i.e., that T acting on m eventually halts—then H passes the input pair (t, m) on to the universal machine
U which then imitates T acting on m. The key observation is that the hybrid
machine UH halts on *every* input pair (t, m). Why? Because H filters out any
nonhalting inputs, and only passes the halting inputs onto U which, imitating
T, must therefore halt.

"Now," Turing writes, "let k be the description number of UH. What does
UH do with (k, k)?"[8] The Russellian answer is that UH halts, and it doesn't
halt. Consider what happens inside the mechanisms of UH when it confronts
its own description number. The filtering component H would say that, yes,

the pair (k, k) is an acceptable input since, as we've just argued, UH halts on every input pair. But then what does U do with (k, k)? It imitates the behavior of UH which, starting over with H, confirms the acceptability of (k, k) and passes it along to U. And we are back where we started! The machine UH, designed to halt on every input pair is very much going round in circles unable to halt—a blatant contradiction. The problem here is not the existence of U, whose instruction book Turing sketched out, but the halting-tester H.

What makes this proof especially interesting is the image of the machine UH struggling, and ultimately failing, to process its own description number. Remembering that Turing would come to view the human mind as an incarnation of one of his abstract universal machines, we start to understand that there is more than an artistic metaphor at stake when we ask what happens when a universal machine encounters its own table of behavior—or one from some other machine of the same genus. The logical paradoxes that arise are inevitable, and just as these paradoxes become the core of the proofs for the limitations on what is theoretically computable, they are also highly suggestive of the limitations we face trying to comprehend ourselves and each other. By virtue of being a playwright, Whitemore is uniquely positioned to recognize this suggestion lurking beneath the surface of Turing's mathematics. The human propensity for empathy—for emulation to the point of animation—lies at the heart of how drama functions as an art form. It is the mechanism by which an audience transforms the illusion perpetrated by actors into a visceral experience, and in *Breaking the Code* it becomes the central motif of the play.

Taking his cues from Turing's proof, Whitemore repeatedly demonstrates the inherent fallibility of our attempts to process the table of behavior of other human beings. When Detective Ross says, "Nothing personal, Mr. Turing, I understand how you must feel," this represents the inflexible, single-tasked police officer's attempt at a superficial kind of sympathy. For the literal-minded mathematician he has just booked, however, it is a personal and painful example of another uncomputable problem:

ROSS: Nothing personal, Mr. Turing. I understand how you must feel.
TURING: No you don't.
ROSS: You're right—I don't.
TURING: How could you?
ROSS: I can't. *(Goes to exit.)* You'll receive official notification of the court proceedings. And keep it simple, sir; all that stuff about mathematics won't go down too well with the local magistrates. *(Ross exits.)* (84)

Alan's Apple

In order to avoid jail time, the historical Turing agreed to a year-long course of estrogen treatments intended to remove any desire he might have to engage in sexual activity. Apart from growing breasts, it's not clear what effect the drugs actually had. As if the drug treatments were not sinister enough, Turing, as a convicted homosexual, was now viewed as a national security risk and kept under surveillance.

There is no shortage of reasons why Turing would decide to take his own life, but neither is there any firm consensus on what ultimately drove him to it. On the night he died, his hormone treatments had been finished for more than a year. Professionally, his research had turned to an interesting question in mathematical biology—a search for the mechanism by which a few initial cells morphed into the complex structures of something like a fir cone. He also maintained a few friendships with colleagues at Manchester and beyond, which ranged from cordial to legitimately warm. Giving no hint to any of these people that he was under duress and leaving no note of explanation, Alan Turing exited the living world with a theatrical flair characteristic of the way he had occupied it. On the morning of June 8, 1954, Turing was discovered dead in his home, and next to his body was an apple missing several bites. The cause of death, it was later determined, was poisoning by potassium cyanide— one of a number of dangerous chemicals Turing kept in his house for various experiments.[9]

What was the point of this final bit of stagecraft? Was it an indication of impending madness, like that which overtook Cantor and Gödel before him? Was Turing assassinated by the government that convicted him before he could decide to take revenge and the murder scene arranged to look like an accident? Or perhaps the staging was a way for Turing's mother to interpret his death as an accident—a theory she did in fact adopt and propagate for the remainder of her life.

There is no sure way to fill in this blind spot in the historical record, which leaves Whitemore free to paint it as he wants. His solution is elegant, in particular because it adds a final exclamation point to the implications Turing's mathematics have to suggest about what it means to be human. "Can the mind exist without the body? Can mental processes take place in something other than a living brain?" Turing asks the audience at the close of the play. He is alone downstage, apple in hand. "How are we to answer that question satisfactorily?"

"Can the mind exist without the body?" *Breaking the Code*;
Mark H. Dold (Turing); Barrington Stage Company, 2014
(photo: Kevin Sprague).

Mixing some of Turing's youthful curiosity in with his current despair makes the moment no less chilling, but at least there is some comfort in seeing the brilliant mathematician back in control of his own fate:

> TURING: Being a practical man as well as a theorist, I tend to look for practical solutions; in this case namely, viz., to dispose of the body and to release what is left. A mind. Or a nothing. Here I have an ordinary apple: red and ripe and English. And here—a tin containing potassium cyanide. (*The ghost of a smile.*) Nothing could be easier, could it? Dip the apple in the brew, let the sleeping death seep through. (105)

Lovesong of the Electric Bear

Have I been saying all that stuff to a man?

—PORGY, FROM *LOVESONG OF THE ELECTRIC BEAR*

Snoo Wilson's 2003 play about Alan Turing starts where Whitemore's ends. In the play's opening moment, Turing lies lifelessly wrapped in a sheet, holding

"Next thing you know, it'll be the morning of your youth." *Lovesong of the Electric Bear*; Tara Giordano (Porgy), Alex Draper (Turing); Potomac Theatre Project NYC, 2010 (photo: Stan Barouh).

a bitten apple. "He could be Christ taken down recently from the cross," the stage directions read. But Turing's death scene is interrupted by Porgy, Turing's stuffed teddy bear who, in addition to being animate himself (herself? itself?), displays a number of other metaphysical talents. Porgy magically restores the half-eaten apple into a perfect red whole and then whisks Turing onto his bicycle where the two fly back through time, crash-landing ungracefully into Turing's childhood years.

Guided by Porgy the bear, Turing embarks on a fanciful romp back through the events of his life. We see him tormented at Sherborne School, heartbroken at Christopher Morcom's passing, running through the Backs at King's College and sauntering with Dilly Knox through dark alleys on his way to a secret rendezvous with a Polish intelligence contact—and all this in just a few minutes of act I. The scenes roll by in a seamless, dreamlike succession, bit by bit filling out a detailed image of the mind and heart of the enigmatic Turing. Throughout the journey, Porgy tags along in the role of narrator and protector, filling in as needed for various characters or pieces of scenery. At Sherborne, for instance, he becomes a toilet seat where his young master can sit and work out calculus problems. There aren't any hard and fast rules for what the bear can and can't do. This causes some confusion for the audience—and occasionally for Porgy—a state of affairs that seems quite deliberate on the part of the playwright. In an early moment from Turing's days as a fellow at Kings, Porgy plays himself. The scene begins with two snarky undergraduates entering Turing's office for a tutorial and finding that Turing hasn't arrived. Porgy is there, however, perched in the corner with a book on his lap:

> UND'GRAD 2: And he keeps his teddy bear in his rooms too. Weird! *(Prods bear)*
> PORGY: Careful how you poke!
> UND'GRAD 1: Makes your average nutty mathematics professor look positively normal.
> PORGY: Oy! Leave my ears alone. And don't try to push your chubby finger up me, I'll have you rusticated and then tarred and feathered! One word to Master is all it takes. Settle down!
> UND'GRAD 1: What's the bear reading?
> UND'GRAD 2: *(looks.) Principia Mathematica*, by Bertrand Russell.
> PORGY: Yes indeed. My dear students, this noble tome attempts to establish the logical truth of mathematics. It was written before the Great War and we now know it is impossible to establish the logical truth of mathematics. So you could say I am reading it in some disappointment. Bertie Russell got it wrong, simple as that. (18)[10]

The undergraduates don't hear Porgy in this particular scene, which is for the best really because Porgy soon heads off into stories about how Russell was a "hopelessly Randy professor" who spent too much time "banging Tom Eliot's poor mad wife" to do any quality work in logic. Turing arrives just in time to take over, but he is an unsightly mess, having come directly from a vigorous run that involved chewing grass for extra energy and spitting out the excess cellulose:

> TURING: Thank you, bear. *(To undergrads)* It's become clear to the next generation of mathematicians that mathematics is no longer classical or

logical. The stuff used to build bombs and bridges with turns out to be unpredictable or ambiguous. Sometimes it stops, and won't go on. Maths is incomplete.

PORGY: You're surely not planning to give this tutorial looking as if a camel has been sick down your front? (19)

Soon enough, Turing is arguing with his stuffed bear about the legitimacy of the universal machine, which prompts the undergraduates to head for the exit. Turing shouts his explanation of the halting problem to the departing students, and when that fails to lure them back, he switches over to the topic of Snow White, which fares no better. Alone with his bear, Turing carries on with his over-the-top impersonation of the wicked queen, eventually feigning his death from a poisoned apple—all to Porgy's deep dismay:

PORGY: O Master, do nothing lightly or presumptive here! At twenty-seven, I know you think you'll live forever, but all too soon you will encounter that interface with the eternal, where even computation stops!

TURING: Who gives a shit about your take on mortality, Porgy? Push your entertainment button now. (*Pause*) Do it! (21)

God's Holy Pantomime

Evident from this short glimpse of Wilson's script is the way that the irreverent and fanciful style of the play meshes with the eccentricities of Turing's personality. Whitemore's traditional realism in *Breaking the Code* gives his play the aura of being authentically biographical, but Wilson's flying bicycles and talking bear liberate him from any rigid expectations in this regard. Ironically, Wilson's whimsical strokes make it possible to gather up a much broader selection of anecdotes that seem farcical but are, in fact, historically accurate. Chewing and spitting out wild herbs while he ran, chanting the evil queen's verse from Snow White in the halls at Kings College, burying silver bullion during the war, running with a gas mask and an alarm clock around his waist, intentionally omitting his signature on his military home guard form so that he would not technically be a soldier, chaining his coffee mug to the radiator, even keeping his teddy bear in his room—all these pieces of Turing lore are woven into Wilson's play. As a rule, whatever might possibly be true in *Lovesong* very likely is. In a scene set in a gay bar, Turing responds to a question from the bartender about what he does for a living by pretending to be a poet:

BARMAN: Hey; lemme hear some!
TURING: Hyperboloids of wonderous light
    Rolling for aye through space and time
    Harbour the waves that somehow might
    Play out God's holy pantomime. (51)

This enigmatic verse was written by the historical Turing on a postcard to his colleague Robin Gandy, three months before he died.

By playing fast and loose with the rules of reality, Wilson is able to create a more colorful portrait of Turing's life and also Turing's ideas, especially his later ideas on artificial intelligence. In *Breaking the Code*, Turing meets Joan (who is Pat in Whitemore's script) during his Bletchley Park interview with Knox, and the pair immediately settle into a discussion of the German Enigma machines. Wilson imagines Turing's first encounter with his future fiancée in a different way. In *Lovesong*, Turing is at his parents' house for Christmas dinner. Porgy is also there, standing center stage, decorated like a Christmas tree while quietly singing *Silent Night* in German. As the family argues over Turing's perpetually dirty fingernails and other moral failings, the stage gradually fills with smoke, ostensibly coming from the burning turkey in the kitchen. Then, from out of the smoke, rises Joan. Awash in glittering lighting, Joan is naked except for a swirl of numbers projected onto her skin from every direction:

TURING: Who else is invited to dinner, father? It seems we are not alone.
TURING SR: You mentioned your readiness for marriage, son, and now
    Christ's glorious natal day brings the gift of spirit clothed in flesh. Fear
    not. (29)

This visually fantastic moment points to a prevalence of religious symbolism in *Lovesong*. Notably, there is a slow gathering of tension between the will of the gods—as interpreted by Porgy—and Turing's endeavors of both the mind and the heart.

In this Christmas moment, the gods seem pleased. Joan has arrived as a gift in exchange for the burnt sacrifice of the Christmas turkey, but the will of the gods is a finicky thing in this morality tale. In exchange for the buried silver—which the historical Turing never found—the gods provide Turing with a second reward, in the companionship of Arnold. This gift, however, comes with a stipulation:

PORGY: How was your gift?
TURING: Pretty brief.

PORGY: You should never bestow a greater affection on anything else, or the gods will be angry. . . . The gods' prohibition against any greater love than the one you have for their gift extends to all created things. I hear there's a powerful computer being built at Cambridge. (57)

And this, in Wilson's telling, turns out to be Turing's fatal sin: his deeply held belief that there is no qualitative divide between minds and machines. "Your ancient gods have to learn humans and animals are no different to computers. We compute," Turing says to Porgy in response to his warnings. Looking back at the birth of Joan moment, Wilson has provided a visual homage to Turing's convictions on this point. The ethereal being that arises out of the mist isn't Joan at the outset but an emotional and mental blank slate, much like the so-called "learning machines" that the historical Turing imagined could be constructed at some point in his not-too-distant future. In Wilson's vision, Joan begins to take shape with Porgy's dutiful assistance.

"Who am I, bear?" she asks, and then, "Tell me, why should I be with this man, and not another?"

"We shall discover in due course, madam, presently," Porgy responds, and instructs Turing to introduce himself to this new creature. Joan is now fully dressed, and at the instant she takes Turing's hand, the lighting abruptly changes, the bustling sounds of the code-breaking business of Hut 6 kick on, and the narrative of Joan and Turing's Bletchley collaboration is taken up in earnest.

Can Machines Think?

Turing's demise at the hands of a host of angry gods is—as far as we know—a narrative device employed by Wilson to connect the protagonist of *Lovesong* to a distinguished line of other mortals who dared to venture out of their designated domain. Like Marlowe's Doctor Faustus, Wilson's Turing follows his ambitions to become too much like the deities in charge, and he suffers the consequences.[11] The historical Turing was an atheist, and his research into what he called "intelligent machinery" did raise some religious and philosophical ire in his day, as it has ever since. Turing addressed these challenges head-on in numerous lectures, in published essays, and once, quite notably, in a public roundtable forum broadcast over the radio. One of the byproducts of this ongoing debate was the so-called Turing test—an eponymous piece of Turing's legacy that displays a curious kinship to improvisational theater.

The following short scene is not from Wilson's play, nor is it from Hugh Whitemore's. It was, in fact, penned by Turing himself and reveals a dry and playful wit:

> INTERROGATOR: In the first line of your sonnet which reads "Shall I compare thee to a summer's day," would not "a spring day" do as well or better?
> WITNESS: It wouldn't scan.
> INTERROGATOR: How about "a winter's day." That would scan all right.
> WITNESS: Yes, but nobody wants to be compared to a winter's day.
> INTERROGATOR: Would you say Mr. Pickwick reminded you of Christmas?
> WITNESS: In a way.
> INTERROGATOR: Yet Christmas is a winter's day, and I do not think Mr. Pickwick would mind the comparison.
> WITNESS: I don't think you're serious. By a winter's day one means a typical winter's day, rather than a special one like Christmas.[12]

From the opening question, we might take the Witness to be Shakespeare—or at least that seems to be the working assumption. The Interrogator, however, is suspicious, and her lawyerly inquisition suggests she is trying to ferret out whether the Witness is really the sonnet writer he claims to be, or whether he is a charlatan. The Witness, thus far at least, is holding his own in this examination, demonstrating a slightly mechanical sense of meter ("a spring day" wouldn't scan) as well as a reasonable understanding of how analogies do and don't work in Elizabethan love poetry.

So how does this banter relate to inciting the ire of the gods in Wilson's play? The dialogue is actually part of an essay Turing published in 1950 in the philosophy journal *Mind*. The article is entitled "Computing Machinery and Intelligence." Its wide accessibility, together with its mixture of insight and provocative jabs, make it the most widely read and debated of all of Turing's writings. It opens with the sentence, "I propose to consider the question, 'Can machines think?'" but Turing then quickly declares the question too ambiguous to be useful. From experience, Turing knew that arguments about the possibility of intelligent machinery very quickly bogged down over attempts to define terms such as "think." When he sat down to write "Computing Machinery and Intelligence," the artificial intelligence debate had been going on for some time, and Turing's agenda in this particular paper was to move the conversation from the abstract philosophical arena to an empirical one. "I shall replace the question by another," he writes, "which is closely related to it and is expressed in relatively unambiguous words."

Turing first describes a game played with three people: a man, a woman, and an interrogator who can be of either gender. The man and the woman are positioned out of the interrogator's sight equipped with a method of communicating that doesn't reveal tone of voice. The game consists of the interrogator asking questions of the two with the goal of determining who is the man and who is the woman. The woman's job is to help the interrogator by answering honestly. The man's job is to try to trick the interrogator into making the wrong identification by fabricating answers he thinks a woman would give.[13] With the parameters of the man-woman imitation game firmly established, Turing gets to the punch line:

> We now ask the question, "What will happen when the machine takes the part of [the man] in this game?" Will the interrogator decide wrongly as often when the game is played like this as he does when the game is played between a man and a woman? These questions replace our original, "Can machines think?"[14]

The imitation game between human and computer is Turing's test. If a computer can pass itself off as a human after, say, ten minutes of questioning from a skeptical interrogator, then, in Turing's view, no matter how one defines "thinking," the term can be safely applied to describe what the computer is doing.

There are no restrictions to the topic of conversation. Nothing is off-limits, and the interrogator should indeed make an attempt to probe the intellectual constitution of each witness in an effort to distinguish the human from the machine. To illustrate this point, Turing provides the following sample dialogue:

> Q: Please write me a sonnet on the subject of the Forth Bridge.
> A: Count me out on this one. I never could write poetry.
> Q: Add 34957 to 70764.
> A: (*Pause about 30 seconds and then give as answer*) 105621.
> Q: Do you play chess?
> A: Yes.
> Q: I have K at my K1, and no other pieces. You have only K at K6 and R at R1. It is your move. What do you play?
> A: (*After a pause of 15 seconds*) R–R8 mate.[15]

By including poetry, arithmetic, and chess in this quick exchange, Turing makes the point that the respondents should be prepared to answer questions about anything from spelling to special relativity. This sample dialogue also makes another point—this one not so explicit. In the addition problem on the fourth line, the answer given after the long pause is actually wrong! What should we make of that?

From the interrogator's point of view, the decision to make is whether the respondent is a human with a tendency for sloppy arithmetic or a machine imitating the kind of mistake a human is liable to make when working with numbers over a certain size. Whichever it might be, we are reminded that the machine is charged with a curious task in Turing's game. Let's return for a second to the man-woman version of the imitation game. For a man to successfully impersonate a woman under close examination requires him to be a performer of sorts, capable of adopting a very different mindset. Because there are no limitations on what questions can be asked, there is no practical way to prepare a ready-made set of responses. The man could get lucky and anticipate a few questions about shoe sizes or recommended daily vitamins, but the vast universe of possible questions makes the prefabricated answers approach untenable. The only way for the man to successfully deceive the interrogator for more than a minute or two of questioning is to try to inhabit the frame of mind of some real or imagined female person and answer the interrogator's questions from that point of view.

A machine attempting to pass itself off as a person is faced with a similar challenge. Despite the extraordinary speed and exponentially expanding memory capabilities of computers, any attempt to pass the Turing test by providing the machine with an enormous lookup table of suitable responses to every conceivable question is doomed to fail. The mathematics is incontrovertible on this point. The universe of question-and-response possibilities is too large by so many orders of magnitude that we can safely discard this possibility both for now and indefinitely into the future. No, a machine that passes the Turing test in the form that Turing proposed it would be a most remarkable entity. There is no reason to suppose that an intelligence supported on a machine should be similar to human intelligence. In fact, there is every reason to expect that the cognitive functions of a thinking machine would have very little in common with human beings, given the stark difference in how these two forms of intelligence would have come about. A thinking machine would likely be capable of a large number of intellectual feats very distinct from anything in the human repertoire. A Turing test winner, however, would list among its talents the ability to replicate a human frame of mind. Whatever else it might do or think or feel, such a machine would have to be a reasonably competent improvisational actor with a rich knowledge of human culture.

Turing was aware he had set a high bar for the computer, but that was by design. His agenda was to bypass the philosophical quagmire around the definition of intelligence and set up a sufficient—but by no means necessary—criterion for demonstrating the presence of intelligence in a machine. Turing predicted the computer would routinely pass this test in "about fifty years' time," an estimate he increased to a hundred years in subsequent discussions. Even this modified estimate appears to be ambitious; Turing had set the bar higher than even he had realized.

Origins of the Hard Problem

The most entertaining section of "Computing Machinery and Intelligence" is the one where Turing sets out a list of argumentative positions opposed to his own, giving each a useful moniker and short summary. Here are the first three:

(1) *The Theological Objection:* Thinking is a function of man's immortal soul. God has given an immortal soul to every man and woman, but not to any other animal or to machines. Hence no animal or machine can think.

(2) *The "Heads in the Sand" Objection:* The consequences of machines thinking would be too dreadful. Let us hope and believe that they cannot do so.

(3) *The Mathematical Objection:* There are a number of results of mathematical logic which can be used to show that there are limitations to the powers of discrete-state machines.[16]

The Mathematical Objection might sound familiar. It is essentially the one that comes up in Stoppard's *The Hard Problem* when Hilary asks her colleague to explain to her "how Gödel's Proof means a brain can't be modelled on a computer." Because formal reasoning is inevitably saddled with incompleteness, the argument goes, and because computers are built on mechanical—i.e., formal—structures, computers will always possess gaps in their capability to respond to certain questions in the imitation game that a human could successfully answer.

As Turing presents the various counterpositions, he simultaneously rebuts them, and his rebuttal to the Mathematical Objection can be summarized as a caution against overestimating human capabilities. "We too often give wrong answers to questions ourselves to be justified in being very pleased at such evidence of fallibility on the part of the machines," he writes. Although this might feel like a case of trying to bring humans down to the level of machines, it's better understood as a reminder that fallibility goes hand in hand with intelligence. Incompleteness is not a *flaw* of PA; it is a necessary side effect of achieving a certain level of sophistication. Along the same lines, "If a machine is expected to be infallible," Turing says, "it cannot also be intelligent."[17]

As we compare the historical Turing's ruminations on machine intelligence with Stoppard's portrayal of this debate in *The Hard Problem*, it is worth remembering that Turing was writing more than sixty years before Stoppard, at a time when the digital computer was in its nascency. Despite the primitive nature of Turing's Manchester computer and others like it, Turing was already many miles down the road considering where this new technology might lead. For example, the next counterposition on Turing's list is the one that most accurately tracks with Hilary's objections in *The Hard Problem*. Indeed, it is the one most commonly adopted by ardent humanists of all sorts when faced with the possibility of having a nonbiological intellectual rival:

(4) *The Argument from Consciousness:* This argument is very well expressed in Professor Jefferson's Lister Oration for 1949, from which I quote. "Not until a machine can write a sonnet or compose a concerto because of thoughts and emotions felt, and not by the chance fall of symbols, could we agree that machine equals brain—that is, not only write it but know that it had written it. No mechanism could feel (and not merely artificially signal, an easy contrivance) pleasure at its successes, grief when its valves fuse, be warmed by flattery, be made miserable by its mistakes, be charmed by sex, be angry or depressed when it cannot get what it wants."[18]

The Professor Jefferson whom Turing quotes was Sir Geoffrey Jefferson, chair of the Department of Neurosurgery at Manchester University, and his Lister Oration entitled "The Mind of Mechanical Man" was directed squarely at Turing.

Turing's rebuttal of this objection is essentially a defense of his proposed imitation game. This is where the previously quoted dialogue between

Interrogator and Witness regarding the "summer's day" sonnet appears. Referring to the insightful repartee of this dialogue, Turing's point is that if a machine provides answers of this nature for a sustained period of time, the only reasonable conclusion would be to concede that the respondent not only wrote the sonnet but, to quote Jefferson, "knew that it had written it." In a sense, the entire argument of "Computing Machinery and Intelligence" comes back to this central claim—that whatever objections people might have to the possibility of artificial intelligence would be overcome if they witnessed a machine winning the imitation game on a regular basis.

As it turns out, statistically based language models are prompting the need for some modifications to Turing's original criteria, but his insights from 1950 are, remarkably, still at the center of the debate.

Swansong of the Electric Bear

When Alan Turing's *Lovesong* journey through time reaches Manchester, Snoo Wilson sets the scene in the dark and dirty bowels of the computer lab. Bronwyn Smith is a graduate student who has sweated away months attempting to implement Turing's famously convoluted flexible programming scheme. For this encounter, Porgy has donned a modest disguise to pass himself off as Sir Porgy Bear, who has come to investigate whether Turing's lab warrants further funding:

> PORGY: It's certainly helped your cause, to go on the radio and stand up for artificial intelligence against the panel of bishops. But I missed the last part of the debate when my wife went into labour. Let me ask you: does the concept "thinking machine" involve awareness? Can you have a machine think which is not conscious of thinking?
> TURING: If you can't tell the difference between human and computer in tests then we have to empirically assign consciousness.
> PORGY: Really? Empiricism has not fared well in human hands, as the measure of things. (66)

Like nearly every other anecdote in Wilson's play, the radio broadcast debate was a historical event from 1950 and included Sir Geoffrey Jefferson as well as Turing and Max Newman. With great effect, Wilson stages his version of this debate among Turing, Bronwyn, and Sir Porgy Bear inside the electronic brain which is the Manchester computer. No help to Turing's cause, the Manchester machine is continually shutting down due to technical problems:

"The test runs stopped for Manchester United taking a penalty kick." *Lovesong of the Electric Bear*; Lilli Stein (Bronwyn), Tara Giordano (Porgy); Potomac Theatre Project NYC, 2010 (photo: Stan Barouh).

BRONWYN: The test runs stopped for Manchester United taking a penalty kick.

PORGY: Bronwyn here is a trifle skeptical about machines' ability to think.

TURING: What does Bronwyn know?

BRONWYN: I knows my own skepticism; surely, you can't challenge that.

PORGY: Skepticism; a state of self-doubt not yet reproducible in any machine.

TURING: Self-doubt can be programmed in, Sir Porgy, I'm sure. It will have a basis in mathematics.

BRONWYN: Not necessarily. Self-programming doesn't feel like consciousness to me. Either this machine is self-aware or it's not. Either the machine is hosting an intelligence, or it's a lot of electrons doing a dance without knowing it. (67)

Wilson gives Bronwyn the upper hand in this exchange and the ones to follow, and this argumentative defeat coincides with the arc of Turing's legal downfall. In the same dazzling way that *Lovesong* whirled us through Sherborne, Cambridge, and Bletchley Park on Turing's way up, we now follow Turing's rapid descent—from his arrest, through his sentencing, to his counseling sessions with an Austrian psychotherapist. Having seen the ending at the opening curtain, the audience knows where events are heading, and soon enough Turing is alone on stage with poisoned apple in hand.

Given the underwhelming report card of AI achievements in the decades since Turing—at least compared to early expectations—it is tempting to view the protagonist of Wilson's play as a dreamer who, despite his substantial cleverness, was ultimately guilty of overstating the possibilities of employing mathematical ideas in areas where they were never meant to be applied. That, at least, is the story that the characters within Wilson's play tell us—but it is not the story the play ultimately shows us. In the closing scene of *Lovesong*, Turing is joined by Porgy, the lifelong companion the historical Turing never had. This is a crucial point where understanding Alan Turing the man is vital to understanding Alan Turing the mathematician and part-time philosopher. Turing's various biographers point to his social isolation as a powerful motivator in Turing's eagerness both to imagine and to bring about a new type of intelligent being. Denied the companion he longed for, Turing resorted to creating one, or at least to defending its right to exist in a world that had proved hostile to people like himself. As a homosexual man living in midcentury England, Turing had to navigate popular assumptions about his own inferiority while simultaneously engaging in a version of the imitation game that compelled gay men to emulate a heterosexual lifestyle. These hard truths bring a deeply poignant quality to Turing's plea for what he termed "fair play" on behalf of the machines.[19]

In giving life to an inanimate bear, Snoo Wilson finds an unobtrusive way to put Turing's grandest dream in the center of the play while also highlighting the reality of Turing's painful isolation. As the title states, Porgy is an "electric bear," justified as a product of Turing's imagination and at the same time an artistic rendering of a universal machine. Porgy successfully imitates humans throughout the performance without the slightest hiccup in his programming. Of the two central characters, in fact, Porgy is the one who comes across as more human in some ways. The unevenly immature Turing fascinates us and earns our respect, but Porgy gets inside our heads:

PORGY: What's this highly poisonous substance doing now right by your bed? Surely it should be placed out of reach, in case of accident during repose, master.

TURING: Leave it there or I'll cut your paws off!

PORGY: Has it come to that, master?

TURING: It's come to that. I'm chucking in the towel. Stop staring Porgy.
(79)

As Porgy pleads with Turing to reconsider, it is the bear's pain the audience feels:

PORGY: To kill yourself makes such a terrible hole in the universe, such a blackness, that the rest of us that love you are going to be carrying it for all our lives.

Unmoved, Turing proceeds to carry out his own execution, and Porgy is helpless to stop him this time. The play, however, is not quite over. Porgy's response to the loss of his master becomes Wilson's most potent homage to the legacy of Alan Turing:

PORGY: What, dead? I'm coming with you. Nothing is stronger than this love, for I am nothing indeed without you, Master. Self-evisceration is swiftest. Paws, lead on! (*Tears straw out of chest.*) There'd be a prickling of eyes now, right up to the gods, if they could see who humbly lays his vital organs at master's feet. See, master! There! (*Tenderly*) There's my heart!
(*End of play.*) (80)

Professor Jefferson's diatribe against Turing's machines appealed directly to the matters of the heart—"no mechanism could feel . . . pleasure . . . grief . . . warmed . . . miserable . . . charmed . . . depressed." In this closing moment, Snoo Wilson crafts Turing's retort in an understated and eloquent way—with an electric bear whose final gesture blunts each prong of Jefferson's argument.

Hearts and Minds

Lovesong of the Electric Bear and *Breaking the Code*, as well as *The Hard Problem*, all take up the debate about the viability of constructing intelligent machines, but rendering a verdict on this question is not paramount to any of these playwrights. The real allure of the AI debate is the way it brings a stark clarity to the challenge of identifying what precisely constitutes human intelligence in the first place. To argue for or against the possibility of a mechanized version of ourselves, we first have to understand what makes us who we are. What the mathematics reveals is that the upward journey toward greater levels of formal sophistication comes with trade-offs—although "trade-offs" is too judgmental a word. A handheld calculator would never botch the addition of two five-digit numbers, but Turing saw that infallibility was not a prerequisite for intelligence—in fact, he realized the two were completely incompatible.

Lovesong, more than the other plays discussed, brings out this subtle implication of the mathematics while not being too heavy-handed about it. Hugh Whitemore's Turing wonders aloud in *Breaking the Code* "if, one day, we could find out just what a machine can feel." Snoo Wilson offers a poetic answer with *Lovesong's* electric bear, and the insights are reassuring. Porgy may have an entertainment button, but the machinations of the heart so dear to Professor Jefferson are not subroutines to be programmed in Porgy any more than they are in us. And not only can they not be programmed, they cannot be avoided. Just as formal systems and Turing machines acquire gaps in their capabilities as they grow in complexity, any conceivable evolution toward cognition is destined to be beset with by-products that may seem like imperfections or extraneous baggage but are really part and parcel of intelligence: fallibility, self-awareness, empathy, charm, misery, pleasure, grief. This is the unspoken suggestion that lingers about in the pell-mell of Porgy's inanimate straw at the end of *Lovesong.*

Half Life

A machine? Are you crazy? . . . I'm not a machine. I'm an artist!

—STANLEY, FROM *HALF LIFE*

Why theater? What particular attributes has theater brought to the task of shaping the legacy of Alan Turing? Cloaked in military secrecy and legal wrangling for many years after he died, Turing's groundbreaking achievements have been methodically brought to light over the last few decades, inspiring writers and artists of all sorts to gravitate to his story. Poets, painters, sculptors, and novelists have each brought their unique insights to bear on Turing's reputation, and in 2017 he became the subject of a major motion picture called *The Imitation Game.* While all these art forms can boast of their specific merits, there is a case to be made that theater possesses the most penetrating set of tools for getting to the core of Turing's life and work.

Turing deftly navigated the boundary between the abstract world and the practical, lived-in world in the same nimble way that theater does. In Turing's day, the most prestigious unsolved problem in pure mathematics was the Riemann Hypothesis. It still is. The problem concerns the location of the roots of the Riemann-zeta function, and when Turing was at Princeton, he made significant headway into acquiring a set of precisely crafted gear wheels for a machine he designed to physically generate the zeros. Unfortunately, the war intervened, but the point is still made. Turing's strategy was to probe abstract ideas by incarnating them into tangible entities. Like a theater practitioner fleshing out a two-dimensional script on a three-dimensional stage, Turing sought to understand the Riemann-zeta function by bringing it to life in much the same way that he revolutionized the notion of computable algorithms by enacting them as imaginary machines. In this latter case, his invented machines were so lifelike that they eventually became so. This propensity to embody the abstract goes hand in hand with the spirit of imitation and performance that pervades Turing's thinking and further highlights the connection to theater. Turing's universal machine is a device that enacts, or performs, the function of other Turing machines. When the debate about artificial intelligence hit an academic roadblock, Turing's solution was to cast a computer in the role of a person. Let the philosophers argue until they are hoarse; in Turing's view, the performance of intelligence would guarantee its existence.

Hugh Whitemore unobtrusively put Turing's mathematics in the center of his play by recognizing that the universal machine's ability to mimic the function of other Turing machines is a model for human empathy. Creating models of other people in the hardware of our own brain is fundamental to the business of being human. It is also how actors become characters in a play and how audiences engage with those characters, and it's why *Breaking the Code* is not just a poignant account of the tragic arc of Turing's life but a potent acknowledgment of its protagonist's deepest ideas. Snoo Wilson's electric bear is a character that can only exist on stage—a place where the inhabitants acquire the concreteness of being three-dimensional, but where the playwright can still tinker with the laws of reality. This is the same type of creative workspace that Turing used to invent his universal machine, and Wilson theatrically brings one to life in a compelling way. "When we create a program that passes the Turing test," Douglas Hofstadter predicted, "we will see a heart even though we know it's not there."[20] This eloquent sentiment, which has its roots anchored in the core of Turing's mathematics, is rendered with stunning beauty in the closing moment of *Lovesong*.

Of the generous handful of other playwrights beyond Whitemore and Wilson to engage Turing, Canadian John Mighton has distinguished himself with his ability to capitalize on the theatrical possibilities embedded in Turing's ideas. Between 1988 and 1995, Mighton penned five full-length plays that collectively established his reputation as an intellectual playwright with a gift for bringing science to bear on unsettling questions about human identity.[21] As his plays engaging astronomy, neuroscience, and the physics of time were performed across Canada and beyond, Mighton was recognized with a steady stream of accolades including the Governor General's Literary Award for Drama in 1992. In 2000, director Robert Lepage adapted Mighton's play *Possible Worlds* into a high-profile motion picture. After a lengthy hiatus from creative writing during which Mighton completed a doctorate in mathematics and became a prominent voice in mathematics education, he found he had another play to write.

Half Life was first performed in 2005. Turing is not a character—the play is set in a nursing home in the present day—but in the second scene the audience is treated to an entertaining rendition of the Turing test. Donald is an AI researcher, and the one time we encounter him outside the nursing home, he is seated next to a closed curtain asking questions to an unseen voice that goes by the name Stanley:

DONALD: I understand you're a mathematician, Stanley.
STANLEY: Who told you that?. . . . I think there's some mistake. I'm an artist.
DONALD: Really?
STANLEY: A painter.
DONALD: Are you sure?
STANLEY: Well, I think I would know what I am.

Donald is clearly enjoying the task of getting to know his new friend. Exploring the relationship between mathematics and art, he goes on to ask about wallpaper designs, tilings, and patterns in general:

DONALD: My phone number has a very unusual pattern in it.
STANLEY: What is it?
DONALD: 314-159-2653.
(*Pause.*)
STANLEY: I said, I'm an artist, not a mathematician.
DONALD: You don't have to be a mathematician to appreciate it.
STANLEY: I'm afraid I can't see any pattern.

So far so good for Stanley the artist. After a few more probing questions about his subject's childhood, however, Donald springs his trap:

DONALD: What's my phone number?
STANLEY: 314-159-2653.
SCIENTIST: Shit. (11)[22]

Stanley has exposed his mechanical nature by forgetting to forget Donald's phone number—which, incidentally, are the first ten digits in the decimal expansion of π—and the frustrated scientist pulls back the curtain to reveal a bank of computers. "A machine? Are you crazy?" the computer babbles on, "I'm not a machine. I'm an artist!"

Mighton makes the initial point of this entertaining excursion a lesson about the importance to human cognition of *not* retaining everything we learn. "We wouldn't survive if we remembered everything," Donald says. Having brought the Turing test into his story, Mighton goes on to have some additional fun with its theatrical nature. A scene midway through the play opens with the same image of Donald beside a closed curtain engaged in a personal Q&A with a voice emanating from behind it. Once the parallels to the earlier scene have firmly taken hold, however, the curtain is pulled back by the nurse on the other side. Momentarily disoriented, the audience quickly recalibrates to see that Donald is in the nursing home. The nurse has been changing Donald's mother behind the curtain, and it is the nurse to whom Donald has been speaking. This clever ambush pays some immediate dividends by piquing the audience's attention, but the playwright has a more ambitious agenda for which this visual trickery is just the setup.

When Stanley is revealed to be a computer, his protestations to the contrary—"I'm not a machine. I'm an artist!" make for a good laugh line. Is Stanley programmed to lie, or does he believe what he is saying? These questions do not really matter in Stanley's case, but they matter a great deal in the case of Donald's mother Clara. Clara is showing signs of cognitive impairment, although it is not clear just how serious this is. The need to answer this question becomes more acute, however, when Clara starts to fall in love with a new resident named Patrick who she may or may not have had an affair with during the war many years before. In many functional ways Clara is like a child—she is forgetful, she needs to be reminded to eat, she is incontinent. So, should she be allowed to make her own decisions about Patrick? Should she be permitted to spend time alone with him? To dance with him? To sleep with him? To marry him, as Patrick eventually requests?

Half Life could appropriately be described as a play that explores how society treats its aging, but the issues it raises are more universal. The one in

"I knew a Patrick once. During the war. He used to take me dancing."
Half Life; Helen Ryan (Clara), Patrick Godfrey (Patrick); Theatre
Royal Bath, 2016 (photo: Simon Annand).

particular that keeps resurfacing is much like the one in Stoppard's *The Real Thing*; specifically, to what degree do people's public presentations of themselves correlate with who they really are? Asked in a different, more technical, way: What role should empirical behavior play in assessing the nature or the degree of human intelligence? This is the real reason why Mighton has imported Turing's imitation game into his play. For Donald, the eroding of his mother's brain function is excruciating because, to a cognitive scientist like himself, this is paramount to the slow and steady loss of his mother. What about to the rest of us? Do we judge adults by their function or is there a soul that shines through even as the mind starts to fail? Mighton ends his play by putting the audience in the uncomfortable position of having to decide for themselves. Simple, understated, and impossible to achieve in a novel or a film, this powerful moment exemplifies theater's distinctive capacity to find human stories inside the mathematical ones.

In the closing scene, the code-breaking Patrick escapes to see his beloved once too often and is escorted away to a higher security floor. A nurse named Diana then kindly tucks Clara into her covers. "Just like a little girl," Diana says and slowly draws the curtain around Clara's bed. "Will I see you again tomorrow?" the unseen Clara asks.

With the curtain closed, the familiar setting of the earlier Turing test is unmistakable:

DIANA: I'll be on Patrick's floor tomorrow.
(*Diana exits.*)
CLARA: I knew a Patrick once during the war . . . You would be too young
    to remember.
(*Pause.*)
Apparently he has a daughter now.
(*Pause.*)
Of course, Dad was on the railroad . . . (81)

For a few more moments Clara's voice continues on in its semicogent per-ambulations, and by now everyone knows how the imitation game is played. With no one else left on stage, it becomes the audience's heart-wrenching job to decide whether they are hearing the intelligent thoughts of a human being coming from behind the curtain.

A Politician's Apology

In *Half Life*, Mighton slips in Turing's name by having the nursing home cler-gyman ask if Patrick worked with Turing during the war. It is not clear how many members of the audience would have known who Turing was in 2005 when *Half Life* was originally performed, but certainly things had improved since 1986 when Hugh Whitemore felt compelled to include Turing's name no less than five times in the opening sixty seconds of his play.

Amid the steady excavation of Turing's seminal contributions, play-wrights, together with artists of all sorts, have made their particular con-tribution to shaping Turing's evolving public reputation. So has the aca-demic community to which Turing belonged. In 2009, computer scientist John Graham-Cumming organized a petition requesting an apology from the

British government for its treatment of Turing at the end of his life. Posted at the beginning of August of that year, the petition generated over thirty thousand signatures in a single month, and on September 10, Prime Minister Gordon Brown published a thoughtful response on the 10 Downing Street website. Brown focuses more on Turing's war heroics than his mathematics, but the overall tone is sincere and somber and represents another step in restoring the humanity to a mathematician whose interest in machines continues to teach us what it means to be human:

> So on behalf of the British government, and all those who live freely thanks to Alan's work I am very proud to say: we're sorry, you deserved so much better.[23]
> —Gordon Brown

7

Auburn: Beautiful Proofs

The book, the math, the dates, the writing, all that stuff you decided with your buddies, it's just evidence. It doesn't finish the job. It doesn't prove anything.

—CATHERINE, FROM *PROOF*

DAVID AUBURN was a relatively unknown young playwright when he penned the script for *Proof* in 1999. A sketch comedy writer and performer during his undergraduate days at the University of Chicago, Auburn spent his twenties moving between Los Angeles and New York, trying to make a living creating theater in various forms. His first modest breakthrough as a writer came in 1997 with *Skyscraper*, an offbeat comedy that enjoyed a brief run off Broadway. Among the companies that took an interest in Auburn's work at the time was the Manhattan Theater Club, and so this is where the thirty-year-old playwright first submitted a draft of *Proof* for a staged reading. When Auburn heard Mary-Louise Parker read the lead part of Catherine, he was deeply impressed. Parker, he thought, had completely nailed it. So, apparently, had Auburn.

Proof opened at the Manhattan Theater Club in May of 2000, where it ran for three months before moving to Broadway. In 2001, it won the Tony Award for Best Play, the Drama Desk Award for Outstanding Play, and the New York Drama Critics Circle Award for Best American Play. It was also awarded a Pulitzer Prize. Its three-year run on Broadway was complemented by a successful London production featuring Gwyneth Paltrow in the lead role. In 2005, Miramax released a film version of *Proof*, also staring Paltrow, to wide acclaim.

Even in the wake of the success of plays like *Arcadia* and *Copenhagen*, the stunning popularity of *Proof*—a play in which three of its four characters are mathematicians—is still hard to reconcile with the larger public's general apprehensions about mathematics. So just how did Auburn manage to write a math play with such extraordinarily wide appeal? One theory is that he did no such thing.

A Mathematical *Proof* or Not?

When *Proof* appeared amid the initial rush of other scientifically themed dramas, reviewers were predictably wary of yet another play with significant quantitative prerequisites. Their concerns, however, were unwarranted. "The story itself is not much more than a highbrow soap opera with painless references to mathematics," wrote Clive Barnes of the *New York Post*. *New York Observer* reviewer John Heilpern actually expressed "relief" that the play wasn't about its "ballyhooed higher mathematics." Far from being critical, Barnes and Heilpern hailed the play as "alive" and "handsomely sustained" and were simply sounding the all-clear siren for math-anxious audiences. Other than a little banter about prime numbers, there really is no substantive mathematics to be found in Auburn's script.

For his part, Auburn is candid about the fact that mathematics was not on his radar when he set out to write *Proof*. In multiple interviews, the playwright has said that the initial germ of the play was a desire to explore the dramatic potential around mental illness and, in particular, the complex dynamics of a child confronted with a sick parent.[1] In the opening moments of *Proof*, Auburn establishes his rendition of such a relationship in a theatrically clever way. At rise, Catherine is startled by the appearance of her father Robert. He offers her a bottle of champagne and, as they talk, we learn that it is Catherine's twenty-fifth birthday, but she has no real friends to help her celebrate. This is because for the past five years Catherine has essentially put her life on hold in order to care for her ill father. Robert is grateful but seems more concerned that Catherine is sacrificing too much on his behalf. Then comes the first of several ambushes: Robert points out that he died a week earlier. Given that Catherine is carrying on an extended conversation with her deceased father, the audience is presented with the play's first proposition in need of a proof, which is that Catherine may have inherited her father's mental instability.

"A very good sign that you're crazy is the inability to ask the question 'Am I crazy?'" *Proof*; Ronald Pickup (Robert), Gwyneth Paltrow (Catherine); Donmar Warehouse, 2002. © Donald Cooper/photostage.co.uk.

The mathematics enters the story via the fact that Auburn based the character of Robert in part on mathematician John Nash. Perhaps not entirely coincidentally, Nash's fame would soar the following year when Sylvia Nassar's biography *A Beautiful Mind* was made into its own award-winning film. Like Nash, Catherine's father is described as a genius who made major contributions to "game theory, algebraic geometry, and non-linear operator theory" before being completely derailed by schizophrenia at the height of his career. In Robert's case, one manifestation of his condition is a deranged form of graphomania. By the time he dies, Robert's upstairs office is stacked with notebooks full of numerological nonsense. One of Robert's former PhD students named Hal gives himself the job of sifting through his mentor's scribbling for any signs of mathematical life—a job that becomes significantly more interesting when, midway through the play, Catherine sends him off with a key that unlocks a particular drawer in her father's desk. Hal returns with a notebook containing a potential proof for a result that we are meant to assume is on par with Fermat's Last Theorem or the Riemann Hypothesis. "It looks like it proves a theorem," Hal announces breathlessly, "a mathematical theorem about prime numbers, something mathematicians have been trying to prove since . . . since there were mathematicians."

This is about as much of a description of the mathematics as we get. In this sense, the proof in the notebook is a "MacGuffin," a Hitchcockian device whereby some shrouded object takes on extraordinary significance without the audience ever knowing precisely what it is. Given this narrative manipulation, it is fair to ask whether the mathematics has any bearing on this play at all. For instance, could we change Robert into a mad composer, hide a brilliant symphony in his cluttered desk, change the title of the play, and still tell the same story?

Well, no, as it turns out.

It's Just . . . Elegant

In a way, Auburn got lucky. The original appeal of mathematics for the playwright was its natural fit with the various plot logistics he needed to work out: mathematics has a reputation for being associated with mental instability, it can be done in solitude on a low budget, it provides an object of extraordinary value that Catherine and her sister Claire must then negotiate over. But as Auburn immersed himself in the culture of mathematics, other benefits began to emerge. On the lighter side, the social eccentricities of mathematicians provided the playwright with an ample supply of quirky flesh for his characters' bones. Auburn arranges for Hal to play in an overly enthusiastic rock band whose set list includes a completely silent tune called "i" (a.k.a. imaginary number). Catherine also has a bit of math geek running through her personality. Although she has dropped out of school, she comfortably holds her own discussing Germain primes with Hal at her father's memorial service-turned-rager. The more interesting way her affinity for mathematics comes through, however, is in subtle turns of phrase she uses such as calling Hal "not boring"—a kind of compliment-by-contradiction. Whatever we make of their respective senses of humor, there is an authenticity to Hal and Catherine's budding courtship that animates the story and spills over into their conversations about mathematics:

HAL: My papers get turned down. For the right reasons—my stuff is trivial. The big ideas aren't there.

CATHERINE: It's not about big ideas. It's work. You've got to chip away at a problem.

HAL: That's not what your dad did.
CATHERINE: I think it was in a way. He'd attack a question from the side, from some weird angle, sneak up on it, grind away at it. He was slogging. He was just so much faster than anyone else that from the outside it looked magical. (37)[2]

Auburn may not have included any theorems, but his intuition about the nature of doing mathematics is compelling, and he is keenly aware of the deeply aesthetic quality of the subject. "Plus the work was beautiful," Hal goes on to say, referring again to his mentor's early results. "You read it for pleasure. It's streamlined: no wasted moves, like a ninety-five-mile-an-hour fastball. It's just . . . elegant."

It is from this vantage point that the deeper significance of the mathematics in *Proof* comes into focus. One reason why the descriptions of mathematical beauty feel so genuine is that they align with the playwright's own artistic sensibilities. The mathematical source that jumps highest off Auburn's stage is G. H. Hardy's *Apology*. Catherine and her father pay homage to Ramanujan with an inside joke about 1729, and Auburn can't resist having Hal invoke Hardy's iconic description of mathematics as a "young man's game." Auburn finds his most common cause with Hardy, however, where other playwrights do—in the shared conviction that their respective creative labors "must be justified as art if they can be justified at all," to use Hardy's phrasing. After declaring that "beauty is the first test: there is no place in the world for ugly mathematics," Hardy devotes multiple sections of his essay to fleshing out just how beauty is manifested in a mathematical theorem. Of his many epiphanies, the one that keeps resurfacing with respect to theater is his observation that in a beautiful proof,

> there is a very high degree of *unexpectedness*, combined with *inevitability* and *economy*. The arguments take so odd and surprising a form; the weapons used seem so childishly simple when compared with the far-reaching conclusions.

This description goes a long way toward explaining the source of *Proof*'s emotional power. The constituent parts of Auburn's play are simple and transparent. Naturalistic in style, *Proof* has just four characters who appear most often in duets on a single back porch in the Hyde Park neighborhood of Chicago. There are no lofty speeches, no extraneous personalities, no wasted moves. The play is full of deep sentiment and humor but never devolves into a sentimental melodrama or stock comedy. One way that it avoids any simplistic taxonomy is its detective quality. Nearly every scene concludes with a revelation that is equal parts unexpected and inevitable. The most potent example of this phenomenon is the gasp-inducing act break when a quivering Hal returns with the notebook from Robert's locked drawer. Catherine's sister Claire tries to get Hal to explain what this discovery means, but he can barely find the words:

> HAL: It means that during a time when everyone thought your dad was crazy . . . or barely functioning . . . he was doing some of the most imp-ortant mathematics in the world. If it checks out, it means you publish instantly. It means newspapers all over the world are going to want to talk to the person who found this notebook.
> CLAIRE: Cathy.
> HAL: Cathy.
> CATHERINE: I didn't find it.
> HAL: Yes you did.
> CATHERINE: No.
> CLAIRE: Well, did you find it or did Hal find it?
> HAL: I didn't find it.
> CATHERINE: I didn't find it.
>   I wrote it.
> *Curtain* (47)

Auburn crafts this moment so that Catherine's declaration is unforeseen, impossible for Hal to accept and yet, somehow, the obvious explanation. To complicate matters, Catherine may be mentally unstable—and Hal may be falling in love with her.

The second act of *Proof* is devoted to sorting out this mystery, which the play does by delving into the past. While the audience's attention is initially on the notebooks, the discoveries that emerge are as much about Catherine's relationships, especially to her father, as they are about how prodigious her latent mathematical talents might be. Arranged in a nonlinear fashion, the scenes are like puzzle pieces revealing disjointed snapshots of the various nar-rative threads, and it is the satisfying way these threads are woven together that imbues *Proof* with the tightly sprung potential energy of one of Stoppard's cricket bats.

On a structural level, then, one compelling reason for Auburn to limit the mathematical details in his play is that they would clutter up the script, ironically working against the economy of the mathematical aesthetic that he is instinctively emulating. With regard to the storyline itself, a more practical reason why Auburn decided to conceal the play's mathematics to some extent is that Catherine is concealing hers. Catherine's reasons are simple enough to understand. To acknowledge that she has inherited her father's mathematical abilities opens the door to asking what else she may have inherited. Vaguely suggested early on and largely at the root of her anxiety, Catherine's inner terror is eventually articulated in stark terms late in the play:

HAL: There is nothing wrong with you.
CATHERINE: I think I'm like my dad.
HAL: I think you are too.
CATHERINE: I'm . . . *afraid* I'm like my dad.
HAL: You're not him.
CATHERINE: Maybe I will be.
HAL: Maybe. Maybe you'll be better. (82)

Stages of Uncertainty

One final argument for the centrality of the mathematics in *Proof*—for why Robert must be a mathematician rather than a composer or a chemist—is that mathematics provides an effective foil for the ambiguity surrounding so many of the events in the play. Is Catherine crazy? Was she right to sacrifice so much of her own life to care for her sick father? What are Hal's motives? And who is the author of the proof in Robert's desk? These are the questions that drive the action in Auburn's play, and for many people—including, quite possibly, the playwright—the rational clarity of mathematics serves as a point of contrast for the kind of certainty we might wish for but that does not exist in the murky negotiations between complicated and compassionate people.

Although there is nothing overtly objectionable about this particular reading of the role of mathematics in *Proof*, the truth about truth in mathematics is much more artfully ambiguous. Yes, mathematics stands apart from the other liberal arts for the precision it wields in declaring what is true and the potential for permanence that this precision engenders. "Archimedes will be remembered when Aeschylus is forgotten," is how Hardy puts it. But the recurring revelation of the collaborations of theater with mathematics through the last century is that mathematics is not exempt from uncertainty, and it is precisely these uncertainties that sparked the imaginations of the playwrights who were bold enough to discover them. The best Tom Stoppard could do with the austerity of Euclidean geometry was invoke it as a passing metaphor for communist orthodoxy in a few political dramas. By contrast, the paradoxes of Zeno and Russell breathe life into *Jumpers*, the uncertainty of Heisenberg's physics is at the heart of *Hapgood*, and the unpredictability modeled by chaos theory animates *Arcadia*. "It's the best possible time to be alive," Valentine says in this latter play, offering the audience a more authentic portrait of a mathematician than they previously had access to, "when everything you thought you knew was wrong."

All avenues of human inquiry have their inherent limitations, but a distinguishing feature of mathematics is the clarity that exists about its own deficiencies. Mathematics is arguably unique in this regard—even when it is drawing attention to its shortcomings, it maintains an unwavering confidence in the efficacy of its methods. Bolyai, Lobachevsky, and those who followed showed that Euclidean geometry held no logical priority among a host of emerging alternatives. Gödel's discovery of the incompleteness of axiomatic systems is expressed in the form of a theorem. Turing's contribution to logic was to prove that there is no algorithmic way to distinguish formally provable statements from unprovable ones. In every case, the conclusions are categorical, restrictive in their assessment of math's authority, and established by exploiting the introspective ability of mathematics to assess its own integrity.

Samuel Beckett may or may not have been familiar with Gödel's work, but he was keenly aware of both the constructive and destructive power in training the scrutiny of his chosen medium back on itself. And Beckett also routinely mined mathematics for its paradoxes, launching *Endgame* with an allusion to Zeno's "impossible heap" and returning multiple times to the heretical discovery of irrational numbers—numbers without names—as a metaphor for the inadequacies of language to capture the human psyche. Although the extent of Beckett's knowledge of twentieth-century logic is up for debate, the role of modern mathematics in Stanislaw Witkiewicz's creative journey is unambiguous. Looking to liberate his art from the confines of realism, Witkacy saw that Euclid no longer had a monopoly on describing reality—if Euclid related to reality at all—and the infinite was no longer forbidden territory. Friedrich Dürrenmatt's attention was snared by the nonorientability of the Möbius strip. For Frayn, it was quantum uncertainty. For McBurney, Cantor's paradoxical infinity of infinities.

This litany is not meant as a rebuttal to the belief of Hardy and others that mathematics occupies the high ground in the hierarchy of universal truths. In fact, it is precisely this reputation that makes the inherent limitations of mathematics so compelling as inspiration for the artists who use them. David Hilbert's declaration, "We must know; we will know," did not suddenly become an irrelevant echo in the wake of Gödel's and Turing's results—quite the opposite in fact. Hilbert's forceful decree had the effect of amplifying the reverberations of these discoveries, and in a way, his edict was met. It wasn't the outcome Hilbert had in mind, nor could it have been if mathematics was to sustain its long-standing kinship to the arts. This is good news all around. When Hugh Whitemore looked at Turing's work on the Entscheidungsproblem, he saw a conclusion he instinctively recognized articulated with bewitching certainty. "Even in mathematics there is no infallible rule for proving what is right and wrong," Whitemore's Turing says. "Each problem—each decision—requires fresh ideas, fresh thought."

The historical Turing's belief that there was no qualitative difference between minds and machines was not an attempt to reduce the hard problem of consciousness to a deterministic mathematical algorithm; it was a declaration that the things we cling to most dearly as defining our humanity—our passions, our poetry, our unpredictability and imperfections—are inevitable and universal. This is what the ubiquity of uncertainty in modern mathematics keeps pointing to, and this is what theater more than any other art form gleans from mathematics and brings to the fore. Gifted with the ability to impersonate characters from his master's life, Snoo Wilson's electric bear is therefore stricken with the capacity for a love stronger than life. Strange indeed, as Beckett's Molloy says, how mathematics helps us to know ourselves, but also strangely reassuring to have the deepest mysteries of life reaffirmed by the theorems of modern mathematics and brought to life on the modern stage.

Auburn did not go as deeply into the foundations of mathematics as some of his counterparts, but he most definitely recognized his art in Hardy's descriptions of beauty, and he had every intent to blur the lines between Catherine's mathematics and her psychological journey. Faced with the proposition that Catherine authored the proof in the notebook he found, Hal encounters first-hand the practical limitations of the correspondence between truth and provability. Initially, Hal fails this test of trust and friendship, and this shortcoming on his part causes real pain. The final offering of mathematics to this story is the insight that "shortcoming" is the wrong word. Incompleteness isn't a defect in the smooth facade of mathematics—it is an intrinsic

"Maybe. Maybe you'll be better." *Proof*; Richard Coyle (Hal), Gwyneth
Paltrow (Catherine); Donmar Warehouse, 2002. © Donald
Cooper/photostage.co.uk.

component of its nature that is responsible for the texture of its intricate landscape. Hal and Catherine, likewise, are rife with bumps and blunders that are
part and parcel of their manifold lives, and Catherine especially must come to
grips with this analogy between life and mathematics. Auburn restrains himself from tying too tidy a bow on his story, but he does offer a glimpse of what it
might look like if Catherine could one day make her peace with all that comes
from being her father's daughter.

"I know . . . it works," Catherine finally confesses to Hal in the play's waning
moments, "but all I can see are the compromises, the approximations, places
where it is stitched together." Perhaps she is talking about her proof, but she
could also be describing so many things about her life, and we seize on this
ambiguity as a means for understanding the implications of the choices she
has made. More than anything else, it is this two-way street between the mathematical and the human that aligns *Proof* with plays like *Arcadia* and *A Disappearing Number* whose mathematical credentials are more self-evident. At its

core, Auburn's play works the same way all the best mathematical plays work, with the inspiration flowing in both directions. The mathematics provides new forms and colors for the storytellers who return the favor by sneaking up from weird angles to portray mathematics as the richly human endeavor that it is.

Despite its absolute character, mathematics is conceived and practiced by flesh and blood people in the finite confines of our stitched-together lives, and in this context it reflects something distinctive, but elusive, about who we are. This is where theater shines its artful light, and beauty is still the first test. From its inception, mathematics has been revered as an unreasonably effective tool for deciphering the laws of the inanimate world; in the hands of gifted playwrights, mathematics becomes an instrument for garnering truths about the soulful, living world. On this more intimate stage, the transcendent power of mathematics is recast into a vehicle for self-discovery, and the image that emerges is no less sublime—not because we see some kind of exalted perfection but precisely because we don't.

ACKNOWLEDGMENTS

IN LARGE measure, this book is the result of two decades of coteaching "Science as Art in Contemporary Theatre" and collaborating on productions of plays with my colleague and friend Cheryl Faraone. In addition to her wisdom as a teacher and theater historian, Cheryl has been a guide and mentor backstage, allowing me to experience theater as the collaborative, creative, three-dimensional event that it is. Engaging with Cheryl and others in the collective endeavor of mounting plays has yielded rewards that go far beyond the scope of this project.

The proliferation of plays engaging mathematics and science in the last few decades has garnered a sizable share of attention from scholars in the humanities. To be clear, this book is intended for a general reader with little or no background in theater, mathematics, or the relevant scholarship. My training as a mathematician provides me with a distinctive perspective on this topic, but to tell a self-contained story, the book retraces ground covered by other authors and draws on the growing body of research being produced. The instances where this dependence is direct are noted in the text and expanded on in the footnotes, but I want to highlight a few colleagues who have been especially helpful to an interloping mathematician.

Kirsten Shepherd-Barr has been a steady source of advice and encouragement from the outset. A central motif of her foundational book *Science on Stage* (Princeton University Press, 2006) is the frequency with which plays about science exploit the merging of scientific content with the form of the play. I adopted this compelling thesis as the theme of chapter 4, where my discussions of the first two plays build on their insightful treatment in *Science on Stage*. The third play, Complicité's *A Disappearing Number*, was produced after Professor Shepherd-Barr's book was published. The company's commitment to weaving mathematics into the structure of *A Disappearing Number* is further confirmation of Shepherd-Barr's original argument.

347

I first read about the non-Euclidean theater of Stanislaw Witkiewicz in Nicolas Salazar-Sutil's *Theatres of the Surd*, which also includes significant forays into the theater of Samuel Beckett and Alfred Jarry. This original exploration of mathematics and theater was a major influence on the contents of chapter 2. Whereas the term "theater of the absurd" is associated with the deliberate undermining of logic and rationality, Salazar-Sutil's clever title points to his view that approaching playwrights like Witkiewicz and Beckett with an eye toward developments in nineteenth- and twentieth-century mathematics reveals how their plays are "not failures of sense-making, but constructions of sense in every way as coherent as the sense which governs realistic theater."[1] My conversations and collaborations with the author have led me to a robust endorsement of this point of view, most especially with respect to Witkiewicz's plays.

The scholar to most thoroughly explore Samuel Beckett's relationship to mathematics is Chris Ackerley, who generously read a draft of this book and provided valuable feedback. In addition to our conversations, I have benefited greatly from Professor Ackerley's meticulous writing on this subject. I am most grateful for *Obscure Locks, Simple Keys: The Annotated Watt*, which chronicles the literary devices at work in *Watt*, proceeding from the original handwritten notebooks up through the published novel. Also, *The Grove Companion to Samuel Beckett: A Reader's Guide to His Works, Life, and Thought*, coauthored with S. E. Gontarski, is a wellspring of insight and information that became a lifeline in my journey through Beckett's novels and plays.

A significant portion of the mathematical focus in the ensuing pages is devoted to the arc of modern logic—from Frege to Russell to Hilbert, with considerable attention devoted to Gödel's contributions. The dominant influence for how this story is rendered here is the writing of Douglas Hofstadter. The version of Gödel's proof running through chapters 3 and 5 is modeled on the approach in *Gödel, Escher, Bach*, Hofstadter's Pulitzer Prize–winning book from 1979. Where Hofstadter's fingerprints are most evident, however, is in the discussions aligning the metamathematics of Hilbert and Gödel with the metatheater of Tom Stoppard. Hofstadter has not written specifically about theater, but he is deeply engaged in probing the significance of reflexive structures in mathematics, music, biology, computing, and cognitive science. Not coincidentally, Stoppard's artistic journey eventually led him to similar questions about brains and minds and the mystery of consciousness. Chapter 5 engages this topic, as does chapter 6 on the mathematics and theater connected to Alan Turing, both adopting a point of view informed by Hofstadter's compelling insights.

My explanations of Turing's mathematics borrow from Martin Davis's approach in *The Universal Computer*, which is notable for the way the author presents mathematical ideas informally but not imprecisely. Of the many books on mathematical philosophy I consulted, the one I returned to most frequently was *Philosophies of Mathematics* by Alexander George and Daniel Velleman. The authors' ability to write concretely and cogently about abstract ideas was deeply appreciated and, as with Hofstadter and Davis, had a measurable influence on my own storytelling.

The research for this project began in earnest over ten years ago during a fellowship at the Center for Research in the Arts, Social Sciences and Humanities (CRASSH) at Cambridge University. I also benefited significantly from a six-week residency at the Harry Ransom Center at the University of Texas in Austin. In between these visiting positions, I have been generously supported by the resources of my home institution of Middlebury College. This includes the inspired help of two student research assistants, Jingyi Wu and Alex Myers.

I am indebted to Tim Spears, Dan Velleman, Rob Cohen, Tony Abbott, and Bill Goettler for their willingness to read and respond to chapter drafts in various stages of disrepair. Steve Kennedy was unreasonably generous in this capacity with his insights and encouragement. Appreciation to Daniel Scharstein and Bruce Torrence for their expert assistance with several of the figures. My thanks also to Princeton University Press—for Susannah Shoemaker's early enthusiasm, Diana Gillooly's wise and generous counsel, and Kiran Pandey's saintly patience that has ushered me to the finish line.

And finally, the most heartfelt gratitude expressible in words to my wife Katy—partner, editor, support team, and everything in between.

Stephen Abbott
Middlebury, Vermont
September 2022

NOTES

Prologue

1. [McB1], p. 9
2. [Har], p. 61
3. [Har], p. 85
4. [Har], p. 84
5. [Har], p. 81
6. [Kit1], p. 176
7. [Els], p. 28
8. [Els], p. 32
9. Kitto makes this same point. See, for example, [Kit1], p. 186.
10. [Els], p. 35
11. [Har], p. 113
12. [Els], p. 32
13. [Tap], p. 22
14. As for the appearance of mathematics as actual content in Greek plays, there is a short scene in *The Birds*, by Aristophanes, where the geometer Meton arrives with his mathematical tools for "measuring the air . . ."

> METON: In truth, the spaces in the air have precisely the form of a furnace. With this bent ruler I draw a line from top to bottom; from one of its points I describe a circle with the compass. Do you understand?
>
> PITHETAERUS: Not in the least.
>
> METON: With the straight ruler I set to work to inscribe a square within this circle; in its centre will be the market-place, into which all the straight streets will lead, converging to this centre like a star, which, although only orbicular, sends forth its rays in a straight line from all sides.
>
> PITHETAERUS: A regular Thales!

15. Rinne Groff's *The Five Hysterical Girls Theorem* is an interesting and ambitious play I could have included. Victoria Gould's *I is a Strange Loop*, cocreated with Marcus du Sautoy, is another innovative mathematical piece which Gould and du Sautoy have performed on multiple occasions but has not been broadly disseminated.

Chapter 1. Stoppard

1. Page numbers for *Arcadia* are from [Sto1].

2. As it appears in [Hea], p. 144.

3. See, for example, Paul Delaney's comment in [Del1], p. 11.

4. As quoted in [Nad], p. 441.

5. As quoted in [Nad], p. 367.

6. [Nad], p. 37

7. [Nad], p. 49

8. As quoted in [Del1], p. 13.

9. Page numbers for *Albert's Bridge* refer to [Sto8].

10. Among other additions to the original script, the screenplay includes a parallel universe where a second version of Albert joins the capitalistic oppression of the working-class by becoming an executive at his father's firm. Some of the fun Stoppard has with this conceit is having the firm invent the eight-year paint that gets the three extraneous painters sacked. Stoppard also makes a conscious attempt to extend the mathematical motif into the new material of the screenplay. Fitch's presentation to the board is redone to include some clever visual math jokes and, in a nod to Kate the maid's untimely pregnancy, the alternate Albert is provided with a parentally approved wife who resorts to differential calculus to predict her ovulation cycle.

11. Page numbers for *Rosencrantz and Guildenstern Are Dead* are from [Sto12].

12. As quoted in [Nad], p. 165.

13. As quoted in [Nad], p. 174.

14. [Sto13], p. 25

15. [Fle], p. 50

16. See, for instance, [Nad], p. 166.

17. [Nad], p. 223

18. [Nad], p. 224

19. Archie's initials are a nod to the nonfictional English philosopher A. J. Ayer, a major twentieth-century thinker associated with the school of logical positivism. In Ayer's framework, only strictly factual statements whose validity can be empirically tested represent meaningful statements. Moral judgments are unambiguously excluded from this category.

20. Page numbers for *Jumpers* are from [Sto6].

21. This anecdote is grounded in historical fact. See [Jen], footnote 12 of chapter 4 on p. 196.

22. Aristotle's rendition refers to a generic moving object that must first traverse half its total distance, and then half of what remains, and so on.

23. One way he achieves this is by naming his protagonist after one of Russell's closest colleagues. George Edward "G. E." Moore was a professor at Cambridge with Russell who wrote extensively about ethics, and in particular about the notion of goodness as an indefinable concept beyond the scope of natural science. The historical George Moore was an atheist, but his arguments are nonetheless quite sympathetic to those of his namesake in *Jumpers*. Russell died in February 1970, two years before *Jumpers* premiered.

24. In multiple interviews, Stoppard has confessed that his personal position on the debate in *Jumpers* is more aligned with George's than with Archie's. "I've always thought the idea of God is absolutely preposterous," Stoppard said to Mel Gussow, "but slightly more plausible than

the alternative proposition that given enough time, some green slime could write Shakespeare's sonnets." [Gus], p. 15

25. "One of the thieves was saved," is part of a longer quotation from Augustine which goes on to point out that one of the thieves was also damned. Samuel Beckett incorporates this same quotation in *Waiting for Godot*. Stoppard, in fact, gives a nod to *Godot* in *Jumpers* by having Archie recast Vladimir's closing speech. Vladimir's line: "Astride of a grave and a difficult birth. Down in the hole, lingeringly the grave digger puts on the forceps," is contorted by Archie into: "At the graveside the undertaker doffs his top hat and impregnates the prettiest mourner. Wham, bam, thank you Sam."

26. From the forward of [Sto2].

27. [Sto2], p. 19

28. [Sto2], p. 26

29. [Sto9], p. 193

30. [Del1], p. 179

31. As quoted in [Del1], p. 193.

32. Page numbers for *Hapgood* are from [Sto3].

33. This quotation originally appears in "Lectures on Physics . . . The Character of Physical Law." See [Fey]

34. The determining factor of whether the waves arrive in or out of phase is the difference in the respective distances between the point on the screen and the two holes on the slide. If the two distances are equal, as they are at the center point of the screen, then the waves arrive in phase and create a bright spot. The same is true if the difference in the two distances is an integer multiple of the wavelength of the light. In between these recurring bright spots are points on the screen where the respective distance to each of the two holes differs by a half wavelength plus some integer multiple of wavelengths (e.g., .5, 1.5, 2.5, etc.). These spots appear dark because the light waves arrive perfectly out of phase, with "peaks" lining up with "troughs" canceling out to create nodes.

35. This is a simplification of what actually happens. With one slit open there are some visible interference effects related to the width of the single slit, but they are quite distinct from the unmistakable pattern of alternating bright spots and dark nodes that characterize two-slit interference.

36. The image is from "Solutio problematis ad geometriam situs pertinensis," *Commentarii Academiae Scientarum Imperialis Petropolitanae*, 8 (1736), pp. 128–140 + Plate VIII.

37. This is because the edges get used two at a time—once upon entering the vertex and once upon leaving. If a vertex is surrounded by an odd number of edges then, eventually, a Königsberg citizen will arrive at that vertex (or land mass to be accurate) but have no edge (bridge) available to depart. If the rules of the game are that the walkers do not have to end up where they started, then it is permissible to have two vertices with an odd number of edges attached—one for the starting location and the other for the ending location.

38. Stoppard makes the admission that this scene may be too short and underwritten in an unpublished essay called "The Hapgood Crib," which he wrote to help actors and directors navigate his play.

39. Paul Delaney points out that, in the original London production, Hapgood "reached up to pull [Ridley] down to her so that she was not so much reciprocating as initiating the kiss." [Del2], p. 137

40. This is essentially a version of Stoppard's summary of the play in his interview with Mel Gussow in [Gus], p. 78–80.

41. Nightengale, Benedict. "The Latest From Stoppard: A Quark and Dagger Thriller," *New York Times*, March 27, 1988.

42. Private letter, from the collection at the Harry Ransom Center, University of Texas at Austin.

43. This framework for viewing *Arcadia* is widely adopted in the scholarship of the play and originates, to some degree, with the playwright himself. "I was thinking about Romanticism and Classicism as opposites in style, taste, temperament, art," Stoppard said to Mel Gussow in response to a question about the origins of *Arcadia*. "I remember talking to a friend of mine, looking at his bookshelves, saying there's a play, isn't there, about the way that retrospectively one looks at poetry, painting, gardening, and speaks of classical periods and the romantic revolution, and so on. Particularly when one starts dividing people up into classical temperaments and romantic temperaments—." [Gus], p. 90

44. Of the copious articles written about *Arcadia*, Prapassaree and Jeffrey Kramer's "Stoppard's *Arcadia*: Research, Time, Loss" had a significant influence over the approach taken here.

45. [Del1], p. 224

46. Jenni Halpin makes a similar observation in [Hal] (p. 67) about this dramatization of the butterfly effect. "[A] relatively small initial error is compounded," she writes in response to both Bernard's flawed Byron hypothesis and Hannah's presumption about a drawing of the Sidley Park hermit, "through the modern characters' developing work and through further revelations in the scenes set in 1809."

47. Mandelbrot actually writes, "Clouds are not spheres. Mountains are not cones."

48. See, for example, [Gle], p. 303.

49. [She1], p. 133

50. Halpin's inventive description of this effect is that "the play shows [the] past unfolding together with their researches, making the past appear mutable." [Hal], p. 67

51. This quotation appears in [Dem2], p. 236, summarizing an argument the author sets out more fully in [Dem1].

52. [Lap], p. 4

53. The full quotation attributed to Newton reads:

I do not know what I may appear to the world; but to myself I seem to have been only like a boy playing on the seashore, and diverting myself in now and then finding a smoother pebble or a prettier shell than ordinary whilst the great ocean of truth lay all undiscovered before me.

This passage does not appear in Newton's writing but is attributed to him in *Memoirs of the Life, Writings, and Discoveries of Sir Isaac Newton*, published in 1855 by Sir David Brewster. An earlier appearance can be found in *The Eclectic Review*, vol. 5, part 1 (London, 1809, p. 232), as part of a preamble to Edmund Turner's "History of Grantham."

54. See [She1], pp. 140–141.

Chapter 2. Jarry and Witkiewicz

1. Appears in [Wit1], p. 3.

2. From *Plato's Republic* (510d)

3. From Aristotle, *Physics B*. This quotation and the previous one from *Plato's Republic* are part of a more in-depth discussion of this topic in [Sha], p. 65.

4. As quoted in [Grb], p. 146.

5. [Boo], p. 330

6. As quoted in [Boo], p. 299.

7. [Str], from the author's preface, p. xii.

8. [Ess2], p. 356

9. As it turns out, Euclid himself was guilty on more than one occasion of unwittingly relying on his figures to justify intuitive steps in a proof. The occasional gaps in his arguments, however, never resulted in asserting faulty propositions.

10. As quoted in [LaPe], pp. 2–3.

11. Although this proposition deals with parallel straight lines, the proof does not require the parallel postulate. Essentially, the argument proceeds by constructing a straight line through the given point such that the alternate interior angles are equal. This has the effect of creating a symmetric situation about the transversal. Specifically, if the straight lines were to intersect on one side, then the standard theorems about congruence of triangles would imply that the lines must also intersect on the other side of the transversal. The net result is a pair of straight lines that intersect twice, which contradicts the assumption that two points determine a unique straight line. We conclude then that the two straight lines do not intersect at all and are therefore parallel.

12. As quoted in [Ess1], p. 360.

13. As quoted in [Bro], p. 161.

14. This opinion of the opening nights of *Ubu Roi* is shaped by Thomas Postlewait's account of this event in [Pos] and Alastair Brotchie's description in [Bro]. Postlewait's agenda is to correct a number of widely circulating misconceptions about the *Ubu* premiere as a case study for bringing a more disciplined and objective approach to historical reconstructions. In my reading, Brotchie's treatment displays the standards Postlewait insists upon while confirming the riotous, and occasionally violent, characterization of the premiere.

15. This viewpoint is the thrust of Shattuck's chapter "Suicide by Hallucination" in [Sht].

16. [Bro], p. 5

17. [Jar1], p. 21

18. Page numbers are from the translation by Connelly and Taylor in [Jar2].

19. As quoted in [Bro], pp. 166–167.

20. This is from Mendès's review of *Ubu Roi*, as quoted in [Bro], p. 165.

21. [ShTa], p. 71

22. [ShTa], p. 84

23. Salazar-Sutil provocatively describes this moment in the play by asserting that "Ubu erupts as violently into Achras's house as non-Euclidean geometry enters the world of modern mathematics." [Sal], p. 109

24. [Jar1], p. 22

25. [Jar1], p. 22

26. This firsthand description from Nobel Laureate André Gide offers a telling portrait of this transformation. "It was the best period in Jarry's life," Gide writes. "He was an incredible figure whom I [met] always with tremendous enjoyment, before he became a victim of frightful attacks of delirium tremens. This plaster-faced Kobold, gotten up like a circus clown and acting a fantastic, strenuously contrived role which showed no human characteristic, exercised a remarkable fascination at the *Mercure*. Almost everyone there attempted, some more successfully than others, to imitate him, to adopt his humor; and above all his bizarre implacable accent—no inflection or nuance and equal stress on every syllable, even the silent ones. A nutcracker, if it could talk, could do no differently." As quoted in [Sht], p. 212.

27. See, for instance, [Bro], pp. 179–182. Shattuck makes a similar argument in [Sht].

28. This sentiment is expressed most clearly in [Sht], p. 202. Salazar-Sutil generalizes this point. "What separates the theatre of the avant-garde from classical theatre," he writes, "is that the former does not subscribe to the view that theatre is a recreation of an event happening somewhere else. Nothingness is a feature of the theatres of the surd, as important in the work of Witkiewicz, as it is in Jarry, Kantor, and of course Beckett, under whose directives nothingness would find perhaps its most powerful theatrical expression." [Sal], p. 47

29. [Bro], p. 240

30. [Jar1], p. 111

31. [Jar1], p. 114

32. In his article on Wells's time machine, Jarry explains that his comments "will apply to all spaces: Euclidean or three dimensional space . . . Riemannian spaces . . . , Lobatchevski's spaces . . . , or any non-Euclidean space identifiable by the fact that it will not permit the construction of two similar figures as in Euclidean space." Salazar-Sutil, meanwhile, points out that Jarry routinely discussed the fact that space can be recast in non-Euclidean terms, concluding that "Riemannian conceptions of affine space and complex manifolds could have held for him a particular significance." [Sal], p. 115

33. As quoted in [Ger1], p. 7.

34. The quoted passage is from [Ger1], p. 135. In [Wit2], p. 33, Gerould goes further by referring to *The Water Hen* as "the best illustration of Witkiewicz's theory."

35. Page numbers for *Tumor Brainiowicz* are from [Wit1].

36. As quoted in [Ger1], p. 15.

37. [Wit1], p. 42

38. [Wit2], from the introduction, p. xxxviii.

39. As quoted in [LaPe], p. 14.

40. As quoted in [LaPe], p. 7.

41. [LaPe], p. 7

42. Legendre employs the phrase in the supplementary notes to a late edition of his *Éléments de Géométrie*; see, for example, the 12th edition from 1823, Note II on p. 279 (as cited in [LaPe], p. 30). Saccheri uses it in Proposition XXXIII of *Euclid Freed of Every Flaw*.

43. As quoted in [Grb], p. 129.

44. This particular version of this image is due to Anton Sherwood and was adopted from https://commons.wikimedia.org/w/index.php?curid=24949640.

45. Among the sources that Witkacy cites in the preface to *Tumor Brainiowicz* are "three orange-colored popular works of Henri Poincaré." It makes perfect sense that the self-taught playwright would be interested in a polymath like Poincaré, whose contributions to mathematics, physics, and philosophy were of the highest order. Late in his career, Poincaré gave a series of lectures to the Psychology Society of Paris that were eventually published as three separate books: *Science and Hypothesis*, *The Value of Science*, and *Science and Method*, which are the likely candidates for Witkacy's reference.

46. [LaPe], p. 8

47. [LaPe], p. 14. When Nicolas Salazar-Sutil finds his way to this quotation, he makes the astute observation that the forthcoming axioms of set theory lead to the whole numbers also being created "out of nothing." [Sal], p. 46

48. [LaPe], p. 14

49. [Grb], p. 145

50. [Grb], p. 146

51. In a 1927 letter, Witkacy referenced *Ubu Roi* in a favorable way, but couldn't recall Jarry's name. [Ger1], p. 165

52. Page numbers for *Gyubal Wahazar* are from [Wit1].

53. Quotes from *An Introduction to the Theory of Pure Form in the Theater* are from the translation that appears in the appendix of [Wit2]. This one is from p. 295.

54. [Sal], p. 69

55. Gauss himself reportedly did this; see [Grb], p. 237.

56. Daniel Gerould makes this argument in much greater detail; see [Ger1], p. 166.

57. Page numbers for *The Water Hen* are from [Wit2].

58. Salazar-Sutil makes the astute observation that in Witkiewicz's incarnation of spherical theater, events can seem normal locally but not from a suitable distance. "Even though it has a center, the play does not have a clear-cut reference point outside of itself," he writes. "Everything in *The Water Hen* seems to be changing as the play unfolds, so things make sense only momentarily. Characters split in two, or else share the same name, often causing a great deal of confusion. The process of ageing is suspended; death becomes reversible and corpses come back to life." [Sal], p. 87

59. [Wit2], pp. 33–39

60. [Wit2], p. 293

61. [Wit2], p. 292

62. [Wit2], p. 294

63. [Wit2], p. 292

64. [Wit2], p. 295

65. The connection between mathematical proofs and Greek plays in *Poetics* is discussed at length in the prologue.

66. [Wit2], p. 296. This point is central to Salazar-Sutil's thesis. Salazar-Sutil notes that Witkiewicz warns his readers: "To pile up absurdities is one thing, and to create formal constructions, which have not been contrived in cold blood, is quite another." [Sal], p. 69

67. The phrase comes from the title of the 1960 article by Eugene Wigner ([Wig]).

68. His companion at the time was Czeslawa Korzenlowska, and the story of his final days is recounted in [Ger1], pp. 20–21. Gerould includes a long footnote about its likely authenticity and related references.

69. [Wit1], p. xxv of the introduction

Chapter 3. Beckett

1. The center point of the square has received its share of attention from scholars. Martin Esslin suggests that the center, which the walkers all conspicuously avoid, represents a point of possible connection or communication, which must therefore be avoided. See also [Sal], p. 205.

2. The choreography of colors looks like this:

$$\ldots YW \to W \to WB \to WBR \to WBRY \to BRY \to RY \to$$
$$Y \to YW \to YWR \to YWRB \to WRB \to RB \to$$
$$B \to BY \to BYW \to BYWR \to YWR \to WR \to$$
$$R \to RB \to RBY \to RBYW \to BYW \to YW \to$$
$$W \to \ldots$$

With four walkers there are fifteen possible combinations (sixteen if we include an empty stage). In order to include all four singleton possibilities via the process of increasing incrementally up to four walkers and then decreasing back to one, it is necessary to repeat a few combinations along the way.

3. [Her], p. 46

4. [Kno], p. 592

5. [Her], p. 52

6. [Kno], p. 105

7. [Bra2], p. 24

8. [Bec1], p. 27

9. In a notable moment late in the novel, the character of Neary offers some geometrically inspired solace to Wylie and Miss Counihan about the shape of the trio's relationship. "Do not despair. Remember there is no triangle, however obtuse, but the circumference of some circle passes through its wretched vertices. Remember also one thief was saved." "Our medians," said Wylie, "or whatever the hell they are, meet in Murphy." "Outside us," said Neary. "Outside us." (213). Confirming Wylie's uncertainty, it is actually the perpendicular bisectors, not the medians, that intersect the center of the circle whose circumference passes through the triangle's wretched vertices. If the triangle is obtuse, then this point is indeed outside the triangle.

10. The Greeks used an indirect line of attack to prove the irrationality of $\sqrt{2}$. The idea is to assume that there *is* a rational number whose square is 2 and then proceed along logical lines until we reach a conclusion that is unacceptable. At this point we will be forced to retrace our steps and reject the erroneous assumption that some rational number squared is equal to 2.

And so assume that there do exist integers m and n satisfying

$$\left(\frac{m}{n}\right)^2 = 2. \tag{1}$$

We may also assume that the fraction m/n is in lowest terms because if m and n happen to have a common factor we could cancel it out. Now equation (1) is equivalent to

$$m^2 = 2n^2. \tag{2}$$

From this we can see that the integer m^2 is an even number because it equals 2 times another integer. Because m^2 is even, m must also be even. (Why? Because squaring an odd number always gives another odd number.) This allows us to write $m = 2p$, where p is another integer. If we substitute $2p$ for m into equation (2), then some modest algebra yields the relationship

$$2p^2 = n^2.$$

But now a contradiction is at hand. This last equation tells us that n^2 is even, and so n must be even as well. Thus, we have that m and n are both even—i.e., divisible by 2—when they were originally assumed to have no common factor. Faced with this logical contradiction, we return to the top of the argument and reject the original premise that there exist integers m and n satisfying equation (1). The conclusion: $\sqrt{2}$ is irrational.

11. Page numbers for *Murphy* are from [Bec5].

12. Adding to the force of this metaphor is the suggestive power of the words themselves—the phrase "irrational number" conflating with the image of "irrational thought." In terms of etymology, Euclid used the term "alogos" to refer to irrational numbers. Since "logos" meant either "word" or "reason," "alogos" could translate as either "unsayable" or "unreasonable." Most likely he intended the former. Because rational numbers were the numbers the Greeks used, an irrational number would be a number without a name. This is exactly the image Beckett is trying to convey by referring to the dark zone of Murphy's mind as a "matrix of surds." It is an uncharted part of the universe, beyond the reach of the descriptive power we currently possess. That "alogos" or "irrational" also suggests a world "beyond reason" is so much the better.

13. There is a debate among scholars about how much relative significance to attach to *Watt* as a component of Beckett's bibliography. There is near universal agreement that his later novels—most notably the trilogy of *Molloy, Malone Dies*, and *The Unnamable*—are superior examples of freestanding works of art that attest to Beckett's great talent. In fact, it wasn't until after Beckett had established his reputation with these later works that he was able to secure a publisher for *Watt*, and even then it was with a small experimental publishing house in Paris that specialized in erotic fiction (which *Watt* most definitely is not).

Watt was written under extraordinarily difficult conditions, with many stops and starts. On various occasions, Beckett has referred to it as an exercise, a diversion, a form of therapy to keep him sane, or simply as a way to "keep in touch" during the war ([Kno], p. 303). Perhaps it is no surprise then that the finished novel feels disjointed. On the other hand, one has to be cautious about suggesting that *Watt* is less than what its author intended. It did undergo a number of drafts and revisions, and Beckett was very eager to have it published, finally signing a contract with Merlin in 1953. "It has its place in the series, as will perhaps appear in time," is how Beckett himself summarized things in a letter to his friend George Reavy in 1947. [But], p. x

14. Watt's singing frogs seem to be descendants of the ones from the chorus of *The Frogs*. In Aristophanes's play, the frog chorus taunts Dionysus with a chant of "Brekekekex koax koax."

15. Translations of this letter appear in several places, including [Bec1]. The version quoted throughout this chapter was translated from the original by Daniel Scharstein and Rebecca Pohl.

16. As quoted in [Wor], p. 68.

17. As quoted in [Bra2], p. 47.

18. As quoted in [Bra2], p. 107.

19. [Gus], p. 6

20. Page numbers for *Endgame* are from [Bec3].

21. The reference here is to Lucky's one and only speech in *Godot*. As a gift to Vladimir, Pozzo offers a performance of Lucky "thinking." Initiated by donning his bowler hat, Lucky's thinking is an uninterrupted five-minute tirade of twisted academic jargon that drives Pozzo and eventually the others into agitated and violent protest. Full of vestiges of intelligent life, the density of philosophical allusions and lack of punctuation in Lucky's verbal spew make it impenetrable in real time and hardly more accessible otherwise. Like all aspects of Beckett's writing, Lucky's speech has been subjected to significant scrutiny. See, for instance, [Atk].

22. After some negotiations and partial recasting, *Waiting for Godot* was performed on Broadway with more successful results. See [Kno], pp. 378–381.

23. The story of the San Quentin performance is beautifully relayed by theater scholar Martin Esslin who explains that "what had bewildered the sophisticated audiences of Paris, London, and New York was immediately grasped by an audience of convicts." Esslin goes on to report the different ways that the San Quentin audience heard their own voices on the stage. "Godot is society." "He's the outside." "If Godot finally came he would only be a disappointment." [Ess1], p. 21

24. [Bea], p. 253

25. [Fre], p. 91

26. [Fre], p. xxiv

27. A useful way to help clarify this distinction is to use the curly brackets from modern set notation to denote extensions. Thus, we are saying the object

Samuel Beckett

should not be confused with the extension

{Samuel Beckett},

which are distinct objects in Frege's formulation.

28. "Your discovery of the contradiction has surprised me and, I should almost like to say, left me thunderstruck," Frege wrote in reply six days after receiving Russell's original letter, "because it has rocked the ground on which I planned to build arithmetic." [Bea], p. 254

29. As quoted in [AlVe], p. 46.

30. "I now begin to feel," Beckett wrote to director Alan Schneider, "the whole idea behind the film... has been chiefly of value on the formal structural level." As quoted in [Ton], p. 99.

31. Page numbers for *Not I* are from [Bec2].

32. [Bai], p. 627

33. This phrase is from Beckett's stage notes for *Not I*.

34. Biographer Dierdre Bair makes this point in [Bai], p. 625. Erik Tonning cites Beckett's notes to director Alan Schneider: "Address less to the understanding than to the nerves of the audience, which should in a sense *share her bewilderment.*" The result, as Toning explains, is that "the audience finds itself intimately involved in the very rhythm of the dilemmas being enacted." [Ton], pp. 117–118

35. The original script has an Auditor, who gestures at this point in the play. Beckett eventually dropped him from the performance because he could not get the effect he wanted and decided that the play held up better without him.

36. [Dut], p. 103

37. [Dut], p. 125

38. This conjecture originates with Hugh Culik and first appears in [Cul2]. It is developed further in [Cul3] and is referenced in [Gib].

39. [PM1], p. 90

40. [PM1], p. 96. The fourth axiom of the list (Associative) was shown to be redundant in the sense that it is possible to derive it as a theorem by starting from the other axioms and using the transformation rules.

41. PM actually uses a system of dots in place of the parenthesis that the authors attribute to Peano.

42. [PM1], p. 94

43. [PM1], p. 96

44. To apply this system of mechanized reasoning to the syllogism stated at the beginning of this section, let p represent the proposition, "Hamm whistles," and let's take q to be the proposition "Clov comes." Premise (1), that "if Hamm whistles Clov comes" translates as $(p \supset q)$. Premise (2) evidently translates as simply $\sim q$. The conclusion, that "Hamm has not whistled," takes the form $\sim p$.

Does this conclusion follow logically from the two premises? Earlier we noted that it instinctively felt like it did, but now we are in possession of a logical calculator designed to verify that our reasoning is sound. Encoding the syllogism into the language of PM, we are asking whether the two premises $(p \supset q)$ and $\sim q$ necessarily imply the conclusion $\sim p$. Thus, what we are really asking is whether it is possible to produce the formula

$$(((p \supset q) \wedge \sim q) \supset \sim p)$$

by starting from the axioms and following the transformation rules. If so, then we are on solid logical ground, provided the machinery of PM is equal to its billing.

45. [PM1], volume I, p. 8

46. [PM1], p. 13. The original notation is modified in the treatment here. In *Principia Mathematica*, the authors mark each asserted proposition with the symbol ⊢, and use a system of dots in place of the grouping parenthesis. The last phrase in the original reads, "i.e., both '⊢ . p' and '⊢ . $\sim p$' cannot legitimately appear."

47. This is known as Goldbach's conjecture and is arguably the most widely known unsolved problem in number theory. It originates from a correspondence between Christian Goldbach and Leonhard Euler from 1742.

48. The notation of PM has not proved so user-friendly to mathematicians over the years. For working with whole number arithmetic, as we are in this case, Peano Arithmetic is more

manageable. This is the axiomatic system used in chapter 5. One version of this proposition in the language of Peano Arithmetic would be:

$$\forall a \, \exists b \, \exists b' \, ((SSa \cdot SS0) = (SSb + SSb') \wedge$$

$$\sim \exists a' \, \exists a'' \, (SSb = (SSa' \cdot SSa'') \vee SSb' = (SSa' \cdot SSa'')))$$

49. At this talk Hilbert only presented a subset of these problems. The full list was published in 1902.

50. The first was to resolve the continuum hypothesis that grew out of Georg Cantor's investigations of infinity.

51. Hilbert's vision was that it could be possible to demonstrate the consistency of a powerful system like PM by using just a portion of its full arsenal of rules. The allowable fraction he suggested was called "finitary methods," and for Hilbert it represented those techniques that fell more firmly within the universally accepted aspects of reasoning and arithmetic. Without going into greater depth about what these finitary methods consist of, it suffices to say that the question became whether the full weight of PM's formal structure could be supported by a carefully specified subset of lighter bricks contained among its rules and axioms.

52. As quoted in 1990 edition of Bair's biography, p. 544.

53. In his in-depth treatment of *Come and Go*, Erik Toning examines the sequence of progressively pared down drafts to illustrate how "Beckett sacrificed the effect of intertwined unfolding ironies in the conversations in order to create a lyrical simplicity of visual, verbal and auditory patterning." [Ton] p. 85

54. In his notes to Billie Whitelaw who was performing the piece Beckett wrote, "The pacing is the essence of the matter. The text: what pharmacists call excipient." As quoted in [Ton], p. 116.

55. [Ton], pp. 271–272

56. With five biscuits, there are $5 \cdot 4 \cdot 3 \cdot 2 \cdot 1 = 120$ ways to arrange them. "But were he to take the final step and overcome his infatuation with the ginger," Murphy thinks to himself as he contemplates his five cookies, "then the assortment would spring to life before him, dancing the radiant measure of its total permutability, edible in a hundred and twenty ways." (96)

57. "Here we find the theatre of the surd summed up as the need to do anything," writes Salazar-Sutil, referring to the title of his thesis, "to perform whatever action is necessary in order to fill the emptiness of a thoughtless time." [Sal], p. 187

58. [Bec6], p. 329

59. [Ack5], p. 17

60. "These two acts are separated by ten thousand years," Beckett explained during the production process. [Kno], p. 593

61. [Bec6], p. 26

62. A little errantly, as it turns out.

63. [Bec6], p. 22

64. [Bec6] p. 170

65. This informal sketch of Gödel's argument adopts a *semantic* approach in that it relies on the intended interpretation of the formulas of PM as meaningful statements of number theory. Among other things, this means formulas can be regarded as either true or false. Godel's original proof was *syntactic* rather than semantic. Avoiding the concept of truth altogether, Gödel

assumed only that the axioms of PM were consistent. He then argued that *G* was *undecidable* in the sense that neither *G* nor $\sim G$ could be derived as a theorem. The semantic argument employed here is informed by the insights of Douglas Hofstadter. See, for example, [Hof1], p. 271. Whereas the semantic approach is more user friendly, there are advantages to taking a purely syntactic approach, which is how Gödel's Theorem is typically treated in formal logic.

Chapter 4. Dürrenmatt, Frayn, and McBurney

1. [Bre], p. 14
2. [Bre], p. 9
3. [Cro], p. 8
4. This point is developed in the essay by Northcott in [Nor].
5. This distinction between Dürrenmatt and Brecht is discussed at length in [Tiu]; see, for example, pp. 217–220.
6. As quoted in the commentary in [Bre], p. xxiii.
7. [Bre], p. 107
8. This quotation is a part of item 21 of Dürrenmatt's "21 Points to The Physicists," all of which appear in [Dur].
9. Page numbers for *The Physicists* are from [Dur].
10. The significance of the Möbius strip to the structure of *The Physicists* is insightfully explored by Kirsten Shepherd-Barr who gravitates to this same moment in the play. "[The Möbius strip] signifies the play's main ideas," she writes, "and the truly remarkable thing is the way in which the play itself enacts or performs the symbol in turn." [She1], p. 80
11. Page numbers for *Copenhagen* are from [Fra].
12. The discussion of *Copenhagen* in this section follows a similar arc to Shepherd-Barr's excellent treatment in [She1]. The quotation is from p. 6.
13. Shepherd-Barr is careful to point out that the term "performativity" does not simply refer to the act of demonstrating or performing an idea. In its original intent, performativity is a quality achieved when the performance *produces* the phenomenon being discussed. "The dialogue does not merely reflect the principle," Shepherd-Barr writes, "it makes it happen, with the audience participating in the act of creation." See the very interesting discussion in [She1], pp. 33–36.
14. One consequence of the attention Frayn's play generated around this historical event was the release of previously unpublished letters, several written by Bohr and one by Heisenberg. Meant to clarify the events that transpired at the Bohr residence in fall 1941, the new information in these letters mainly served to exacerbate the confusion. A thoughtful analysis of how these new letters reinforce the thesis of Frayn's play can be found in [She1], pp. 105–110.
15. To be accurate, the histogram would approximate the square of the modulus of the wave function.
16. [Fra], p. 99
17. Shepherd-Barr draws attention to this same quotation from the postscript, adding her own italics to indicate the emphasis she feels it deserves. She goes on later to quote the playwright again in a *New York Review of Books* article called "*Copenhagen* Revisited," published in 2002: "The epistemology of intention is what this play is about."

18. This anecdote was relayed by Frayn as part of his remarks during a conference on science and theater held at the University of Lincoln, April 2014. Shepherd-Barr also discusses Blakemore's contributions to *Copenhagen*. See [She1], p. 5 and 104 for a discussion of Blakemore's argument for how *Copenhagen* demonstrates the distinctive role of theater among the arts.

19. The quotation originally appeared in [Bla]. This point, and this quotation in particular, is discussed in [She1], p. 5.

20. Frayn made this comment during his 2014 University of Lincoln presentation.

21. The sentiment attributed to Hans-Peter Durr, as well as Frayn's personal agreement with it, was expressed by Frayn as part of his presentation at the conference in Lincoln. Peter Langdal's belief that Heisenberg's trip was an attempt "to know yourself by knowing each other" is expressed by Shepherd-Barr in [She1], p. 108.

22. In his introduction, C. P. Snow mentions that novelist Graham Greene makes this very point in an early review.

23. [Har], p. 61

24. [Har], p. 65

25. [Har], p. 139

26. This is the opinion of Victoria Gould who was part of an initial read through of *A Mathematician's Apology* that McBurney arranged.

27. Page numbers for *A Disappearing Number* are from [McB2].

28. This quotation, as well as the ensuing ones by Gould and du Sautoy are from a personal interview with the author.

29. Gould reports that these sessions included a number of improvisations about imagined conversations between Hardy and Ramanujan that in the end did not yield very satisfying results. This is how it gradually came to be that the two historical figures would say only lines we know they actually said.

30. An account of some of these exercises are described in the resource materials for the play in [AFG], p. 4.

31. McBurney writes warmly about the contributions of du Sautoy and Gould in the published play's introduction (despite getting the color of du Sautoy's eyes wrong). The two were instrumental with regard to contributing mathematical insights about the infinite, and by all accounts Gould was the creative force behind the character of Ruth. After collaborating on *A Disappearing Number*, Gould and du Sautoy went on to create their own math play, *I is a Strange Loop* (originally *X & Y*).

32. [Har], p. 85

33. [McB1]

34. In this passage, Ruth is quoting from *And Our Faces, My Heart, Brief as Photos*, by John Berger.

35. [Har], p. 84

36. For an examination of how the process of devised theater is amenable to exploring science more generally, see the interesting piece by Mike Vanden Heuvel in [She3].

37. [Har], p. 113

38. [AFG], p. 4

39. [Mur], p. 105

40. He also sounds a great deal like Jacques Lecoq who founded the influential theater school in Paris where McBurney trained. See, for instance, [Mur], p. 101.

41. From Frayn's 2014 University of Lincoln presentation. With this comment, Frayn was emphasizing that a playwright does something similar to what the theoretical physicists in *Copenhagen* were doing, but it applies even more aptly to mathematicians.

42. See, for instance, the quotation from [Wor], p. 68: "When I was working on *Watt*, I felt the need to create for a smaller space, one in which I had some control of where people stood or moved, above all, of a certain light. I wrote *Waiting for Godot.*"

43. [Har], p. 84

Chapter 5. Stoppard

1. I was comforted to find that Nicolas Salazar-Sutil took a similar position in his thesis, *Theatres of the Surd.* Acknowledging that Stoppard "has devoted himself to the exploration of mathematical thinking in the theatre in ways that are just as paradoxical as Pirandello's," Salazar-Sutil argues that "modern theatre dramatizes how somethingness emerges from self-perception, how one comes out of the empty set, and finally, how an interior-to-interior participatory presence finally asserts itself over the absent externality of Nature and Truth." [Sal], p. 142

2. [Sto9], p. 193

3. As quoted in [Nad], p. 351.

4. Page numbers for *The Real Inspector Hound* are from [Sto10].

5. Page numbers for *Artist Descending a Staircase* are from [Sto8].

6. Page numbers for *Dogg's Hamlet* are from [Sto10].

7. The play is eventually the thing, and when the schoolboys do finally take the stage it is Stoppard's *15 Minute Hamlet*. In this performance, Shakespeare's original script is trimmed to the marrow, but all the famous quotations remain and the central narrative is miraculously intact. The scripted encore for this performance is then a two-minute attempt of the same trick.

8. This discussion closely follows [Hof1], pp. 267–272, where, for a host of clever reasons, Hofstadter actually uses three-digit codons for each PA symbol. Gödel employed a different method. He assigned numbers to each symbol of the formal system similar to the way we've started, but then the Gödel number for a string of symbols is obtained in the following way: If the string consisted of, say, twelve symbols, then the Gödel number of the string is formed by taking the product of the first twelve primes, each one raised to a power equal to the Gödel number of the corresponding symbol in the string. The fact that every number has a unique factorization into a product of primes implies that no two strings get mapped to the same Gödel number.

9. One unimportant difference is that the concept of theorem number depends on the particular Gödel-numbering algorithm in place. Different Gödel-numbering schemes result in different sets of theorem numbers, whereas even numbers are a fixed subset of the whole numbers.

10. One way to generate every even number is to start with 0 and then successively add 2. Given any even number, adding 2 always produces another even number, and a little thought reveals that every even number can be generated in this way. Because this process of generating

even numbers has a familiar arithmetic description, it feels very natural that it falls within the domain of PA's expressive powers.

To expand on the analogy of theorem numbers with even numbers, recall that PA consists of a collection of axioms—starting points or "free" theorems in a sense. There is also a list of transformation rules for generating new theorems from old ones. Now each of these axioms has a Gödel number, and the collection of Gödel numbers that correspond to each of the axioms becomes our initial set of theorem numbers. These are "free" theorem numbers, so to speak. Relating this to the example of even numbers, this initial set of theorem numbers plays the same role as the number 0 we started with as our first "free" example of an even number. One axiom of PA we have seen a few times is

$$\forall a \,(a + 0) = a.$$

Computing its Gödel number gives

$$141317132110271113$$

and thus 141317132110271113 is a theorem number.

The transformation rules for PA then become a way of generating new theorem numbers from old ones in the same way that adding 2 is a way to generate new even numbers from previously established ones. Here we have to wave our hands a bit, but not too egregiously. In the previous proof, we saw how the Rule of Specification could be invoked so that the axiom

$$\forall a \,(a + 0) = a \qquad \text{is transformed into} \qquad (S0 + 0) = S0$$

In terms of our new Gödel-numbering notation, this transformation gets rephrased as saying that the number

$$141317132110271113 \qquad \text{is transformed into} \qquad 172010211027112010.$$

Note that when we phrase this using Gödel numbers in place of the PA formulas we are converting one whole number into another one, and what should be at least plausible is that there is a way to simulate this conversion with a fixed *arithmetic* algorithm—i.e., a finite sequence of steps that involves operations like addition and multiplication rather than syntactic symbol-shifting. Because the transformation rules are a bit involved, the corresponding arithmetic algorithms are too. Trying to write out the arithmetic steps is not much fun and not important for this discussion. What matters is that we recognize that every typographical manipulation that the PA transformation rules carry out on a given formula has a corresponding arithmetic analog that can be applied to the formula's Gödel number to achieve the corresponding outcome. As an example, the typographical step of appending the parenthesis ")" to the right end of a formula can be simulated by multiplying the formula's Gödel number by 100 and adding 27. Generalizing from this simple example, the takeaway is that manipulating symbols according to PA's formal rules becomes indistinguishable from doing arithmetic on the corresponding Gödel-numbered copy of PA.

11. Following the earlier example, we encode the equivalent statement, "There exists an integer b that, when doubled, gives a result of 101110." In PA notation, this can be written as

$$\exists b (2 \cdot b) = SSS \ldots SSS0,$$

where there are precisely 101110 S symbols before the 0.

12. [Göd], p. 39

13. Airbrushed out of this description of G is the puzzle of how a formula can contain its own Gödel number. It is evident from the simplest examples that a formula's Gödel number is excessively large. The PA numeral for that Gödel number will then contain a similarly excessive string of S symbols, which has no chance of somehow being wedged back into the original formula. To say it more succinctly, a formula of PA can't explicitly contain its own Gödel number—it won't fit. The work-around that Gödel devised was to embed a description, or recipe, for how to construct a particular number inside G that when followed yields the Gödel number of G itself.

There are various levels of precision from which to appreciate this part of the proof. Ernest Nagel and James Newman's 1958 book *Gödel's Proof* provides an authentic feel for the structure of the argument that requires only a modest background in formal logic. Douglas Hofstadter's *Gödel, Escher, Bach* manages the same trick with more detail but also more logic required. In *I am a Strange Loop*, Hofstadter describes a fascinating linguistic analogy to Gödel's proof due to W.V.O. Quine that requires no formal logic at all.

14. This outline of Gödel's proof is essentially the one in [NaNe], p. 85; see also [Hof1], p. 271.

15. [Gus], p. 72

16. Page numbers for *The Real Thing* are from [Sto11].

17. [Gus], p. 41

18. This observation originates, in my experience, with Douglas Hofstadter. Here is how he explains it: "Kurt Gödel's bombshell. . . revealed the stunning fact that a formula's *hidden meaning* may have a peculiar kind of 'downward' causal power, determining the formula's truth or falsity (or its derivability or nonderivability inside *Principia Mathematica* or any other sufficiently rich axiomatic system.)." [Hof2], p. 169. Hofstadter's name and contributions are introduced later in the chapter.

19. "I'm much happier working on 'Shakespeare in Love' which I think suits me better," Stoppard wrote to friend and film executive Barry Isaacson in September 1992 as a way of breaking the news that he was clearing his desk of Isaacson's other requested projects. "The previous plan of doing a quick three or four weeks on 'Shakespeare in Love' hasn't worked out, in a way it has worked out more happily from my point of view. It has turned into more of a take-over than a short interruption. . . . I have had to strip things back so that I am using almost none of the original superstructure, and not much of the foundations either." This quotation is from a private correspondence between Stoppard and Isaacson, which is part of the larger collection of Stoppard papers at the Harry Ransom Center.

20. Page numbers for *The Hard Problem* are from [Sto4].

21. The most careful articulation of this argument is due to J. R. Lucas, who published it in a 1961 paper called "Minds, Machines, and Gödel." To get a sense of the debate see, for example, [Hof1], pp. 471–479, and also Turing's "Computing Machinery and Intelligence," where he discusses the "Mathematical Objection." This latter article can be found in [Cop1].

22. Hofstadter's 1979 Pulitzer Prize-winning book *Gödel, Escher, Bach* uses Gödel's Theorem as the launching point for an exploration of the reflexive structures that permeate such diverse fields as painting, music, biology, the theory of computation, and cognitive science.

David Chalmers, the cognitive scientist responsible for the phrase "the hard problem of consciousness," was Hofstadter's PhD student at Indiana University in the early 1990s.

23. Here is the full quotation from [Hof1], p. 709:

> My belief is that the explanations of 'emergent' phenomena in our brains—for instance, ideas, hopes, images, analogies, and finally consciousness and free will—are based on a kind of Strange Loop, an interaction between levels in which the top level reaches back down toward the bottom level and influences it, while at the same time being itself determined by the bottom level. In other words, a self-referencing 'resonance' between different levels—quite like the Henkin sentence which, by merely asserting its own provability, actually becomes provable. The self comes into being at the moment it has the power to reflect itself.

24. [Sto].

25. It is possible to give a reasonably specific description of this "sufficiently powerful" threshold for formal systems. If we start with the goal of building a formal system of arithmetic from scratch, we might first create the axioms and rules to express basic facts about addition. If we stopped here, our system could be considered complete in the sense that it could prove all the truths it could articulate—it just wouldn't be able to articulate anything very interesting. A statement like "$2 + 3 = 5$" could be encoded into the language of our low-power formal system and it might indeed be a theorem, but a statement like "5 is prime" would be inexpressible without access to multiplication. And so we augment the symbols and the axioms and the transformation rules to increase the reach of our system. Soon enough, we can prove the statement "5 is prime," then "a is prime" for any choice of a that yields a true statement. With a few more rules for handling quantifiers, it is not long until we can prove the true instances of "a is a theorem number." It is at this point that we have crossed the threshold. Whether or not we are paying attention to this fact, at each stage of development, the architecture of the system we are building can be reflected inside the world of whole numbers via Gödel-numbering. Because we are building a system that is progressively describing larger amounts of terrain within the world of whole numbers, there is a critical moment where this described terrain expands enough to engulf the system's reflection. At this point, like it or not, our system has the capacity to talk about itself.

26. Stoppard all but acknowledges this sufficiently powerful threshold for plays in the same piece where he asserts their potential for expressing metanarratives. "I did say that all narratives have some such capacity," he writes, "but the plays that are important to the advancement of art (as opposed to plays that are merely good in many different ways) are those that suggest this capacity to a very high degree." He goes on to cite *Waiting for Godot* and *The Birthday Party* as two important examples. See [Sto].

27. Page numbers for *The Invention of Love* are from [Sto5].

28. From *New York Review of Books*, September 21, 2000, vol. 47, no. 14.

29. As quoted in the *Invention of Love* program from the performance at Lincoln Center.

30. This aspect of Gödel's work is at the heart of Hofstadter's analysis. "In my many years reflecting about what Kurt Gödel did in 1931," Hofstadter writes in *I Am a Strange Loop*, "it is this insight of his into the roots of meaning—his discovery that, thanks to a mapping,

full-fledged meaning can suddenly appear in a spot where it was entirely unexpected—that has always struck me the most."

31. And not so odd that he keeps inserting himself into the chapter.

Chapter 6. Whitemore, Wilson, and Mighton

1. Page numbers for *Breaking the Code* are from [Whi].

2. As quoted in [Lea], p. 16.

3. As quoted in [Cop1], p. 48.

4. [Lea], pp. 19 and 115

5. [Har], p. 140. Leavitt makes the same observation in [Lea], p. 133, musing that "Hardy was so utterly wrong in his affirmation that one cannot help wondering how much of what he wrote was wishful thinking."

6. What is particularly poignant about this passage is that the historical Knox was dying of lymphatic cancer at this time. It had been diagnosed as early as 1939 and he died in February 1943.

7. In using the term "mutual masturbation," Turing might have been aware of the legal definition of the more serious crime involving anal penetration, a consequence of which was capital punishment not many years earlier.

8. [Cop1], p. 73. The notation is slightly altered.

9. There is no record that the apple was ever tested for cyanide or anything else. The autopsy concluded death by cyanide poisoning but did not speculate on the agent that brought it into his body. Copeland has an extended discussion of the problems with the autopsy and inquest. He concludes that it was possible Turing inhaled cyanide fumes accidently.

10. Page numbers for *Lovesong of the Electric Bear* are from the original script cited in [Sno].

11. This observation is from a program note written by Steve Kennedy for a production of *Lovesong* at Carleton College in spring 2007.

12. [Cop1], p. 452

13. It is fair to say that, writing in 1950, Turing was operating on a different set of assumptions about gender identity than would be in play currently.

14. [Cop1], p. 441

15. [Cop1], p. 442

16. [Cop1], p. 450

17. This is from a lecture Turing gave in 1947 addressing the same issue. The transcript appears in [Cop1], and this particular quotation is on p. 394.

18. [Cop1], p. 451

19. An in-depth analysis of the relationship between Turing's sexual orientation and his ideas about artificial intelligence can be found in [Lea]. See, for example pp. 205–206.

20. [Hof1], p. 679

21. Mighton's first play, *Scientific Americans*, explores the changing terrain of ethical predicaments brought on by the military possibilities for modern research in fields like computer science. In *A Short History of Night*, Mighton recounts the plight of sixteenth-century astronomer and mathematician Johannes Kepler in a way that suggests some cautionary lessons for

contemporary society. In *Possible Worlds*, Mighton borrows ideas from neuroscience to create a work of science fiction that explores questions of human identity in an original and unsettling way. The budding technology of virtual reality was Mighton's vehicle for probing human nature in the surreal and steamy comic-drama *Body and Soul*, and the physics of time is at the core of *The Little Years*.

22. Page numbers for *Half Life* are from [Mig4].

23. In 2017, the United Kingdom passed the "Alan Turing Law" that pardoned over 50,000 men convicted under historical legislation that outlawed homosexual acts.

Chapter 7. Auburn

1. For example, Auburn mentions this in his *Talk of the Nation: Science Friday* interview on NPR in October 2000.

2. Page numbers for *Proof* are from [Aub].

Acknowledgments

1. [Sal], p. 213

BIBLIOGRAPHY

[Abb1]   Abbott, Stephen. "A Disappearing Number." *Math Horizons*, vol. 16, no. 4 (2009): pp. 17–19.

[Abb2]   _____. "Simon McBurney's Ambitious Pursuit of the Pure Math Play." In "New Directions in Theatre and Science," special issue, *Interdisciplinary Science Reviews*, vol. 39, no. 3 (February 2014): pp. 224–237.

[Abb3]   _____. "The Dramatic Life of Mathematics: A Centennial History of the Intersection of Mathematics and Theater." *A Century of Advancing Mathematics*, Mathematical Association of America, Washington, DC, 2015.

[Abb4]   _____. "A Mathematical Proof or Not?" *Math Horizons*, vol. 13, no. 3 (February 2006): pp. 26–27.

[Abb5]   _____. "A Beautiful Mind." *Math Horizons*, vol. 9, no. 4 (April 2002): pp. 5–7.

[Abb6]   _____. "Turning Theorems into Plays." *The Edge of the Universe: Celebrating Ten Years of Math Horizons*, Mathematical Association of America, Washington, DC, 2006, pp. 113–119.

[Ack1]   Ackerley, C. J. *Obscure Locks, Simple Keys: The Annotated Watt.* Journal of Beckett Studies Books, Tallahassee, FL, 2005.

[Ack2]   _____. "Samuel Beckett and Mathematics." *Cuadernos de Literatura Inglesa y Norteamericana*, vol. 3, no. 1–2 (1998): pp. 77–102.

[Ack3]   _____. "Samuel Beckett and Science." In S. E. Gontarski, *A Companion to Samuel Beckett*, pp. 143–163.

[Ack4]   _____. "Monadology: Samuel Beckett and Gottfried Wilhelm Leibniz." *Sofia Philosophical Review*, vol. 5, no. 1 (2011): pp. 122–145.

[Ack5]   _____. "Samuel Beckett and the Mathematics of Salvation." In "Beckett and the Afterlife," edited by Christopher Conti, Matthijs Engelberts, and Sjef Houppermans. Special issue, *Samuel Beckett Today/Aujourd'hui* 33, no. 1 (2021): pp. 15–29.

[AcGo]   Ackerley, C. J., and Gontarski, S. E. *The Grove Companion to Samuel Beckett: A Reader's Guide to his Works, Life, and Thought.* Grove Press, New York, 2004.

[Alb]   Albers, Don et al. (editors). *The G. H. Hardy Reader.* Cambridge University Press and Mathematical Association of America, Cambridge, 2015.

[AFG]   Alexander, Catherine; Freedman, Natasha; and Gould, Victoria. *A Disappearing Number Resource Pack.* www.complicite.org.

[AlVe]   Alexander, George, and Velleman, Daniel. *Philosophies of Mathematics.* Blackwell Publishing, Malden, MA, 2002.

[Art]   Artmann, Benno. *Euclid: The Creation of Mathematics*. Springer, New York, 1999.

[Atk]   Atkins, Anselm. "Lucky's Speech in Beckett's 'Waiting for Godot': A Punctuated Sense-Line Arrangement." *Educational Theatre Journal*, vol. 19, no. 4 (1967): pp. 426–432.

[Bai]   Bair, Deirdre. *Samuel Beckett: A Biography*. Harcourt Brace Jovanovich, New York, 1978.

[Bar]   Barrow, John. *The Infinite Book*. Vintage, London, 2005.

[Bea]   Beaney, Michael (editor). *The Frege Reader*. Blackwell Publishing, Malden, MA, 1997.

[Bel]   Bell, E. T. *Men of Mathematics: The Lives and Achievements of the Great Mathematicians from Zeno to Poincaré*. Simon and Schuster, New York, 1965.

[Ber]   Berger, John. *And Our Faces, My Heart, Brief as Photos*. Pantheon Books, New York, 1984.

[Bil]   Billington, Michael. "Stoppard's Secret Agent." In Paul Delaney, *Tom Stoppard in Conversation*, pp. 193–198.

[Bla]   Blakemore, Michael. "From Physics to Metaphysics and the Bomb." *New York Times*, April 9, 2000.

[Bln]   Blansfield, Karen. "Atom and Eve: The Mating of Science and Humanism." *South Atlantic Review*, vol. 68, no. 4 (2003): pp. 1–16.

[Boo]   Booth, Michael R. "Nineteenth-Century Theatre." In John Russell Brown, *The Oxford Illustrated History of the Theatre*, pp. 299–340.

[Bra1]   Brater, Enoch. "Mis-takes, Mathematical and Otherwise in *The Lost Ones*." *Modern Fiction Studies*, vol. 29 (1983): pp. 93–109.

[Bra2]   _____. *Why Beckett*. Thames and Hudson, London, 1989.

[Bra3]   Brater, Enoch (editor). *Beckett at 80: Beckett in Context*. Oxford University Press, New York, 1986.

[Bri1]   Brits, Baylee. "Beckett and Mathematics." In Rabaté, *The New Samuel Beckett Studies*, pp. 215–230.

[Bri2]   _____. "Ritual, Code, and Matheme in Samuel Beckett's *Quad*." *Journal of Modern Literature*, vol. 40, no. 4 (2017): pp. 122–133.

[Bro]   Brotchie, Alister. *Alfred Jarry: A Pataphysical Life*. MIT Press, Cambridge, MA, 2011.

[Brw]   Brown, John Russell (editor). *The Oxford Illustrated History of the Theatre*. Oxford University Press, Oxford, 1995.

[But]   Büttner, Gottfried. *Samuel Beckett's Novel* Watt. University of Pennsylvania Press, Philadelphia, 1984.

[Cam1]   Campos, Liliane, "Searching for Resonance: Scientific Patterns in Complicité's *Mnemonic* and *A Disappearing Number*." *Interdisciplinary Science Reviews*, vol. 32, no. 4 (December 2007): pp. 326–334.

[Cam2]   _____. *The Dialogue of Art and Science in Tom Stoppard's* Arcadia. Presses Universitaires de France, Paris, 2011.

[Cam3]   _____. "Mathematics and Dramaturgy in the Twentieth and Twenty-First Centuries." In Englehardt, *The Palgrave Handbook of Literature and Mathematics*, pp. 243–262.

[CaSh]   Campos, Liliane, and Shepherd-Barr, Kirsten. "Science and Theatre in Open Dialogue: *Biblioetica, Le Cas de Sophie K.* and the Postdramatic Science Play." *Interdisciplinary Science Reviews*, vol. 31, no. 3 (July 2006): pp. 245–253.

[Car]   Cardullo, Bert, and Knopf, Robert (editors). *Theater of the Avant-Garde: 1890–1950; A Critical Anthology.* Yale University Press, New Haven, CT, 2001.

[Cas]   Case, Sue-Ellen. *Performing Science and the Virtual.* Routledge, New York, 2007.

[Cop1] Copeland, Jack (editor). *The Essential Turing: Seminal Writings in Computing, Logic, Philosophy, Artificial Intelligence, and Artificial Life.* Clarendon Press, Oxford, 2004.

[Cop2] _____. *Turing: Pioneer of the Information Age.* Oxford University Press, Oxford, 2014.

[FeOv2] Craig, George; Fehsenfeld, Martha Dow; Gunn, Dan; and Overbeck, Lois More (editors). *The Letters of Samuel Beckett, Volume II, 1941–1956.* Cambridge University Press, Cambridge, 2012.

[FeOv3] _____ (editors). *The Letters of Samuel Beckett, Volume III, 1957–1965.* Cambridge University Press, Cambridge, 2014.

[FeOv4] _____ (editors). *The Letters of Samuel Beckett, Volume IV, 1966–1989.* Cambridge University Press, Cambridge, 2016.

[Cro]   Crockett, Roger. *Understanding Friedrich Dürrenmatt.* University of South Carolina Press, Columbia, SC, 1998.

[Cul1] Culik, Hugh. "Science and Mathematics." In Uhlmann, *Beckett in Context*, pp. 348–357.

[Cul2] _____. "Mathematics and Metaphor: Samuel Beckett and the Esthetics of Incompleteness." *Papers on Language and Literature*, vol. 29, no. 2 (Spring 1993): pp. 131–150.

[Cul3] _____. "Raining & Midnight: The Limits of Representation." *Journal of Beckett Studies*, vol. 17, no. 1–2 (May 2009): pp. 127–152.

[DaSe] Davis, Chandler, and Senechal, Marjorie Wikler. *The Shape of Content.* A. K. Peters, Wellesley, MA, 2008.

[Dav]   Davis, Martin. *The Universal Computer: The Road from Leibniz to Turing.* W. W. Norton and Company, New York, 2000.

[Del1] Delaney, Paul (editor). *Tom Stoppard in Conversation.* The University of Michigan Press, Ann Arbor, 1994.

[Del2] _____. *Tom Stoppard: The Moral Vision of the Major Plays.* St. Martin's Press, New York, 1990.

[Dem1] Demastes, William. *Theatre of Chaos: Beyond Absurdism, into Orderly Disorder.* Cambridge University Press, New York, 1998.

[Dem2] _____. "Portrait of an Artist as Proto-Chaotician: Tom Stoppard Working His Way to *Arcadia*." *Narrative*, vol. 19, no. 2 (May 2011) pages 229–240.

[DoMa] Doxiadis, Apostolos, and Mazur, Barry. *Circles Disturbed: The Interplay of Mathematics and Narrative.* Princeton University Press, Princeton, NJ, 2012.

[Dra]   Drain, Richard (editor). *Twentieth-Century Theatre, A Sourcebook.* Routledge, London, 1995.

[Dun]   Dunham, William. *Journey Through Genius.* Wiley, New York, 1990.

[Dut]   Duthuit, Georges (editor). "Three Dialogues." From *Transition* no. 5, Paris, 1949.

[Ell]   Ellis, Reuben J. "'Matrix of Surds': Heisenberg's Algebra in Beckett's *Murphy*." *Papers on Language and Literature*, vol. 25, no. 1 (January 1989): pp. 131–150.

[Els]   Else, Gerald. *Aristotle: Poetics.* Translated, with an introduction. University of Michigan Press, Ann Arbor, 1970.

[EJT] Englehardt Nina; Jenkins, Alice; and Tubbs, Robert (editors). *The Palgrave Handbook of Literature and Mathematics*. Palgrave Macmillan (Springer Nature), Cham Switzerland, 2022.

[Ess1] Esslin, Martin. *The Theatre of the Absurd*. Third edition. Penguin Books, London, 1980.

[Ess2] _____. "Modern Theatre: 1890–1920." In John Russell Brown, *The Oxford Illustrated History of the Theatre*, pp. 341–379.

[Fat] Faticoni, Theodore. *The Mathematics of Infinity*. Wiley, Hoboken, NJ, 2006.

[FeOv1] Fehsenfeld, Martha Dow, and Overbeck, Lois More (editors). *The Letters of Samuel Beckett, Volume I, 1929–1940*. Cambridge University Press, Cambridge, 2009.

[Fey] Feynmann, Richard. *Lectures on Physics/The Character of Physical Law*. Published lectures by the BBC, 1965.

[Fle] Flemming, John. *Stoppard's Theatre: Finding Order amid Chaos*. University of Texas Press, Austin, 2001.

[Fur] Furlani, Andre. "Earlier Wittgenstein, Later Beckett." *Philosophy and Literature*, vol. 39, no. 1 (2015): pp. 64–86.

[Ger1] Gerould, Daniel Charles. *Witkacy. Stanislaw Ignacy Witkiewicz as an Imaginative Writer*. University of Washington Press, Seattle, 1981.

[Ger2] _____ (editor). *Twentieth-Century Polish Avant-Garde Drama: Plays, Scenarios, Critical Documents*. Cornell University Press, Ithaca, NY, 1977.

[Ger3] _____ (editor). *The Witkiewicz Reader*. Northwestern University Press, Evanston, IL, 1992.

[Gib] Gibson, Andrew. *Beckett and Badiou: The Pathos of Intermittency*. Oxford University Press, Oxford, 2006.

[Gle] Gleick, James. *Chaos*. Viking, New York, 1987.

[Gon] Gontarski, S. E. (editor). *A Companion to Samuel Beckett*. Blackwell Publishing, West Sussex, 2010.

[Goo] Goodall, Jane. *Performance and Evolution in the Age of Darwin*. Routledge, New York, 2002.

[Gra] Gratton-Guinness, I. "Russell and G. H. Hardy: A Study of Their Relationship." *Journal of the Bertrand Russell Archives*, n.s. 11 (Winter 1991–92): pp. 165–179.

[Grb] Greenberg, Marvin Jay. *Euclidean and Non-Euclidean Geometries: Development and History*. W. H. Freeman and Company, San Francisco, 1974.

[Gre] Greene, Brian. *The Fabric of the Cosmos: Space, Time, and the Texture of Reality*. Vintage Books, New York, 2004.

[Gus] Gussow, Mel. *Conversations with Stoppard*. Grove Press, New York, 1995.

[Hal] Halpin, Jenni G. *Contemporary Physics Plays: Making Time to Know Responsibility*. Palgrave Studies in Literature, Science, and Medicine, Cham, Switzerland, 2018.

[Har] Hardy, G. H. *A Mathematician's Apology*. Cambridge University Press, Cambridge, 1993.

[Hea] Heath, Sir Thomas. *Diophantus of Alexandra: A Study in the History of Greek Algebra*. Second edition. Cambridge University Press, Cambridge, 1910.

[Her] Herren, Graley. "Samuel Beckett's *Quad*: Pacing to Byzantium." *Journal of Dramatic Theory and Criticism* (Fall 2000): pp. 43–59.

[Hod] Hodges, Andrew. *Alan Turing: The Enigma*. Simon and Schuster, New York, 1983.

[Hof1] Hofstadter, Douglas. *Gödel, Escher, Bach: An Eternal Golden Braid*. *(20th Anniversary Edition)*. Basic Books, New York, 1999.

[Hof2] _____. *I Am a Strange Loop*. Basic Books, New York, 2007.

[Hor] Hornby, Richard. *Drama, Metadrama and Perception*. Associated University Presses, Cranbury, NJ, 1986.

[How] Howard, Alane. "The Roots of Beckett's Aesthetic: Mathematical Allusions in *Watt*." *Papers on Language and Literature*, vol. 30, no. 4 (Fall 1994): pp. 346–351.

[Hug] Hugill, Andrew. *'Pataphysics: A Useless Guide*. MIT Press, Cambridge, MA, 2015.

[Hun] Hunter, Jim. *A Faber Critical Guide: Tom Stoppard* (R&G, Jumpers, Travesties, Arcadia). Faber and Faber, London, 2000.

[Jam] James, Clive. "Tom Stoppard: Count Zero Splits the Infinite." *Encounter* (November 1975): pp. 68–75.

[Jar] Jarrett, Joseph. *Mathematics and Late Elizabethan Drama*. Palgrave Macmillan, London, 2019.

[Jen] Jenkins, Anthony. *The Theatre of Tom Stoppard*. Second edition. Cambridge University Press, Cambridge, 1989.

[Jer] Jernigan, Daniel. "Tom Stoppard and Postmodern Science: Normalizing Radical Epistemologies in *Hapgood* and *Arcadia*." *Comparative Drama*, vol. 37, no. 1 (2003): pp. 3–35.

[Kie] Kiebozinska, Christine. "Witkacy's Theory of Pure Form: Change, Dissolution, Uncertainty." *South Atlantic Review*, vol. 58, no. 4 (November 1993): pp. 59–83.

[Kit1] Kitto, H.D.F. *The Greeks*. Penguin Books, Baltimore, MD, 1956.

[Kit2] _____. *Greek Tragedy: A Literary Study*. Third edition. Routledge, London, 1961.

[Kno] Knowlson, James. *Damned to Fame: The Life of Samuel Beckett*. Grove Press, New York, 1996.

[Kra] Kramer, Prapassari, and Kramer, Jeffrey. "Stoppard's *Arcadia*: Research, Time, Loss." *Modern Drama*, vol. 40, no. 1 (1997): pp. 1–10.

[LaNu] Lakoff, George, and Nunez, Rafael. *Where Mathematics Comes From*. Basic Books, New York, 2000.

[Lap] LaPlace, Pierre Simon. *A Philosophical Essay on Probabilities*. Translated from the sixth French edition by Frederick Wilson Truscott and Frederick Lincoln Emory. John Wiley & Sons, New York, 1902.

[LaPe] Laubenbacher, Reinhard, and Pengelley, David. *Mathematical Expeditions: Chronicles by the Explorers*. Springer-Verlag, New York, 1999.

[Lea] Leavitt, David. *The Man Who Knew Too Much: Alan Turing and the Invention of the Computer*. W.W. Norton, New York, 2006.

[Lee] Lee, Hermione. *Tom Stoppard: A Life*. Alfred A. Knopf, New York, 2021.

[Lin] Lindley, David. *Where Does the Weirdness Go?* Basic Books, New York, 1996.

[Mac] Macaskill, Brian. "The Logic of Coprophilia: Mathematics and Beckett's *Molloy*." *Substance*, vol. 17, no. 3, issue 57 (1988): pp. 13–21.

[Maz] Mazur, Joseph. *The Motion Paradox*. Dutton, New York, 2007.

[McB1] McBurney, Simon. "Note from the Director." Program for National Theatre Live broadcast of *A Disappearing Number*, 2010.

[Mur] Murry, Simon. *Jacques Lecoq*. Routledge, London, 2003.

[Nad] Nadal, Ira. *Tom Stoppard: A Life*. Palgrave Macmillan, New York, 2002.

[NaNe] Nagel, Ernest, and Newman, James. *Gödel's Proof*. New York University Press, New York, 1986.

[Nor] Northcott, Kenneth (editor). *Friedrich Dürrenmatt: Selected Writings, Volume I, Plays*. Translated by Joel Agee. University of Chicago Press, Chicago, 2006.

[Pil] Pilling, John (editor). *The Cambridge Companion to Beckett*. Cambridge University Press, Cambridge, 1994.

[PoRo] Polster, Burkard, and Ross, Marty. *Math Goes to the Movies*. Johns Hopkins Press, Baltimore, MD, 2012.

[Pos] Postlewait, Thomas. *The Cambridge Introduction to Theatre Historiography*. Cambridge University Press, Cambridge, 2009.

[Pou] Pountney, Rosemary. *Theatre of Shadows: Samuel Beckett's Drama 1956–1976*. Colin Smythe, Gerrards Cross, Buckinghamshire, 1988.

[Rab] Rabaté, Jean-Michel (editor). *The New Samuel Beckett Studies*. Cambridge University Press, Cambridge, June 2019.

[Rus] Russell, Bertrand. *Autobiography of Bertrand Russell*. Routledge, London, 1998.

[Sal] Salazar-Sutil, Nicolas. *Theatres of the Surd: Study of Mathematical Influences in European Avant-Garde Theatre*. Doctoral thesis, Goldsmiths College, University of London, 2011.

[Sch] Schmitt, Natalie Crohn. *Actors and Onlookers: Theatre and Twentieth-Century Scientific Views of Nature*. Northwestern University Press, Evanston, IL, 1990.

[Sha] Shapiro, Stewart. *Thinking about Mathematics*. Oxford University Press, Oxford, 2000.

[Sht] Shattuck, Roger. *The Banquet Years: The Origins of the Avant-Garde in France, 1885 to World War I*. Revised edition. Jonathan Cape London, 1969.

[ShTa] Shattuck, Roger, and Taylor, Simon Watson. *Selected Works of Alfred Jarry*. Eyre Methuen, London, 1980.

[She1] Shepherd-Barr, Kirsten. *Science on Stage: From Doctor Faustus to Copenhagen*. Princeton University Press, Princeton, NJ, 2006.

[She2] _____. *Theatre and Evolution from Ibsen to Beckett*. Columbia University Press, New York, 2015.

[She3] _____ (editor). *The Cambridge Companion to Theatre and Science*. Cambridge University Press, Cambridge, 2020.

[Skl] Sklar, Jessica, and Sklar, Elizabeth (editors). *Mathematics in Popular Culture: Essays on Appearances in Film, Fiction, Games, Television and Other Media*. McFarland & Company, Jefferson, NC, 2012.

[Smi] Smith, Peter. *An Introduction to Gödel's Theorems*. Cambridge University Press, Cambridge, 2013.

[Smt] Smith, Russell. "Childhood and Portora." In Uhlmann, *Samuel Beckett in Context*, pp. 7–18.

[Sno]   Snow, C. P. *The Two Cultures and the Scientific Revolution* (The Rede Lecture, 1959). Cambridge University Press, New York, 1961.

[Spe]   Spector, Don. "Distinguishing the Multiverse from an Echo: Quantum Mechanics and *Waiting for Godot*." *Samuel Beckett Today/Aujourd'hui*, vol. 20, no. 1 (2008): pp. 245–257.

[Ste]   Stevens, Brett. "A Purgatorial Calculus: Beckett's Mathematics in *Quad*." In S. E. Gontarski, *A Companion to Samuel Beckett*, pp. 164–181.

[Sti1]   Stillwell, John. *The Four Pillars of Geometry*. Springer UTM series, New York, 2005.

[Sti2]   _____. *Roads to Infinity: The Mathematics of Truth and Proof*. A. K. Peters, Natick, MA, 2010.

[Sti3]   _____. *Mathematics and Its History*. Third edition. Springer, New York, 2010.

[Sto]   Stoppard, Tom. "Pragmatic Theater." *The New York Review*, September 23, 1999.

[Tap]   Taplin, Oliver. "Greek Theater." In John Russell Brown, *The Oxford Illustrated History of the Theater*, pp. 13–48.

[Teu]   Teuscher, Christof (editor). *Alan Turing: Life and Legacy of a Great Thinker*. Springer, New York, 2004.

[Tiu]   Tiusanen, Timo. *Dürrenmatt: A Study in Plays, Prose, and Theory*. Princeton University Press, Princeton, NJ, 1977.

[Ton]   Tonning, Erik. *Samuel Beckett's Abstract Drama*. Peter Lang International Academic Publishing, Oxford, 2007.

[Tur]   Turing, Alan. "Computing Machinery and Intelligence." Originally published in *Mind* in 1950. In Copeland, *The Essential Turing*, pp. 433–464.

[Uhl1]   Uhlmann, Anthony (editor). *Samuel Beckett in Context*. Cambridge University Press, New York, 2013.

[Uhl2]   _____. "Staging Plays." In Uhlmann, *Samuel Beckett in Context*, pp. 173–182.

[Van]   Vanden Heuvel, Mike. "Devised Theatre and the Performance of Science." In Shepherd-Barr, *The Cambridge Companion to Theatre and Science*, pp. 131–145.

[Wat]   Waters, Steve. *The Secret Life of Plays*. Nick Hern Books, London, 2010.

[Wee]   Weeks, Jeffrey. *The Shape of Space: How to Visualize Surfaces and Three-Dimensional Manifolds*. Marcel Dekker, New York, 1985.

[Wig]   Wigner, Eugene. "The Unreasonable Effectiveness of Mathematics in the Natural Sciences." *Communications on Pure and Applied Mathematics*, vol. 13 (1960): pp. 1–14.

[Wit]   Wittgenstein, Ludwig. *Philosophical Investigations*. Third edition. Macmillan Publishing Company, New York, 1958.

[Wol]   Wolf, Robert. *A Tour through Mathematical Logic*. Mathematical Association of America, Washington, DC, 2005.

[Wor]   Worton, Michael. "*Waiting for Godot* and *Endgame*: Theatre as Text." In Pilling, *The Cambridge Companion to Beckett*, pp. 67–87.

[Zeh]   Zehelein, Eva-Sabine. *Science: Dramatic. Science Plays in America and Great Britain, 1990–2007*. Universitätsverlag Winter, Heidelberg, 2009.

Creative Works (Plays, Films, Novels, Mathematics)

[Aub]   Auburn, David. *Proof*. Faber and Faber, New York, 2001.

[Bar]   Barfield, Tanya. *Blue Door*. Dramatists Play Service, New York, 2006.

[Brw]   Barrow, John. *Infinities*. Directed by Luca Ronconi for Piccolo Teatro, Milan, 2002.

[Bax]   Baxter, Craig. *Let Newton Be*. Directed by Patrick Morris for the Menagerie Theatre Company. Performed at Trinity College, Cambridge, 2009.

[Bec1]   Beckett, Samuel. *Disjecta: Miscellaneous Writings and a Dramatic Fragment*. Grove Press, New York, 1984.

[Bec2]   _____. *The Collected Shorter Plays*. Grove Press, New York, 1984.

[Bec3]   _____. *Endgame and Act without Words I*. Grove Press, New York, 1957.

[Bec4]   _____. *Happy Days*. Grove Press, New York, 1961.

[Bec5]   _____. *Murphy*. Grove Press, New York, 1957.

[Bec6]   _____. *Three Novels: Molloy, Malone Dies, The Unnamable*. Grove Press, New York, 1958.

[Bec7]   _____. *Waiting for Godot, a Tragicomedy in Two Acts*. Grove Press, New York, 1954.

[Bec8]   _____. *Watt*. Grove Press, New York, 1953.

[Bre]   Brecht, Bertolt. *Life of Galileo* Commentary by Hugh Rorrison. Translated by John Willet. Methuen Drama, London, 1986.

[DeC]   De Cari, Gioia. *Truth Values: One Girl's Romp through the MIT Male Math Maze*. Directed and performed by the author. Originally performed at the New York International Fringe Festival, 2009.

[Dje]   Djerassi, Carl. *Calculus (Newton's Whores)*. In Djerassi and Pinner, *Newton's Darkness*. Imperial College Press, London, 2003.

[Dox]   Doxiadis, Apostolos. *Seventeenth Night*. (Formally titled *Incompleteness*.) Directed by Tony Stevens. Original performance at the Epi Kolonos Theater, Kolonos, Athens, 2004.

[Dur]   Dürrenmatt, Friedrich. *The Physicists*. Translated by James Kirkup. Grove Press, New York, 1964.

[Euc]   Euclid. *Elements*. Edited by Dana Densmore. Translated by Thomas Heath. Green Lion Press, Santa Fe, NM, 2013.

[Fra]   Frayn, Michael. *Copenhagen*. Anchor Books, New York, 2000.

[Fre]   Frege, Gottlob. *The Foundations of Arithmetic: A Logico-Mathematical Inquiry into the Concept of Number*. Second revised edition. Translated by J. L. Austin. Harper and Brothers, New York, 1953.

[Geo]   George, Madeleine. *The Curious Case of the Watson Intelligence*. Samuel French, New York, 2014.

[Göd]   Gödel, Kurt. *On Formally Undecidable Propositions of Principia Mathematica and Related Systems*. Introduction by R. B. Braithwaite. Translated by B. Meltzer. Dover Publications, New York, 1992.

[GoSa]   Gould, Victoria, and du Sautoy, Marcus. *I Is a Strange Loop*. (Formally titled *X&Y*.) Performed by the authors. Original production at the Latitude Festival, Suffolk, England, 2012.

[Gra]   Granville, Jennifer, and Granville, Andrew. *Math Sciences Investigation: The Anatomy of Integers and Permutations.* https://dms.umontreal.ca/~andrew/PDF/MSRIProgram.pdf.

[Gro]   Groff, Rinne. *The Five Hysterical Girls Theorem.* Playscripts, New York, 2006.

[Gun1]   Gunderson, Lauren. *Deepen the Mystery: Science and the South Onstage.* iUniverse, New York, 2005.

[Gun2]   _____. *Emilie: La Marquis Du Châtelet Defends Her Life Tonight.* Samuel French, New York, 2010.

[Gun3]   _____. *Ada and the Engine.* Dramatists Play Service, New York, 2018.

[Gun4]   _____. *Silent Sky.* Dramatists Play Service, New York, 2015.

[Hau]   Hauptmann, Ira. *Partition.* Playscripts, New York, 2003.

[Hna]   Hnath, Lucas. *Isaac's Eye.* Dramatists Play Service, New York, 2013.

[Jar1]   Jarry, Alfred. *Exploits and Opinions of Doctor Faustroll, Pataphysician.* Translated and annotated by Simon Watson Taylor. Exact Change, Cambridge, MA, 1996.

[Jar2]   _____. *The Ubu Plays.* Translated by Cyril Connelly and Simon Watson Taylor. Grove Press, New York, 1968.

[Joh]   Johnson, Terry. *Insignificance.* Heinemann Educational Books, London, 1988.

[Keh]   Kehlmann, Daniel. *Ghosts in Princeton.* Directed by Robert Lyons at the Workshop Theater, New York, 2012.

[LeRo]   Lessner, Joanne Sydney, and Rosenblum, Joshua. *Fermat's Last Tango.* Original production at the York Theatre (NYC) directed by David Stern, 2001. Produced by and available through the Clay Mathematics Institute (www.claymath.org/).

[Mad]   Maddow, Ellen. *Delicious Rivers.* Directed by Paul Zimet at La Mama Theater Club, New York, 2006.

[McB2]   McBurney, Simon. *A Disappearing Number.* (Conceived and directed by Simon McBurney, devised by the company.) Oberon Books, London, 2007.

[Mig1]   Mighton, John. *A Short History of Night.* Playwrights Canada Press, Toronto, Ontario, 1989.

[Mig2]   _____. *Possible Worlds.* Playwrights Canada Press, Toronto, Ontario, 1988.

[Mig3]   _____. *Body and Soul.* Coach House Press, Toronto, Ontario, 1994.

[Mig4]   _____. *Half Life.* Playwrights Canada Press, Toronto, Ontario, 2005.

[Mos]   Moses, Itamar. *Completeness.* Samuel French, London, 2011.

[New]   Newton, Isaac. *Mathematical Principles of Natural Philosophy, Volume I: The Motion of Bodies.* University of California Press, Berkeley, 1962.

[NoSt]   Norman, Marc, and Stoppard, Tom. *Shakespeare in Love.* Philipp Reclam jun., Stuttgart, 2000.

[Pae]   Patera, Valeria. *La Mela di Alan: Hacking the Turing Test.* Di Renzo Editore, Rome, 2007.

[Pat]   Patterson, Kevin. *A Most Secret War.* Samuel French, London, 1988.

[Pey]   Peyret, Jean-Francois, and Steels, Luc. *Le Cas de Sophie K.* Directed by Jean-Francois Peyret. Performed at the Festival d'Avignon, 2005.

[Pin]   Pinner, David. *Newton's Hooke.* In Djerassi and Pinner, *Newton's Darkness.* Imperial College Press, London, 2003.

[PM1]   Russell, Bertrand, and Whitehead, Alfred North. *Principia Mathematica, Volume I.* Second edition. Cambridge University Press, London, 1925.

[PM2]   _____. *Principia Mathematica, Volume II*. Second edition. Cambridge University Press, London, 1927.

[PM3]   _____. *Principia Mathematica, Volume III*. Second edition. Cambridge University Press, London, 1927.

[Sta]   Stan's Café. *Simple Maths*. Devised by Stan's Café company members Sarah Dawson, Amada Hadingue, Jake Oldershaw, Craig Stephens, and Nick Walker. First performed at Belgrade Studio Theatre, Coventry, UK, 1997.

[Ste]   Stephens, Simon. *The Curious Incident of the Dog in the Night-Time*. (Based on the novel by Mark Haddon.) Methuen Drama, London, 2012.

[Sto1]   Stoppard, Tom. *Arcadia*. Samuel French, London, 1993.

[Sto2]   _____. *Every Good Boy Deserves Favor and Professional Foul*. Grove Press, New York, 1978.

[Sto3]   _____. *Hapgood*. Faber and Faber, London, 1988.

[Sto4]   _____. *The Hard Problem*. Grove Press, New York, 2015.

[Sto5]   _____. *The Invention of Love*. Grove Press, New York, 1997.

[Sto6]   _____. *Jumpers*. Grove Press, New York, 1972.

[Sto7]   _____. *The Plays for Radio, 1964–1983*. Faber and Faber, London, 1990.

[Sto8]   _____. *Plays, Volume 2*. Faber and Faber, London, 1990.

[Sto9]   _____. *Plays, Volume 3*. Faber and Faber, London, 1998.

[Sto10] _____. *The Real Inspector Hound and Other Plays*. Grove Press, New York, 1996.

[Sto11] _____. *The Real Thing*. Faber and Faber, New York, 1984.

[Sto12] _____. *Rosencrantz and Guildenstern Are Dead*. Grove Press, New York, 1967.

[Sto13] _____. *Rosencrantz and Guildenstern Are Dead: The Film*. Faber and Faber, London, 1991.

[Sto14] _____. *Travesties*. Grove Press, New York, 1975.

[Str]   Strindberg, August. *Miss Julie*. Dover Publication, Mineola, New York, 1992.

[Wal]   Walet, Kathryn. *Victoria Martin: Math Team Queen*. Samuel French, New York, 2007.

[Wel]   Wellman, Mac. *Hypatia, or the Divine Algebra*. Alexander Street Press, Alexandria, VA, 2005.

[Whi]   Whitemore, Hugh. *Breaking the Code*. Samuel French, New York, 1987.

[Wil]   Wilson, Snoo. *Lovesong of the Electric Bear*. Bloomsbury Methuen Drama, London, 2015.

[Wit1]   Witkiewicz, Stanislaw Ignacy. *Seven Plays*. Translated and edited by Daniel Gerould. Martin E. Segal Theatre Center Publications, New York, 2004.

[Wit2]   _____. *The Madman and the Nun and Other Plays*. Translated and edited by Daniel Gerould and C. S. Durer. University of Washington Press, Seattle, 1968.

[Wit3]   _____. *Tropical Madness: The Winter Repertory*. Translated by Daniel Gerould and Eleanor Gerould. Winter House, New York, 1972.

[Zac]   Zacarias, Karen. *Legacy of Light*. Directed by Molly Smith at the Arena Stage, Washington, DC, 2009.

INDEX

A NOTE ON THE TYPE

This book has been composed in Arno, a serif typeface in the classic Venetian tradition, designed by Robert Slimbach at Adobe.

CREDITS

Chapter 1

Chapter 2

Chapter 3

Chapter 4

Chapter 5

Chapter 6